STRY

Modern Chemical Analysis and Instrumentation

UNDERGRADUATE CHEMISTRY

A Series of Textbooks

edited by
J. J. LAGOWSKI
Department of Chemistry
The University of Texas at Austin

Volume 1: Modern Inorganic Chemistry, *J. J. Lagowski*
Volume 2: Modern Chemical Analysis and Instrumentation, *Harold F. Walton and Jorge Reyes*

MODERN CHEMICAL ANALYSIS AND INSTRUMENTATION

Harold F. Walton and Jorge Reyes

Department of Chemistry
University of Colorado
Boulder, Colorado

Department of Chemistry
National University of Trujillo
Trujillo, Peru

MARCEL DEKKER, INC. New York 1973

MARCEL DEKKER, INC.
95 Madison Avenue, New York, New York 10016

LIBRARY OF CONGRESS CATALOG CARD NUMBER: 72-90967
ISBN: 0-8247-6033-6

PRINTED IN THE UNITED STATES OF AMERICA

PREFACE

This is a textbook of instrumental analysis (or, as we prefer to call it, Modern Chemical Analysis) for the undergraduate student. It is assumed that he knows the fundamentals of titration and stoichiometry and how to handle volumetric glassware and the analytical balance. Generally this material is taught in the freshman year.

The incentive to write this book came from the course which the authors taught at the University of Trujillo, Peru. We taught this course under some handicaps. Space and facilities were limited and only a few instruments were available. At the same time we had advantages and opportunities that, unfortunately, are not always present in the United States. Because the laboratory was small, we could have only a few students at a time; and because we had no teaching assistants, one or both of us was in the laboratory all the time. We could see difficulties as they arose and immediately offer our personal help. Another advantage was that we taught the whole analytical process. Wherever possible, we used samples from the real world: samples of soil from the desert or water from the pipes, fertilizer, paints, ores, products from the pharmacy or grocery. Though the final measurements were made with instruments, chemical separations and sample preparation went before.

This book is primarily a practical textbook, containing some of the valuable features of a laboratory manual. Enough experiments are described for a full year course and more. Some of them are described in detail, step by step, and in others the problem is indicated in general terms and the student is asked to work out his own procedure. Some are "open-ended" and an imaginative student or instructor can use them to lead into research projects.

The theory of each method is discussed, as is the construction of each instrument. A few instruments are described in detail, with a "conducted tour" of the control panel that explains what each button or switch is for. It would be impossible to do this for all commercial instruments, of course, and most instruments are described in more general terms. If, in selecting certain models for detailed description, we seem to have favored certain manufacturers, the reason is simply that we had these models in our laboratory and that the instruments are in common use. In no case is any endorsement of one instrument maker over another implied or intended.

We prefer short books, and to keep this one within bounds, we have included only the most common instrumental techniques. This is not an encyclopedia. Several excellent texts on instrumental analysis are available that are more comprehensive than ours, and bulkier. We omitted mass spectrometry, for instance, because very few institutions have mass spectrometers that undergraduate students use personally. The omission of nuclear magnetic resonance is less defensible, because moderate-priced NMR spectrometers are available in many small colleges. Their use, however, is for identification of pure compounds and for structure determination rather than for the analysis of mixtures. The same comment could be made about infrared spectrometers, which we have included, but infrared spectrometers are relatively cheap and very common, and they can be used for a wider variety of materials than NMR spectrometers. Radioactivity is omitted, because radiotracers are used for research rather than for the analysis of "things as they are."

When a student has learned a few instruments really well, so that he knows what calibrations and checks are necessary and what the read-out on the dial or recorder really says, he will easily be able to learn to use a new instrument when he meets it for the first time.

Many of the experiments in this book present the analysis of "things as they are," and include separations, titrations, and other chemical operations as well as the actual instrumental measurements. We feel that this aspect of analysis is neglected in many instrumental analysis texts.

We acknowledge with gratitude the help and encouragement of students and staff of the National University of Trujillo, Peru, the Pedagogical Institute of Caracas, Venezuela, and the University of Colorado.

CONTENTS

Modern Chemical Analysis and Instrumentation

INTRODUCTION

Today, chemical analysis means instrumental analysis. The days when the analytical balance was the only instrument in the laboratory have long gone. To solve new and old problems of chemical analysis the modern analytical chemist has at his disposal a great variety of instruments, and his work would be impossible without them. It follows, therefore, that the student of chemistry must learn about analytical instruments and have some practical experience in their use.

Analytical instruments are expensive, and some of them are very expensive. Their cost poses a real problem for colleges and universities, and the problem is especially acute in the developing countries. Fortunately, it is not necessary for a student to work with every kind of instrument there is. What is important is for him to get a "feeling" for instruments, to know how they work and what they will do. He must see for himself, by his own practical experience, that they give useful information for solving important practical problems.

In this book we have taken a simple approach to teaching instrumental analysis. We have described only the simpler and commoner instruments and have not included the very expensive, though very powerful, techniques of mass spectroscopy and nuclear magnetic resonance. Other omissions will be noted; we have not included any methods that depend on radioactivity, nor have we included x-ray methods, emission spectroscopy, or thermal analysis. On the other hand we have included modern methods of chemical separation that are very powerful and can be performed with modest equipment, namely, paper and thin-layer chromatography and ion exchange.

The experiments in this book make considerable use of chemical reactions, in the preparation of samples for instrumental measurements, in titrations, and in other ways. A common misconception is that instruments have eliminated the need to know any chemistry. Needless to say this is not the case. Analytical chemists solve problems of chemical composition, and they solve them by the simplest and most effective means at their disposal. To tell whether a certain liquid is ethyl bromide or diethyl ether, for instance, one could use a mass spectrometer or scan the infrared spectrum, but he can get the answer within minutes by measuring the density, using a micropipet and a chemical balance. Simple problems should be solved by simple means. There are many occasions when a titration or a simple qualitative test performed in a test tube will give the information one needs. The classical methods of volumetric and gravimetric analyses are not obsolete; however, they are often far too slow to meet the needs of modern industry, and they may not give the kind of information that one needs. When classical methods are inadequate the chemist turns to his instruments.

PREPARATION FOR THIS BOOK

We have assumed that the student who uses this book has learned the elements of classical qualitative and quantitative analyses. He must know how to form and filter an analytical precipitate and how to wash, dry, and weigh it. He must know how to use pipets, burets, and volumetric flasks; he must know the techniques of titration, quantitative transfers, and quantitative dilutions. He must have a grasp of experimental error, precision, and accuracy. He must know some of the commoner reagents and their uses. He must have more than an empirical or "cookbook" knowledge; he must know why certain things are done and must understand that volumes and concentrations must be measured carefully and accurately in some cases, while in others, like making a solution acid for a thiosulfate titration or diluting a sample to a convenient volume to go into a titration flask, a very rough estimate of volume or concentration will do. This point may seem trivial, but many times we have watched students in the laboratory carefully measuring 50 ml of distilled water to dissolve a sodium carbonate sample for titration, because the textbook said, "Dissolve in 50 ml of water."

The student must understand chemical equations and how to balance them. He must understand stoichiometry, gravimetric factors, and chamical equivalents, molar and normal concentrations, and how to standardize a volumetric solution. He should also have been introduced to chemical equilibrium and understand solubility products, ionization constants, buffer solutions, and pH.

To help in reviewing this basic material we have included a set of problems at the end of this chapter.

It is essential to understand the basic theory of the instrumental methods that we shall discuss. Each chapter has a section on theory and a general description of the instrument before the experiments are described. To read these chapters it will be a help if the student has had an introduction to physics and physical chemistry. One can use a pH meter without being an expert in thermodynamics or electrochemistry, but one should know what is meant by free energy, what is a reversible process, what is electrical potential, and what is the distinction between chemical equilibrium and rate of reaction. Likewise one need not master quantum mechanics to use a spectrophotometer in chemical analysis, but one should have an idea of energy levels and quantum numbers.

Nearly all analytical instruments depend for their operation on electronics and electronic circuits. To understand his instruments properly the chemist should study electronic circuitry; several excellent textbooks and courses are available in this area. We shall not emphasize this aspect in this book. The average analytical chemist knows enough to make minor adjustments and repairs and even to adapt his instruments to special uses, but he is usually helpless if anything goes seriously wrong. Thanks to the partnership between the electrical engineer and the chemist, commercial instruments are simple to operate and highly reliable, but they do occasionally need repairs. Large laboratories and universities have their own service men who can maintain their instruments in good working condition, but the problem of maintenance is sometimes serious, particularly in the developing countries. One should take into account the availability of service personnel before deciding which make of instrument to buy.

DESCRIPTIONS OF INSTRUMENTS

In a few cases we have described specific instruments in considerable detail, explaining how the controls are labeled and how they should be operated. These are instruments that are very common and widely used. It would, however, be an impossible task to describe every commercial instrument of its kind, and it would be self-defeating, because new models and new instruments are constantly appearing. We have therefore given a general description of each type of instrument and tried to explain how it works. When one is ready to use the instrument that one has in one's laboratory, one must first *read the instructions*. Every instrument maker supplies a booklet with his instruments that explains in detail how to install the instrument and get it ready for use, how to operate it, and what to do if something goes wrong. *Keep this booklet in a safe place where it can be consulted by everybody who uses the instrument.*

THE EXPERIMENTS

Instructions are given in this book for many experiments, more than could be performed in a semester or a year. The instructor thus has a wide choice. Some of the experiments measure quantities of intrinsic chemical interest, like ionization constants. Many of them describe the analysis of natural materials like soils, minerals, and fruit juices, or of commercial products like flavoring essences, fertilizers, and drugs. Experiments like this lend realism to a course in chemical analysis. The examples we have given in our text are intended to show the possibilities. An instructor or student with imagination and energy can multiply these examples tenfold. He will find materials to analyze in any pharmacy, hardware store, grocery, or a store where agricultural supplies are sold; he can go to the market or the fields and collect fruit to analyze for its amino acids or carboxylic acids; he can test soils for their mineral nutrients and their trace metals; he can measure the fluoride content of public water supplies; he can investigate environmental pollution. Many times in trying to analyze "real-life" samples he will meet unexpected complications that will test his knowledge of chemistry and his ability to solve problems.

This book is addressed to beginners, and most of the laboratory directions are given in considerable detail, even to the point of repetition. Each student should keep his own laboratory notebook in which he records all that he does, the things he observes and the measurements he takes, his calculations and graphs, and the conclusions he draws. Learning to keep a good laboratory record is a vital part of the student's training.

ONE FINAL WORD

Scientific instruments are delicate precision instruments and must be treated with care. If a lever or switch does not move easily, never force it. Stop and think whether you are doing the right thing. Re-read the instructions if necessary. Never slam a lever or switch; move it smoothly and gently. The time you save by slamming it is a small fraction of a second; the damage you do may shorten the instrument's life by years.

REVIEW QUESTIONS

1. A solid sample of a water-soluble salt weighs 0.264 g. It gives 0.185 g of silver chloride. What is the percentage of chloride in the sample?

2. Sea water contains 18.98 g of chloride ions per kilogram of sea water. Its density is 1.022 g/ml. What weight of AgCl would be produced from 0.500 ml of sea water after treating with excess silver nitrate?

3. Sea water contains sulfate ions as well as chloride. 5.00 ml of sea water yields 33.0 mg of barium sulfate. How many grams of sulfate ions, SO_4^{2+}, are present in 1 kg of sea water? What is the molar concentration of sulfate (moles per liter)?

4. What weight of $BaSO_4$ would be produced from 0.100 g of iron pyrites, FeS_2?

5. A solution of hydrochloric acid is standardized by titration against pure sodium carbonate, Na_2CO_3. Titration is done to the methyl orange end point at pH 4. 0.160 g of Na_2CO_3 = 15.0 ml HCl. What is the normality of the acid?

6. Constant-boiling hydrochloric acid is used as a standard for preparing volumetric solutions. The acid that boils at 760 mm contains 20.22 percent of hydrochloric acid by weight. What is the normality of a solution prepared by dissolving 15.00 g of constant-boiling acid in water and making the volume up to 500 ml?

7. You wish to prepare a dilute solution of copper sulfate, $CuSO_4 . 5H_2O$, containing 10.0 mg of Cu per liter. You start by weighing out 1.00 g of copper sulfate and dissolving it in water in a 250-ml volumetric flask, and adding water to the mark. Call this solution "A". Now, what volume of Solution A must you add to a <u>second</u> flask, also 250 ml, to give a solution which, when made up to volume, will contain 10.00 mg of copper per liter?

8. (a) Write a balanced half-reaction equation (that is, an equation in which electrons are shown as participants in the reaction) for the reduction of the permanganate ion, MnO_4^{2-}, in acid solution to form manganous ion, Mn^{2+}.
   (b) A solution contains 7.88 g of $KMnO_4$ in 1 l. What is its molarity? What is its normality as an oxidizing agent in acid solution?

9. (a) When sodium thiosulfate is oxidized by iodine, what are the products of the reaction?
   (b) What is the normality of a 0.100 M sodium thiosulfate solution?
   (c) How is a standard sodium thiosulfate solution used to titrate copper? What volume of 0.100 M sodium thiosulfate is equivalent to 10.00 mg of copper?

10. How many milligrams of magnesium ions, Mg^{2+}, are there in 100 ml of (a) pure water, (b) a solution buffered at pH 11.0, if both solutions are saturated with magnesium hydroxide, $Mg(OH)_2$?

11. Calculate the pH of each of the following solutions: (a) 0.0010 M HCl; (b) 0.0010 M $Ca(OH)_2$; (c) 0.20 M HOAc (acetic acid); (d) 0.20 M sodium acetate; (e) a solution made by mixing 50 ml 0.10 M HOAc with 25 ml 0.30 M NaOAc.

12. What volume of 0.10 M sodium hydroxide must be added to 100 ml 0.050 M acetic acid to yield a solution of pH 5.00?

\- - - - - - - - -

DATA NEEDED FOR THESE PROBLEMS

Atomic weights: Ag 107.87, Ba 137.34, C 12.00, Cl 35.45, Cu 63.54, Fe 55.85, H 1.01, K 39.10, Mg 24.31, Mn 54.94, Na 22.99, O 16.00, S 32.06.

Solubility product: $Mg(OH)_2$, 1.8×10^{-11}
Ionization constant: Acetic acid (HOAc), 1.75×10^{-5}

\- - - - - - - - -

ANSWERS

1, 17.3%; 2, 39.5 mg, or 0.0395 g; 3, 2.78 g/kg, or 0.0277 M; 4, 0.389 g; 5, 0.201 N; 6, 0.166 N; 7, 2.46 ml; 8(b) 0.050 M, 0.250 N; 9(b) 0.100 N, (c) 1.57 ml; 10(a) 0.40 mg, (b) 0.043 mg; 11(a) 3.0, (b) 11.3, (c) 2.73, (d) 9.02, (e) 4.93; 12, 31.8 ml.

Chapter 1

POTENTIOMETRY AND THE pH METER

Next to the analytical balance, the pH meter is the commonest instrument to be found in chemical laboratories. Its usefulness goes beyond the measurement of pH. Essentially the pH meter is a voltmeter, a device for measuring electromotive force, and it can be used for any analytical determination that depends on the measurement of electromotive force of differences in electrical potential. The reading can be applied directly to the measurement of the quantity desired, as in the measurement of pH, or it can be used to follow the progress of a titration, in which case the changes in the reading are of interest. In this chapter we shall review the various ways of using a pH meter in chemical analysis, but before doing so, we must make clear what is meant by the terms electrical potential and electromotive force and show how they are related to chemical quantities.

I. POTENTIAL AND ELECTROMOTIVE FORCE

Electrical energy is measured by the product of two factors, quantity of electricity times potential drop. The quantity of electricity ultimately means the number of electrons, or elementary electric charges, that change places. It is found by multiplying the current, which gives the rate of flow of electrons, by the time. If the time is measured in seconds and the current in amperes, the quantity of electricity is given in coulombs. The potential drop may be likened to pressure, and one may think of the electrical potential at a particular point as being the pressure of electrons at that point. Differences in potential are expressed in volts. It is the difference in electrical potential that decides which way electrons will flow, just as it is the difference in temperature that decides which way heat will flow. There is one complication, however. Back in the eighteenth century, when Benjamin Franklin called one kind of electricity positive and the other kind negative, he picked the wrong names. For we now know that electrons are the things that move in a metal wire or in an evacuated space, and electrons carry, according to Franklin's notation, a negative charge. So we have to explain that electrons move from a point that is more negative in potential to a point that is less negative, or more positive. We assume, of course, that no external force enters to oppose or reverse that movement.

Electromotive force is any non electric force that starts electrons moving. It comes from another form of energy, which may be mechanical work, as in the familiar electric generator; radiant energy, as in a photoelectric cell; heat, as in thermocouples; or chemical energy from a chemical reaction. It is this last source of electromotive force that is used in electrochemical methods of analysis.

Consider a simple process of oxidation and reduction, the reaction of zinc metal with hydrogen ions to form hydrogen gas and zinc ions. Electrons are transferred from the zinc atoms to the hydrogen ions, and we can catch these electrons and turn them into an electric current by using the arrangement shown in Fig. 1-1. This is a "voltaic cell." It has two compartments separated by a porous partition. The purpose of this parti-

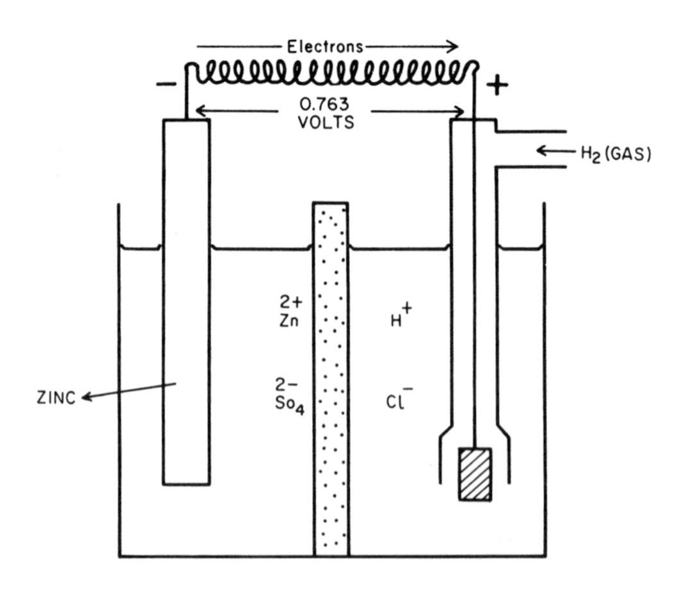

Fig. 1-1. Voltaic cell.

tion is to prevent the hydrogen ions and the zinc from coming into direct contact. Instead of the arrangement shown we could use an inverted U-tube filled with potassium chloride solution; such "salt bridges" are described in textbooks of physical chemistry. In one compartment is a piece of zinc metal surrounded by zinc sulfate solution, and in the other a piece of platinum foil or gauze, covered with a catalytic coating of platinum black, over which is bubbled hydrogen gas. The solution on this side is dilute hydrochloric or sulfuric acid.

At the two metal surfaces the following reversible reactions occur:

$$Zn^{2+} + 2e = Zn, \quad 2H^+ + 2e = H_2$$

If the metals are not connected to each other these reactions are each in equilibrium. It happens, however, that the equilibrium represented by the first equation is relatively far over on the left, that is to say, the pressure of electrons in the zinc is greater than the pressure of electrons in the platinum. (Remember that metals, by their nature, act as "solvents" for electrons and pathways for electrons in motion.) If now we connect the zinc to the platinum by a wire going outside the cell, the electrons flow of their own accord from where they are concentrated (on the zinc) to where they are more dilute (on the platinum). In other words an electric current flows, and the electrons flow through the wire from the negative terminal of the cell, which is the zinc, to the positive terminal, which is the platinum. At the surface of the zinc this process occurs: $Zn \rightarrow Zn^{2+} + 2e$. At the (catalytic) surface of the platinum this process occurs: $2H^+ + 2e \rightarrow H_2$, and the net result is the well-known chemical reaction: $Zn + 2H^+ \rightarrow Zn^{2+} + H_2$. It is the energy of this reaction that drives the electric current and creates the electromotive force.

The electromotive force depends on the temperature and on the concentrations of the reactants. If the zinc is pure and the hydrogen is at one atmosphere, and if the solutions of acid and zinc salt are ideal one-molar solutions (what is called "unit activity") the electromotive force is 0.763 volt. Because the hydrogen electrode is taken as the standard of reference, we call this the <u>standard potential</u>, and because the zinc is the negative pole of the cell, we write

$$Zn^{2+} + 2e = Zn, \quad E^o_H = -0.763 \text{ volt.}$$

Writing the zinc ions as the reactants, rather than the products, gives the minus sign of the potential another meaning; it indicates that the reduction of zinc ions to zinc metal is <u>not spontaneous</u> when hydrogen gas is used as the reducing agent. By contrast, we write

$$Cu^{2+} + 2e = Cu, \quad E^o_H = +0.337 \text{ volt.}$$

The plus sign before the electromotive force tells us two things; that in a cell like that of Fig. 1-1, but with copper metal and copper ions instead of zinc, the copper metal would be the positive pole, and that copper ions are spontaneously reduced to copper metal by hydrogen gas.

The subscript $_H$ means that the potentials are referred to the hydrogen electrode as reference. We are reminded that we can only measure differences in potential in electric circuits, not absolute potentials. The subscript is usually omitted because it is generally understood that the hydrogen electrode is taken as the level of reference. The superscript ^{o} means that the potentials are standard, that is, measured or calculated for pure metals, gases at one atmosphere or rather unit fugacity, and solutions of unit activity. To understand the "activity" concept requires an understanding of thermodynamics and free energy, and it will suffice for the present to say that the activity of a dissolved substance becomes equal to its molar concentration in the limit of very dilute solutions.

A table of standard electrode potentials is given in the Appendix. From it we can calculate the electromotive force of any combination of electrodes. For example, a cell with copper and copper sulfate solution on one side and zinc and zinc sulfate solution on the other would have an electromotive force of 0.763 + 0.337 = 1.100 volt, and the zinc would be the negative pole.

II. CONVENTIONS OF SIGNS

It is a common practice to write electrode reactions with electrons shown as products, rather than as reactants. When this is done, the sign of the standard potential is reversed also. Thus the zinc metal - zinc ion reaction is shown as

$$Zn = Zn^{2+} + 2e; \quad E^o{}_H = +0.763 \text{ volt}$$

Again a plus sign goes with a spontaneous reaction. We all know that zinc metal will react spontaneously with hydrogen ions (a dilute acid solution) to form hydrogen gas and zinc ions. By the same token, copper metal does not liberate hydrogen gas from a dilute acid, and the sign of the standard potential, written with copper metal as the reactant, is negative:

$$Cu = Cu^{2+} + 2e; \quad E^o{}_H = -0.337 \text{ volt}$$

Applying this system to represent the electromotive force of a cell, a cell made by combining the copper-cupric ion electrode with the hydrogen gas-hydrogen ion electrode would have the hydrogen, or reference electrode, as the negative electrode. The cell electromotive force would be 0.337 volt with the hydrogen (reference) electrode negative. This is exactly what we said in the preceding section. There we said "the copper metal would be the positive pole."

The system of writing electrode reactions with the electrons as the products was used extensively by the great physical chemist G. N. Lewis and by his school, and it is sometimes called the "California convention" or the "American convention" of signs. The system of writing the electrons as reactants, and showing the sign of the electrode potential as the sign of the electrode in question with respect to the reference electrode, was adopted by the International Union of Pure and Applied Chemistry, meeting in Stockholm, and is known as the "Stockholm convention." We shall use the Stockholm convention in this book.

III. EFFECT OF CONCENTRATION: THE NERNST EQUATION

We noted that the electromotive force of a cell depends on the energy of its chemical reaction, and this in turn depends on the concentrations. In the example of $Zn + 2H^+ \rightarrow Zn^{2+} + H_2$, the electromotive force is greater the more concentrated the acid, and less the more concentrated the zinc salt (because concentrated zinc ions oppose the reaction). The electromotive force depends on the logarithms of the concentrations, for this reason: the work needed to move a mole of dissolved substance from a dilute solution of concentration c_1 to a more concentrated solution of concentration c_2 equals $RT \ln(c_2/c_1)$, where R is the gas constant, T the absolute temperature, and "ln" means the natural logarithm, or logarithm to the base e. This relation holds for "ideal" solutions. Its proof is given in textbooks of physical chemistry. If we transfer n moles the work is n times as great, and we recall that $n \log x = \log(x^n)$.

We are now ready to write the Nernst equation. For zinc metal immersed in a solution of zinc ions, the potential of the zinc metal is

$$E = E^o + \frac{RT}{2F} \ln [Zn^{2+}] = E^o + \frac{2.303RT}{2F} \log [Zn^{2+}]$$

The square bracket around the symbol for zinc ion means molar concentration, and again we are considering the solutions to be ideal. The symbol F stands for the faraday, the charge on one mole of univalent ions, and since the zinc ion has two charges, a mole of zinc ions carries a charge of two faradays; hence the factor 2 in the denominator.

Let us take a couple of more complicated cases. Consider a platinum wire (platinum is used because it is chemically inert and yet catalytic) in a solution of ferric and ferrous ions mixed together; the reaction at the electrode surface is

$$Fe^{3+} + e = Fe^{2+}$$

and the potential of the platinum wire is

$$E = E^{o} + \frac{RT}{F} \ln \frac{[Fe^{3+}]}{[Fe^{2+}]}$$

Common sense tells us which way to write the fraction; raising the concentration of ferric ions, Fe^{3+}, would draw electrons out of the metal and make it more positive; raising the concentration of ferrous ions, Fe^{2+}, would drive electrons into the platinum metal and make it more negative.

As another case, consider a platinum wire dipped into a solution that contains uranyl ions, $UO_2{}^{2+}$, uranium(IV) ions and hydrogen ions; the reduction reaction is

$$UO_2{}^{2+} + 4H^{+} + 2e = U^{4+} + 2H_2O$$

and the corresponding Nernst equation is

$$E = E^{o} + \frac{RT}{2F} \ln \frac{[UO_2{}^{2+}][H^{+}]^4}{[U^{4+}]}$$

In making calculations it is more convenient to use logarithms to base 10, and moreover we are usually working at room temperature; the standard temperature for chemical work is 25°C or 298.1°K. The factor in front of the logarithm thus becomes

$$\frac{2.303\ R \times 298.1}{nF} = \frac{0.05915}{n}$$

(The gas constant, R, equals 1.99 calories mole^{-1} deg^{-1}; the faraday, using consistent units, is 23,060 calories volt^{-1} equiv^{-1})

IV. PROBLEMS USING THE NERNST EQUATION

1. 50 ml 0.10 M $FeSO_4$ is titrated with 0.10 M $Ce(SO_4)_2$. A platinum wire is placed in the solution to indicate the potential. What is the

potential of the wire, referred to the standard hydrogen electrode, when (a) 25 ml, (b) 49.5 ml, (c) 50.0 ml ceric sulfate have been added? The reaction is $Fe^{2+} + Ce^{4+} = Fe^{3+} + Ce^{3+}$;
the potentials are: $Fe^{3+} + e = Fe^{2+}$, $E^o = +0.770$ V;
$Ce^{4+} + e = Ce^{3+}$, $E^o = +1.44$ V.

(a) At this point, just half of the Fe^{2+} has been converted to Fe^{3+}; the concentrations of the two ions are equal, and $E = E^o = \underline{0.77\ V}$.

(b) Here, 99% of the Fe^{2+} has been oxidized to Fe^{3+}, and very nearly, $[Fe^{3+}] = 100[Fe^{2+}]$; $E = 0.77 + 0.059 \log[Fe^{3+}]/[Fe^{2+}]$ $= 0.77 + 0.059 \log 100 = 0.77 + 0.12 = \underline{0.89\ V}$.

(c) Here we are at the point of exact equivalence. We have added exactly 1 mol of Ce^{4+} for every mole of Fe^{2+}, and they have reacted in the ratio 1:1. Thus, $[Ce^{4+}] = [Fe^{2+}]$, $[Ce^{3+}] = [Fe^{3+}]$ and

$$E = 0.77 + 0.059 \log \frac{[Fe^{3+}]}{[Fe^{2+}]} = 1.44 + 0.059 \log \frac{[Ce^{4+}]}{[Ce^{3+}]}$$

(We note that the indicating wire can only have one potential at a time!) Combining these equations, we get

$$1.44 - 0.77 = 0.67 = 0.059 \log \frac{[Fe^{3+}][Ce^{3+}]}{[Fe^{2+}][Ce^{4+}]} = 0.059 \log \frac{[Fe^{3+}]^2}{[Fe^{2+}]^2}$$

and, $[Fe^{3+}]/[Fe^{2+}] = 10^{5.6} = 4 \times 10^5$; $E = 0.77 + (0.059 \times 5.6)$ $= 1.10$ V. This is just midway between the two standard potentials, 0.77 and 1.44 V. This happens because the electron transfer is symmetrical, that is, one electron gained by each Ce^{4+} and one electron lost by each Fe^{2+}.

2. A solution 0.001 $\underline{M}$ in Fe^{3+} is passed through a tube packed with granulated silver, which reduces Fe^{3+} to Fe^{2+}. Reduction is facilitated by having the solution 0.10 $\underline{M}$ in hydrochloric acid, which forms silver chloride, AgCl, which is sparingly soluble. If the solubility product of AgCl is 1.6×10^{-10}, and E^o for $Ag^+ + e = Ag$ is 0.80 V, what is the final (equilibrium) ratio of Fe^{3+} to Fe^{2+} in the solution?

Again, the solution inside the tube can have only one potential at a time, and whatever potential we calculate for the silver ions - silver couple must also apply to the ferrous-ferric ions couple. First, calculate the concentration of silver ions, Ag^+. From the solubility product and the chloride ion concentration, $[Ag^+] = 1.6 \times 10^{-10}/0.10 = 1.6 \times 10^{-9}$. Putting this into the Nernst equation,

$$E = 0.80 + 0.059 \log 1.6 \times 10^{-9} = 0.80 - (0.059 \times 8.80)$$
$$= 0.80 - 0.52 = \underline{0.28 \text{ V}}.$$

Now let us write the equation for the ferrous-ferric couple;

$$E = 0.28 = 0.77 + 0.059 \log \frac{[Fe^{3+}]}{[Fe^{2+}]}$$

The ratio of ferric to ferrous ions, Fe^{3+}/Fe^{2+}, is $10^{-8.3}$, or $\underline{1 \text{ to } 2 \times 10^8}$.

If we look in tables of data we find that the standard reduction potential, E^o, for silver ions is quoted to four significant figures, +0.7996 V. We used only two figures in this calculation, and we also ignored the slight drop in the chloride ion concentration that occurs when silver reacts with ferric ions in hydrochloric acid to form ferrous ions and silver chloride. There is little point in carrying the calculations to high accuracy as long as we ignore the activity coefficients. For highly charged ions these coefficients may be very different from 1.

V. MEASUREMENT OF pH; HYDROGEN AND GLASS ELECTRODES; REFERENCE ELECTRODES

The electrode of hydrogen gas at one atmosphere pressure, placed in a hydrochloric acid solution of unit activity, has by definition a potential of zero. If we change the acid concentration the electrode potential changes according to the Nernst equation:

$$E = E^o + \frac{2.303RT}{F} \log[H^+] = 0 + 0.059 \log[H^+] \text{ at } 25^oC$$
$$= \underline{-0.059 \text{ pH}}$$

for we recall that $\text{pH} = -\log[H^+]$. We can thus use the hydrogen electrode to find the pH of a solution; all we have to do is measure its potential. However, this immediately raises the question, "potential with respect to what?" We have noted that we cannot measure absolute potential, only differences in potential, and a cell whose electromotive force we wish to measure must always have two electrodes. We could compare the hydrogen electrode in the sample solution with another hydrogen electrode in a one-molar hydrochloric acid solution, connecting the two solutions by a salt bridge, but there is an easier way. We use for the second electrode a saturated calomel electrode, as shown in Fig. 1-2. The actual electrode material, the metal that serves as the pathway for electrons, is liquid mercury. It is in contact with solid mercurous chloride ("calomel") and a saturated solution of potassium chloride; these, between them, maintain a fixed concentration of mercurous ions, and the equilibrium $Hg_2^{2+} + 2e = 2Hg$ fixes the "pressure of electrons," or the potential of the mercury. The saturated potassium chloride solution makes contact with the test solution through a very small hole which usually consists of a tiny asbestos fiber fused into the glass; see Fig. 1-2. This contact is the "salt bridge." The test solution can be changed at will, but the potential of the mercury with respect to the surrounding potassium chloride solution stays constant. Arrangements such as this are called "reference electrodes." The saturated calomel electrode has a potential of +0.244 volt with respect to the standard hydrogen electrode. There is a small potential (of the order of 0.002 volt) across the fiber junction, so it is hard to give a very precise value for the calomel electrode potential.

Problem. If the standard potential for $Hg_2^{2+} + 2e = 2Hg$ is +0.796 V, and saturated potassium chloride is 4.0 M, what is the solubility product,

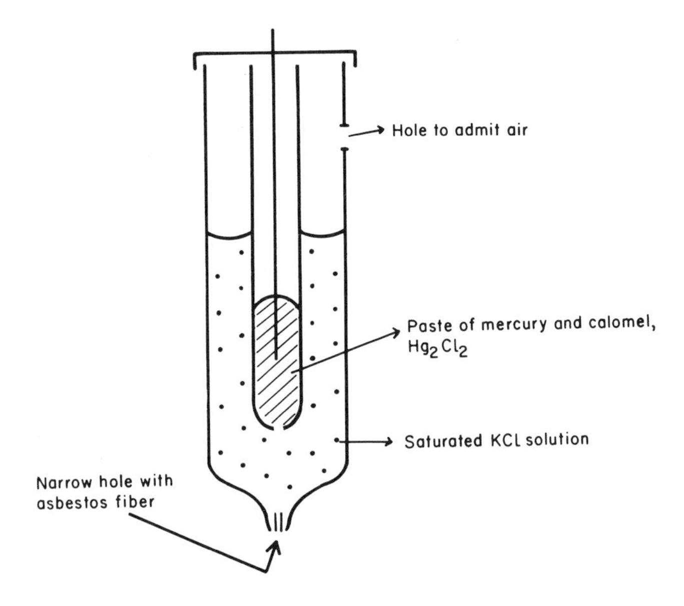

Fig. 1-2. Saturated calomel electrode.

$[Hg_2^{2+}][Cl^-]^2$, of mercurous chloride? Answer: $[Hg_2^{2+}] = 1 \times 10^{-18}$, $K_{sp} = 1.6 \times 10^{-17}$. (Of course this calculation cannot be very accurate; the activity coefficient of KCl in its saturated solution is about 0.5.)

We could, therefore, make a "pH meter" by combining a hydrogen electrode with a saturated calomel reference electrode. But the hydrogen electrode is very cumbersome, and it is affected by oxidizing agents. A much more convenient arrangement is the glass electrode.

In 1909 Haber and Klemensiewicz found that a thin bulb of glass conducted electricity and that, if one put two different solutions of different acidities inside and outside the bulb, an electrical potential was produced across the glass whose value depended on the logarithm of the ratio of the two hydrogen ion concentrations:

$$E_{glass} = 0.059 \log \frac{[H^+]_{inside}}{[H^+]_{outside}}$$

The side of the glass where the hydrogen ion concentration was smaller was was the side that was positive.

Glass is an ion-exchanging material (see Chapter 10) and has a pronounced preference for hydrogen ions. What happens when the glass bulb conducts electricity is this. Hydrogen ions are adsorbed on one surface; the positive charge of these hydrogen ions is transmitted through the glass by displacement of the sodium ions that the glass contains; then hydrogen ions are desorbed on the other side. The process is shown in Fig. 1-3. Effectively, the thin film of glass is like a filter, or membrane, that lets hydrogen ions pass but no other ions.

The hydrogen ions move of their own accord from the side where these ions are more concentrated (usually the inside of the bulb) to the side where they are more dilute. When they move they carry their positive charge with them, and the side of the glass toward which they move - the "dilute" side - becomes charged positively. This charge, of course, opposes the movement of more hydrogen ions. Very soon an equilibrium is set up. At equilibrium the free energy change for the movement of hydrogen ions becomes zero; in thermodynamic notation,

$$\Delta G = RT \ln \frac{[H^+]_{dilute}}{[H^+]_{conc.}} + EF = 0$$

and

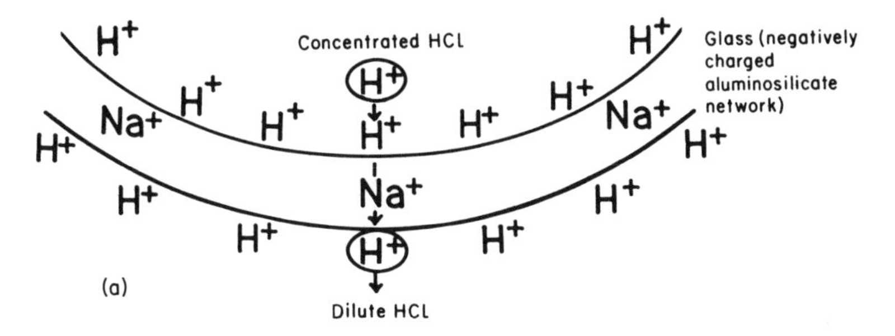

Effectively,

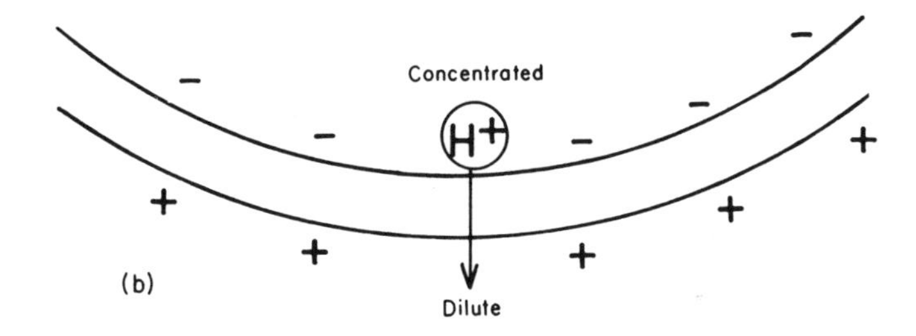

Fig. 1-3. Glass electrode.

$$E = \frac{RT}{F} \ln \frac{[H^+]_{conc.}}{[H^+]_{dilute}}$$

Converting to common logarithms,

$$E = \frac{2.3RT}{F} \log [H^+]_{conc.} + \frac{2.3RT}{F} \text{ pH}$$

where "pH" is the pH of the "dilute" solution, outside the glass bulb. The outside of the bulb is considered positive.

Before we can measure the electromotive force E we need reference electrodes. The usual arrangement is shown in Fig. 1-4. Inside the bulb is a solution of hydrochloric acid, and dipping into this solution is a silver wire coated with solid silver chloride. Outside the bulb is the test solution, and dipping into this is a saturated calomel electrode. In normal operation the calomel electrode is positive, and we can express the electromotive force of the whole cell as follows:

$$E = E^o + 0.059 \text{ pH} \quad (\text{at } 25^oC)$$

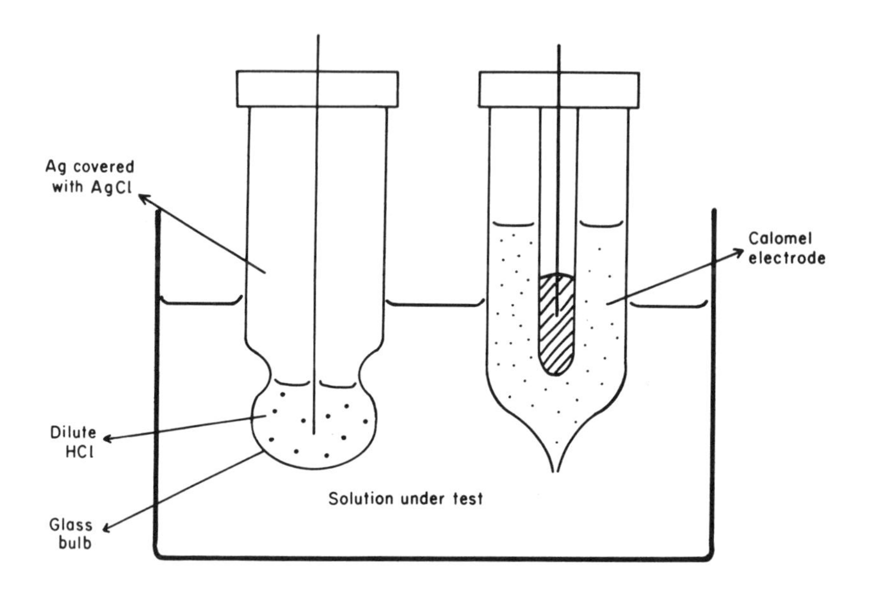

Fig. 1-4. Cell used in pH meter.

We can calculate the value of E^o if we know the concentration of hydrochloric acid inside the bulb.

Problem. Inside a glass electrode is 0.10 M HCl and a silver wire coated with silver chloride; outside is a solution of pH 4.00 and a saturated calomel electrode; the temperature is 25°C. What is the EMF of the cell?

The EMF is made up of three components or "steps." The first is the potential of the silver metal with respect to the 0.1 M hydrochloric acid solution. We have already calculated this; it is 0.281 V (see page 14). The silver is positive with respect to the solution (more correctly, with respect to a standard hydrogen electrode placed in the solution).

The second "step" is the potential difference across the glass. Inside the hydrogen ion concentration is 10^{-1}; outside it is 10^{-4} M. The difference (that is, the ratio) is three powers of 10; for every power of 10 the potential difference is 0.059 V; the total is $3 \times 0.059 = 0.177$ V. The outside of the bulb is positive.

The third step is the potential of the calomel electrode with respect to the test solution. This is 0.244 V, with the mercury positive. Summing up all the potentials with proper regard to sign (see Fig. 1-5), the total EMF is $-0.281 + 0.177 + 0.244 = \underline{0.140\ V}$, with the calomel electrode positive.

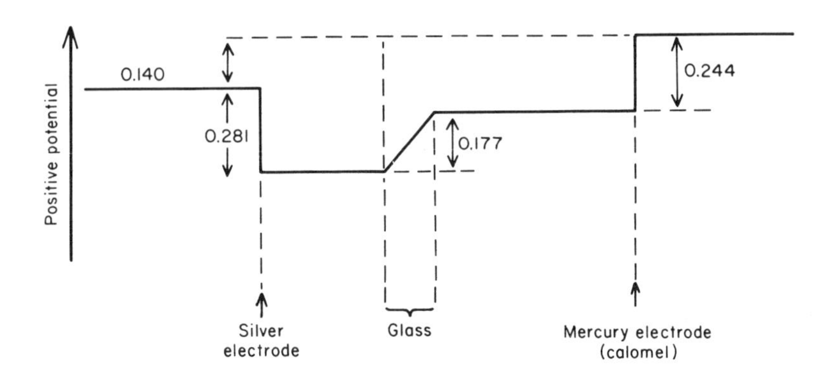

Fig. 1-5. Potentials in glass-electrode cell.

In using a pH meter one does not normally make calculations like this. For one thing, there is a small potential across the glass, of the order of a few millivolts, called the "asymmetry potential," which is due to differences between the two glass surfaces and is hard to evaluate. For another, commercial glass electrodes are sealed, and one does not know the concentration of the hydrochloric acid they contain. One is not particularly interested in the actual value of the cell electromotive force; one only wants to know how this electromotive force depends on the pH.

One always calibrates a pH meter with a standard buffer solution of known pH. Every pH meter has a control knob that allows one to set the indicator needle wherever one wants to set it on the scale. To use the instrument one turns it on and warms up the electronics, places the glass and calomel electrodes in a standard buffer solution, then adjusts the needle to read correctly the known pH of the solution. The meter should now read correctly over the entire scale, provided one other adjustment has been made, the temperature control.

As we said at the beginning of this chapter, the pH meter is simply a voltmeter. We could take the reading in volts or millivolts (1 millivolt = 0.001 volt) and use the relationship

$$\text{difference in pH} = \frac{\text{difference in voltage}}{0.05915}$$

to find the pH. It is easier to make the dial read directly in pH units, with divisions such that 1 pH unit = 0.05915 volt, and this is done. The scale can be read in pH units or in millivolts, whichever is desired. Conversion from one scale to the other is made by turning a switch. However, the proportionality factor 0.05915 only holds at 25°C. The factor is actually 2.303RT/F; it is proportional to the absolute temperature. Instruments therefore have a temperature adjustment which must be set within a degree or two of the sample temperature, if the pH scale is to be used. For millivolt readings, of course, the temperature adjustment does not apply.

Problem. A pH meter is set to read correctly in a standard buffer of pH 4.00 at 25°C. However, the temperature control is erroneously set for 40°C. A test solution shows a pH reading of 8.50. What is its true pH?

Between the standard and the sample, the pH difference as read is 4.50 units. However, this does not mean 4.50 x 0.05915 V, for the temperature dial was wrongly set for 40°C. At 40°C (313°K) 1 pH unit = 0.05915 x 313/298 = 0.06213 V. The difference in EMF between the sample and the standard is therefore 0.06213 x 4.50 V, and the true difference in pH is (0.06213 x 4.50)/0.05915 units, or more simply 4.50 x 313/298 = 4.73 units. The actual pH of the test solution is 4.00 + 4.73 = 8.73.

To verify that a pH meter is calibrated correctly and that the temperature control is properly set, it is wise to check it with two standard buffers, rather than one, and to use a standard whose pH is not too far from that of the solutions being tested. In very careful work the standard buffer reading should be checked several times during a series of measurements.

VI. STANDARD BUFFERS; THE pH SCALE

The proper definition of pH is: pH = $-\log a_{H^+}$, where a_{H^+} is the activity of the hydrogen ion, not its concentration. Activities and concentrations are only equal in the limit of an ideal solution. Activities are defined thermodynamically through the free energy of transfer: $\Delta G_{1\to 2} = RT \ln(a_2/a_1)$ where $\Delta G_{1\to 2}$ is the free energy or reversible work needed to transfer one mole of the dissolved substance from a solution having an activity a_1 to one having an activity a_2. Now, it is experimentally impossible to transfer weighable quantities of positive ions from one place to another without at the same time transferring negative ions. Thus the term "activity of hydrogen ions" is thermodynamically meaningless. One gets over the difficulty by using theoretical models of ions in solutions, and finally has to make an arbitrary choice. Many years of investigation and thought have gone into the choice of a consistent pH scale, largely at the National Bureau of Standards in Washington under the direction of R. G. Bates. The National Bureau of Standards tables appear in standard reference works; a highly abbreviated version appears in Table 1.

TABLE 1

Standard pH Buffers

Temperature, $^{\circ}C$	Potassium hydrogen tartrate, sat. at $25^{\circ}C$	Potassium hydrogen phthalate, 0.050 M	A solution 0.025 M in KH_2PO_4, 0.025 M in Na_2HPO_4	Borax, 0.010 M
0	—	4.01	6.98	9.46
10	—	4.00	6.92	9.33
20	—	4.00	6.88	9.22
25	3.56	4.01	6.86	9.18
30	3.55	4.01	6.85	9.14
40	3.54	4.03	6.84	9.07

For most purposes the potassium hydrogen phthalate standard is sufficient. Its pH varies little with concentration; 0.02 M solution has pH 4.04 at $25^{\circ}C$; thus, for rough calibration of a pH meter it is sufficient to stir some potassium hydrogen phthalate crystals into water, without weighing, and set the meter to read pH 4.0. A more-or-less saturated solution of sodium bicarbonate has pH 8.3. There are many purposes for which a pH reading to $\pm$ 0.1 is quite adequate, and a rough check saves laboratory time. Also, instrument companies sell concentrated buffer solutions which, when diluted according to instructions, give pH values valid to 0.01-0.02 unit. Here again, we should remember that, to a first approximation, the pH of a buffer solution does not change with dilution, and if high accuracy is not needed, the dilution of a buffer concentrate can be quite rough.

VII. pH-TITRATION CURVES

The experiments described in this chapter include the plotting and interpretation of titration curves, so we now present a brief review. The topic is treated in detail in many textbooks. Here we offer an approximate treatment which ignores activity coefficients but is adequate so long as the pH values are within the range 3-11, roughly, and the pK values within the range 4-10.

The ionization of a weak acid, HA, may be represented: $HA = H^+ + A^-$. The equilibrium constant of this reaction is called the ionization constant and given the symbol K_a. Its negative logarithm, by analogy with pH, is called $pK_a = -\log K_a$. The anion of the weak acid, A^-, is a base, because it attracts protons: $A^- + H_2O = HA + OH^-$. The ionization constant of A^- as a base is

$$K_b = \frac{[HA][OH^-]}{[A^-]} = \frac{K_w}{K_a} ,$$

K_w being the ionic product of water: $K_w = [H^+][OH^-]$. The concentration of water itself is not included in the equations because it is considered to be constant. $K_w = 1.0 \times 10^{-14}$ at 25^oC.

A very important weak acid is the ammonium ion, NH_4^+, whose ionization constant is

$$K_a = \frac{[NH_3][H^+]}{[NH_4^+]} = 6.0 \times 10^{-10} \text{ at } 25^oC.$$

Ammonia, NH_3, as a base has

$$K_b = \frac{[NH_4^+][OH^-]}{[NH_3]} = \frac{K_w}{K_a} = 1.7 \times 10^{-5}.$$

A dibasic acid like carbonic acid has two ionization constants, K_1 and K_2. Schematically, for an acid H_2A,

$$K_1 = \frac{[H^+][HA^-]}{[H_2A]} , \quad K_2 = \frac{[H^+][A^{2-}]}{[HA^-]}$$

Again, $pK_1 = -\log K_1$; $pK_2 = -\log K_2$.

Similar expressions are used for weak bases.

Figure 1-6 shows the titration curve of a weak acid whose pK_a is 1×10^{-5}, titrated with a strong base like sodium hydroxide. It has an in-

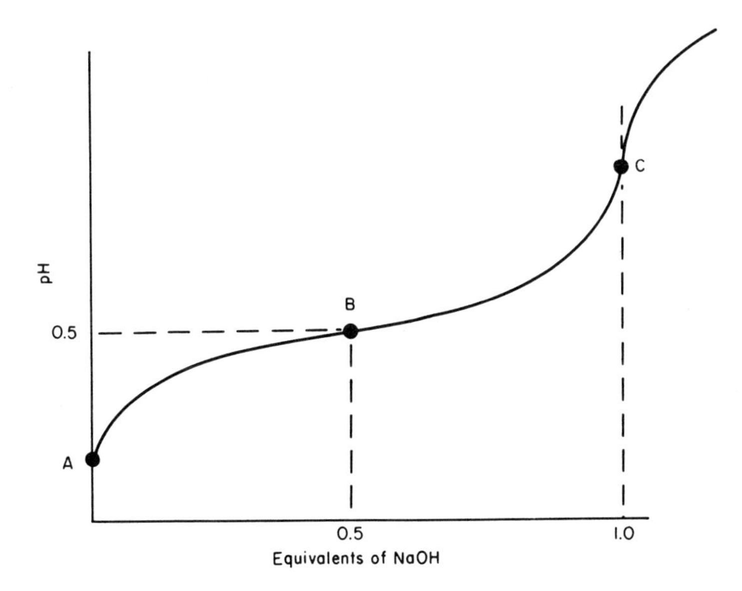

Fig. 1-6. Titration of a weak acid ($pK_a = 1 \times 10^{-5}$).

flection, a point of maximum slope, at "C." This coincides almost exactly with the point of chemical equivalence, where one mole of NaOH is added for every mole of HA.

At point A, the starting point,

$$[H^+] = [A^-] = \sqrt{[HA]\,K_a}, \quad pH = \frac{1}{2}(pK_a - \log a),$$

where a is the concentration of HA, or approximately the stoichiometric concentration of the acid present at the start.

At point B, the half-neutralization point,

$$[HA] = [A^-], \quad [H^+] = K_a, \quad pH = pK_a$$

Locating the half-neutralization point is a quick way to find the ionization constant of HA.

At point C, the equivalence point,

$$[HA] = [OH^-] = \sqrt{[A^-]\,K_b} \text{ or } pOH = \frac{1}{2}(pK_b - \log b),$$

where b is the concentration of the salt NaA. This comes about because the "principal equilibrium" is the hydrolysis of A^-: $A^- + H_2O = HA + OH^-$; HA and OH^- are formed together and their concentrations are equal. We recall that $K_b = K_w/K_a$. It may be more familiar, though more complicated, to write

$$[H^+] = \sqrt{\frac{K_a K_w}{b}} .$$

A weak dibasic acid, H_2A, is neutralized in two stages, forming first HA^-, then A^{2-}. An intermediate inflection point is seen, corresponding to HA^-, provided the two ionization constants are sufficiently far apart; K_1/K_2 must be at least 100. Figure 1-7 shows the titration curve of sodium carbonate, which will be obtained in Experiment 1-1. This is the titration curve of a bifunctional base, CO_3^{2-}, and it is like that of carbonic acid turned backward. Point B is the half-neutralization point for CO_3^{2-} to HCO_3^-; pH = pK_2 for carbonic acid, approximately, for the pH is high, the activity coefficient of the doubly charged ion CO_3^{2-} is low, and the errors are great. In addition, there is an error in the glass electrode reading itself. When the hydrogen ion concentration is very low and the concentration of sodium ions is high, some sodium ions are adsorbed on the glass, as well as hydrogen ions, and the pH reading is lower than it should be. This "sodium ion error" must be considered when using glass electrodes at high pH. At point A, Fig. 1-7, the errors are so great the actual pH reading has little meaning.

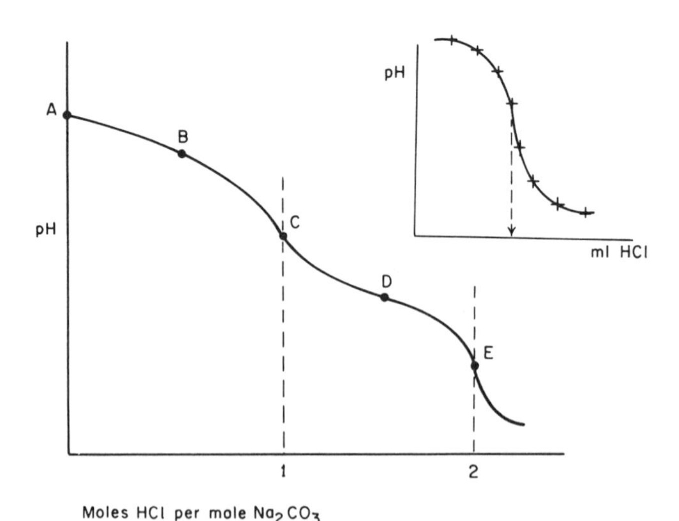

Fig. 1-7. Titration of sodium carbonate with HCl.

At point D, pH = pK_1 very nearly. At point C, the first inflection point, pH = $\frac{1}{2}(pK_1 + pK_2)$; $H^+ = \sqrt{K_1 K_2}$. This very important relation arises because the solution is really a solution of sodium bicarbonate, and bicarbonate ions react with themselves to produce H_2CO_3 and CO_3^{2-} in equal quantity: $HCO_3^- + HCO_3^- = H_2CO_3 + CO_3^{2-}$. It is evident that $K_1 K_2 = [H^+]^2$.

In Experiment 1-1 you are asked to use the titration curve to find the normality of the hydrochloric acid, assuming the sodium carbonate to be pure. You must find as precisely as possible the number of milliliters of hydrochloric acid required to reach one of the two inflection points. The second inflection, corresponding to formation of H_2CO_3, is much sharper than the first, and one may locate it very exactly by taking a number of readings close together and plotting a separate graph of the equivalence region on an expanded scale; see Fig. 1-7.

Experiment 1-2 requires the titration of a solution that contains both hydrochloric acid (strong) and phosphoric acid (weak). The titration curve of the mixture is indicated in Fig. 1-8. The inflection points corresponding to the ions $H_2PO_4^-$ and HPO_4^{2-} stand out well (points A and C), and the volume of standard base required to go from one to the other gives the quantity of phosphoric acid in the solution. However, we see no inflection to mark the neutralization of HCl, because the first ionization of phosphoric acid is too strong; it obscures any inflection. Also, we see no inflection to mark the formation of PO_4^{3-}, because the third stage of ionization of H_3PO_4 is too weak; the pH is too high.

The three ionization constants of H_3PO_4 can be deduced from Fig. 1-8 by using the points A, B, and C.

If an acid or base is very weak, the pH-titration curve shows no inflection, or such a poor one that one cannot locate the equivalence point with any accuracy. The reason is that the water itself is both a weak acid and a weak base, and the water competes with the weak acid or base being titrated. The solution to this difficulty is to get rid of the water. Very weak bases may be titrated in glacial acetic acid, using perchloric acid as the titrant; see Experiment 1-3. Very weak acids may be titrated in solvents like dimethylformamide or a benzene-methanol mixture, with a very strong base like tetramethylammonium hydroxide as the titrant.

The titration of acids and bases in nonaqueous solvents is very important in practice, especially in the organic chemical industry. Most of the solvents have low dielectric constants, and this increases the attractive force between ions of opposite charge. In methyl ethyl ketone, for example, sulfuric acid can be titrated as a dibasic acid that shows two distinct in-

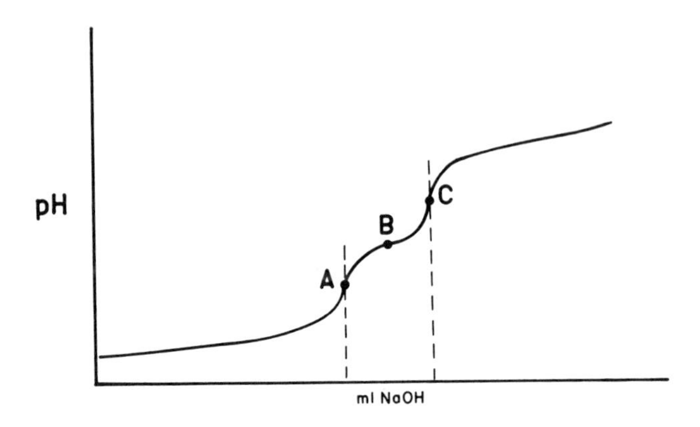

Fig. 1-8. Titration of a mixture of HCl and H_3PO_4.

flection points, because the attraction between H^+ and SO_4^{2-} is so great. Charged and uncharged acids, like NH_4^+ and C_6H_5OH, which have about the same strength in water, can be distinguished in a solvent of low dielectric constant, because the attractive force between $C_6H_5O^-$ and H^+ is increased considerably.

Let us now look at some of the ways to use a pH meter for other purposes than the measurement of pH. The pH meter is just a voltmeter, a device for measuring differences in electrical potential. It is easy to take out the glass electrode and replace it by another electrode (an adapter must be used; however, see Experiment 1-4), and we can substitute other electrodes for the calomel reference electrode if we wish.

VIII. THE SILVER ELECTRODE

Replacing the glass electrode by an electrode of metallic silver we can titrate chloride, bromide, iodide, and thiocyanate ions with silver nitrate (Experiment 1-4). The theory has been introduced already (pages 11-14). Chloride and the other anions mentioned form sparingly soluble salts with silver, and in the presence of solid silver chloride, for example, the silver ion concentration (which determines the potential of the electrode) is linked with the chloride ion concentration through the solubility product: $K_{sp} = [Ag^+][Cl^-]$. We can calculate the potential of the electrode in a solution of known chloride ion concentration by first calculating the concentration of silver ions, then using the standard potential of $Ag^+ + e = Ag$ and the Nernst equation to calculate the electrode potential. An easier way is to combine the silver ion standard potential with the solubility product to get a new standard potential, as follows:

$$Ag + e = Ag, \quad E^o = +0.7996 \text{ volt}$$

$$[Ag^+][Cl^-] = 1.7 \times 10^{-10}$$

$$AgCl + e = Ag + Cl^-, \quad E^o = +0.2223 \text{ volt}$$

The value 0.2223 volt is an experimental value determined with high accuracy. It is the potential of a silver electrode in contact with silver chloride and a hydrochloric acid solution of activity 1.000.

We can apply the Nernst equation directly to this potential; it is:

$$E^o = 0.2223 - \frac{2.303RT}{F} \log [Cl^-]$$

or, more correctly, a_{Cl^-}. The metal becomes more negative if the chloride ion concentration is raised; hence the minus sign before the log term.

In Experiment 1-4 we titrate chloride ions (or bromide, or thiocyanate) with silver nitrate and use the curve to calculate the solubility product of the precipitated salt. To do this calculation we need only remember that every time the silver ion concentration changes by a factor of 10, the electromotive force changes by 0.059 volt at 25°C. The actual value of the electromotive force does not matter and need not be known. Consider the following:

Problem. Suppose 100.0 ml of a solution of potassium thiocyanate, KCNS, is titrated with 0.100 M silver nitrate, and a curve is obtained like that of Fig. 1-9. Between the points P and Q the EMF changes by 350 mV (0.350 V). What is the solubility product of AgCNS?

Equivalence is reached where the curve is steepest, at 10.0 ml. At 11.0 ml $AgNO_3$ (point Q) there is 1.0 ml $AgNO_3$ in excess; the total volume is 111 ml; $[Ag^+] = 0.100 \times 1.0/111 = 9.0 \times 10^{-4}$ M. By the same reasoning, at 9.0 ml $AgNO_3$ (point P) thiocyanate is in excess, and $[CNS^-] = 0.100 \times 1.0/109 = 9.1 \times 10^{-4}$ M. However, we can calculate $[Ag^+]$ at point P, too, using the Nernst equation. The potential difference between points P and Q is 350 mV; this corresponds to a factor of 350/59 = 5.93 powers of 10; the silver ion concentration at point P is thus 9.0×10^{-4}

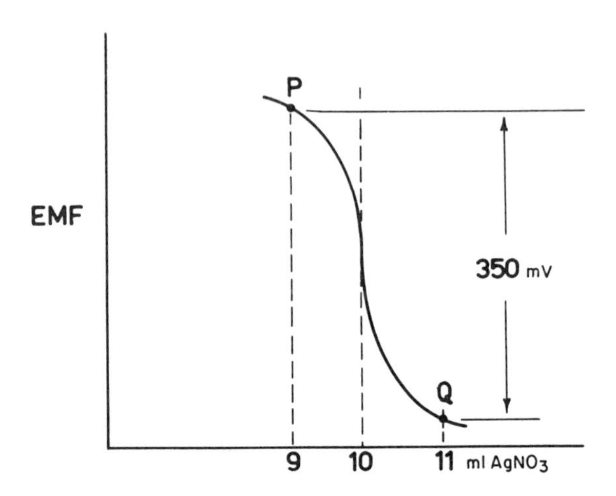

Fig. 1-9. Titration of CNS^- with Ag^+ (end-point region).

divided by $10^{5.93} = 1.05 \times 10^{-9}$. Multiplying the silver concentration by the thiocyanate concentration at point P we get the solubility product: $K_{sp} = 1.05 \times 10^{-9} \times 9.0 \times 10^{-4} = \underline{1.0 \times 10^{-12}}$, rounding off the number to two significant figures.

Mixtures of two ions, like chloride and bromide, can be titrated (see Fig. 1-10), but the problem arises that chloride ions are strongly absorbed by the freshly precipitated silver bromide and separate out before the inflection point is reached, making it look as if the solution contains more bromide than it actually does. The error can be reduced by adding, at the start of the titration, a large quantity of a salt, such as calcium nitrate, that coagulates the silver bromide as it forms and prevents it from absorbing so much chloride. Even so, this method of simultaneous titration is only practical when the concentrations of chloride and bromide are not very different.

A possible source of error in titrating halides with silver nitrate, using a pH meter, arises if we use a calomel electrode as reference. Inside this electrode is a concentrated solution of potassium chloride which is bound to leak out into the solution during the titration. Is this leakage sufficient to cause an error in the titration? This question can only be answered by experiment, and we invite the reader to devise a simple experiment to find out how bad the leakage is. It will differ from one calomel electrode to another. Most commercial electrodes have such a small hole at the liquid junction that the leakage of potassium chloride is negligible, provided the

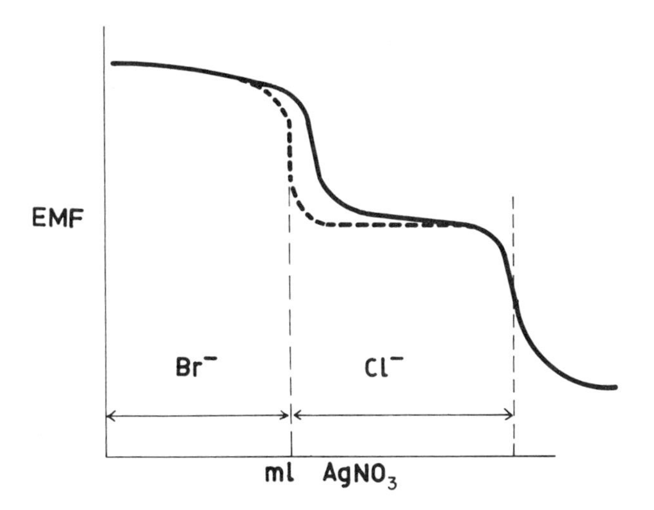

Fig. 1-10. Mixture of Br^- and Cl^- titrated with $AgNO_3$. Full curve: without added salt; dashed curve: with $Ca(NO_3)_2$ added.

titration is done quickly and the solution being titrated is not too dilute. There is an easy way out of this problem, however. Instead of using the calomel electrode for reference, use the glass electrode; take out the calomel electrode and put the silver electrode in its place, then, before titrating, add enough dilute nitric acid to the solution that the hydrogen ion concentration does not change significantly and abruptly during the titration. The curve of the electromotive force versus volume of silver nitrate will of course go in the opposite direction to what it would if calomel were taken as reference.

IX. ION-SELECTIVE ELECTRODES: THE FLUORIDE ION ELECTRODE

Since 1966 a new class of electrodes has been developed, a type that bears some resemblance to the glass electrode. The glass electrode is a membrane that is selectively permeable to hydrogen ions. Membranes are now available that are selectively permeable to other ions and can be used to measure the concentrations of these ions through measurements of electromotive force.

The most successful of these electrodes measures fluoride ions. Its construction is shown in Fig. 1-11. The membrane is a disk cut from a single crystal of lanthanum fluoride. The fluoride ions in this crystal are

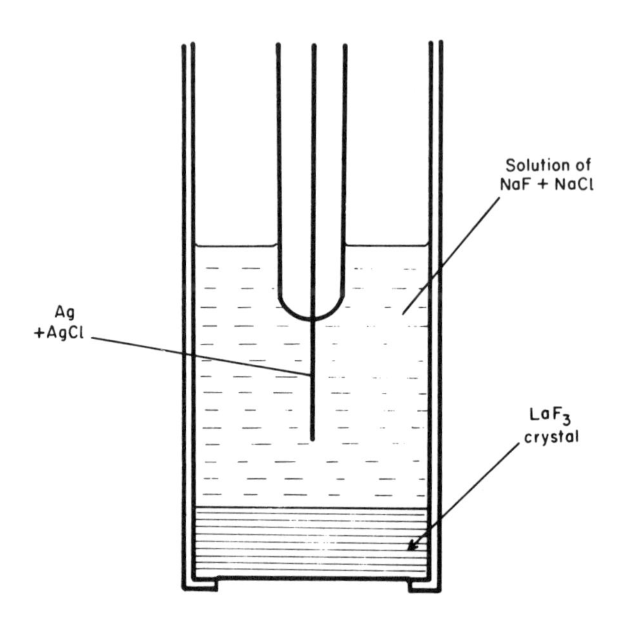

Fig. 1-11. Fluoride ion electrode.

relatively mobile, and fluoride ions can enter on one side of the crystal and come out of the other. The only other ion that can get into the crystal is the hydroxyl ion, OH^-; it has almost the same radius as the fluoride ion and, as is well known, replaces fluoride in minerals. So the lanthanum fluoride electrode cannot be used at high pH. It cannot be used at very low pH either, for hydrogen ions and fluoride ions combine to form hydrofluoric acid, a weak acid with $K_a = 3.5 \times 10^{-4}$.

The way the electrode acts is shown in Fig. 1-12. The tube holding the membrane is filled with a sodium fluoride solution that is relatively concentrated, say 0.1 molar. The solution also contains chloride, so as to give a stable potential at the internal reference electrode, which is a silver wire coated with silver chloride. When the electrode is placed in a more dilute fluoride solution, fluoride ions pass from the inside of the membrane to the outside, building up a negative charge on the outside. The charge displacement is opposite to what happens in the glass electrode; see Fig. 1-12. Again the potential across the membrane depends on the ratio of the two fluoride concentrations. Ignoring activity effects,

$$E_{\text{lanthanum fluoride}} = 0.059 \log \frac{[F^-]_{\text{inside}}}{[F^-]_{\text{outside}}}$$

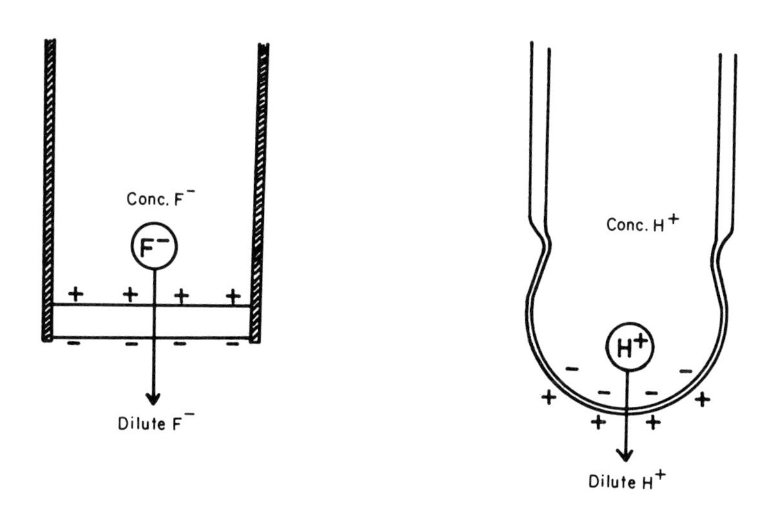

Fig. 1-12. Comparison of polarity, LaF_3 and glass electrodes.

This is like the equation for the glass electrode (page 17) except that the side where the fluoride concentration is smaller is the side that is negative.

To use the lanthanum fluoride electrode in a pH meter one takes out the glass electrode and replaces it by the lanthanum fluoride electrode, keeping the calomel reference electrode in place. The electromotive force of the cell thus formed depends on the reference electrode as well as the electromotive force across the lanthanum fluoride crystal itself, and the cell must be calibrated with solutions of known fluoride ion concentration, just as the pH meter is calibrated with known pH buffers. Such calibration is illustrated in Experiment 1-5. In using the fluoride electrode it is very important to distinguish between concentrations and activities. This distinction is less important in glass-electrode measurements, because we use the glass electrode to measure pH, which is the hydrogen ion activity. Using the lanthanum fluoride electrode and others like it, we generally want to know the concentration. The correct equation for the potential across the lanthanum fluoride membrane is

$$E_{LaF_3} = 0.059 \log \frac{a_{F^-}\ \text{outside}}{a_{F^-}\ \text{inside}} = 0.059 \log\left(\begin{matrix}\text{concentration}\\ \text{ratio}\end{matrix}\right) \times \frac{\gamma_{out}}{\gamma_{in}}$$

where γ is the activity coefficient. One can calculate γ approximately from this equation (valid for 25°C):

$$\log \gamma = -\frac{0.5z^2\sqrt{\mu}}{1 + \sqrt{\mu}}$$

where z is the valence of the ion and μ is the ionic strength $= \frac{1}{2}\sum_i c_i z_i^2$. c_i is the molar concentration and z_i the valence, or charge, of ions of the kind i. We note that $\mu = c$ for salts of singly charged ions like sodium chloride.

Two solutions that have the same concentration of fluoride ions but different total salt concentrations will give different potential readings with the lanthanum fluoride electrode. The higher the total salt concentration, the lower the fluoride ion concentration will appear to be. To overcome this effect in routine analysis one often adds to the sample an equal volume of a concentrated solution containing sodium citrate and sodium chloride, plus acetic acid to give a pH of 5. This "total ionic strength adjustment buffer" does three things: it provides a uniform ionic strength, hence constant activity coefficients; it controls the pH to avoid interferences by hydrogen or hydroxyl ions; and the citrate ions form complexes with any iron(III) or aluminum present, which would otherwise combine with some of the fluoride ions.

The accuracy with which concentrations can be found by direct readings of electrical potential is only moderate. Even if there were no uncertainties in the activity coefficients, a small error in the electromotive force measurement causes an appreciable uncertainty in the concentration. Thus, a difference of 2 millivolts in electromotive force means a difference in the logarithm of the concentration of $2/59 = 0.03$ unit; this corresponds to a factor 1.07 or an error of 7 percent. (This error is doubled if a doubly charged ion, like Ca^{2+}, is being determined.) Highest accuracy is obtained by using specific-ion electrodes to follow the course of a titration. Thus, fluoride ions can be titrated in 60 percent alcohol with a solution of lanthanum or thorium nitrate, using the fluoride ion electrode to plot a titration curve in the same way that the silver electrode was used in the last section.

Ion-selective electrodes are available for several ions, including Cl, Br, I, S^{2-}, CN, NO_3, Ca, and Cu. They are of different types, the calcium-selective and nitrate-selective electrodes using a layer of a "liquid ion exchanger" as the selective membrane. Few of these are as selective as the fluoride electrode. The bromide ion electrode, for example, will not measure the bromide content of sea water, because there is 500 times as much chloride as bromide (on a scale of chemical equivalents) in sea water, and the chloride interferes with the bromide response.

X. THE PLATINUM ELECTRODE

A platinum electrode is used in conjunction with a calomel reference electrode to measure the oxidation-reduction potential of solutions and to perform oxidation-reduction titrations potentiometrically. The practical importance is not great, but an example of the use of a platinum electrode is given in Experiment 1-6.

XI. POLARIZED PLATINUM ELECTRODE PAIRS; "DEAD-STOP" TITRATIONS

Another use for a pH meter is illustrated by Experiment 1-7. Most meters have a connection for a "polarizing current," and this connection permits one to pass a very small direct current, about 10 microamperes, between two electrodes placed in a solution. The electrodes can be two platinum wires or small pieces of platinum foil mounted in glass. The dial of the pH meter shows the drop in potential across these two electrodes. If current passes easily between them the potential drop will be small. If there is any obstacle to electron transfer at one of the electrodes, however, the potential will rise. In certain titrations the end point is marked by a sudden rise in potential or a sudden fall. These are sometimes called "dead-stop" titrations."

An example is the titration of arsenite by iodine in neutral solution. The reaction is

$$I_2 + H_2AsO_3^- + H_2O = HAsO_4^{2-} + 3H^+ + 2I^-$$

The reduction and oxidation of arsenic by an electric current at platinum electrodes are slow and difficult because of what is called an "overvoltage." Reduction of iodine and oxidation of iodide ions, on the other hand, are fast and easy. When iodine and iodide ions are present together, iodine is reduced to iodide ions at one electrode while iodide ions are oxidized to iodine at the other; no net chemical work is done, current passes easily, and the potential drop is small. When arsenite is in excess, however, and there are iodide ions but no iodine, current passes with great difficulty and the potential drop is large.

A pair of polarized platinum electrodes is used to detect the end point in the Karl Fischer titration of water, which is a very important titration. Water is titrated in methyl alcohol solution by a reagent that contains iodine, sulfur dioxide, and pyridine. The reaction (somewhat simplified) is

$$SO_2 + I_2 + H_2O = SO_3 + 2HI$$

then,

$$SO_3 + CH_3OH + C_5H_5N = CH_3SO_4^-C_5H_5NH^+$$

$$2HI + 2C_5H_5N = 2C_5H_5NH^+ I^-$$

As long as water is in excess the electrodes are "polarized" and the potential between them is high. As soon as iodine is in excess, the iodine-iodide combination "depolarizes" the electrodes, current passes easily, and the potential drops to nearly zero.

XII. THE MEASUREMENT OF ELECTRICAL POTENTIAL AND ELECTROMOTIVE FORCE

We have left to the last a description of how electrical potential differences are actually measured. The common "voltmeter" that is used to test batteries cannot be used, for this is really an ammeter that draws a small current through a high resistance. In measuring the electromotive force of cells we must draw as little current as possible and approach, as closely as we can, the ideal of thermodynamic reversibility. The same is true if we want to measure the potential drop across a high resistance.

The most accurate way to measure the electromotive force of a cell is with a potentiometer. Figure 1-13 shows the basic circuit of a potentiometer. This is a very important circuit that every chemist should know and understand. A battery A sends a steady current through a wire PR which is highly uniform. The voltage drops uniformly from one end of this wire to the other, and the voltage drop is proportional to the distance along the wire. The rate of voltage drop can be adjusted by the variable resis-

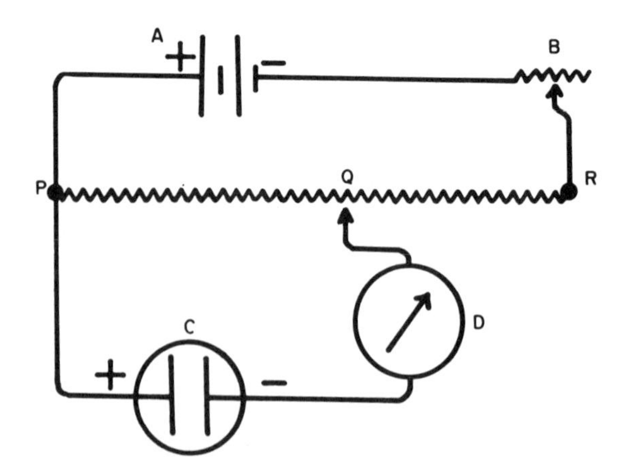

Fig. 1-13. Basic circuit of a potentiometer.

tance B. The cell to be measured is shown as C; the positive pole of this cell and the positive pole of the battery A are both connected to the same end of the slide wire PR. Q is a sliding contact that can be moved at will along the length of PR. The object is to find a position for Q such that no current passes through the sensitive galvanometer D. When this is the case, the electromotive force of the cell C is exactly balanced by the potential drop along PQ, and this is a fraction PQ/PR of the total potential drop along PR. The position of Q is read from a scale or dial. To find exactly how many scale divisions correspond to how many volts one puts a standard cell (not shown) into the circuit in place of C, and finds the point at which the sliding contact balances.

The galvanometer D is purely a null instrument. The current that corresponds to each galvanometer division need not be known, and it need not be uniform along the scale. The only requirement is that D be sensitive. The accuracy of the potentiometer is determined by the uniformity of the slide wire PR and the accuracy of the standard cell used for calibration.

Even a very sensitive galvanometer, however, will barely detect the tiny current that passes through a glass electrode or a lanthanum fluoride disk. The resistance of these membranes may be ten million ohms or more. This is done by connecting the negative pole of C to the grid of an electronic vacuum tube. The potential of C controls the flow of electrons across this tube from its cathode to its plate. The more positive is the grid, the more electrons can flow, and vice versa. The plate current is passed through the galvanometer and is detectable even though the sensitivity of the latter is quite moderate.

Vacuum tube amplification can be used in two ways. One way, and the more accurate, is to retain the basic potentiometer circuit and use the galvanometer (carrying the amplified current) as a null instrument. This principle is used in pH meters like the Beckman Model G. The other method is to do away with the potentiometer circuit and use the electromotive force of the test cell to control the value of the current flowing across a series of vacuum tubes, which is then measured by a milliammeter having a linear scale. The current is directly proportional to the cell electromotive force. This principle is used in direct-reading pH meters, which are by far the more popular. They are, of course, easier to use than the null-balance potentiometer type and less subject to drift. The drawback for precise work is that the scale reading may not be strictly proportional to the electromotive force.

All glass electrodes come with a shielded lead. This is necessary because the resistance is so high and the currents that pass through the electrode are very minute. In the socket where the glass electrode lead enters the pH meter, the outside shield is grounded. The socket is designed for glass electrodes, and when another electrode of low resistance, like a silver electrode, is used an adapter must be used to fit into the socket.

XIII. OPERATION OF A DIRECT-READING pH METER

The directions that follow apply to the Beckman Zeromatic pH meter. Other models have their controls arranged differently, but basically all pH meters have much in common. The manuals supplied by the instrument makers should always be consulted.

1. Plugging in. There is no "on-off" switch; when the instrument is plugged into the electric outlet, current passes immediately and the tubes are warmed up. As is the case with most electronic equipment, it is a good idea to leave the instrument plugged in and warmed up, unless it will be left unused for several days. In humid climates it is specially desirable to leave equipment warmed up.

2. "Standby" and "Read." The Zeromatic has two buttons on the right-hand side of the case, in front, called "Standby" and "Read." When "Standby" is depressed the electronics are warmed but the electrodes are not in circuit. The instrument should always be on "Standby" when electrodes are being changed or taken out of the solution. The "Read" button is depressed only when one is actually ready to take a reading. Other instruments have switches to be turned instead of buttons to be pressed, but the function is the same.

3. Temperature compensation. On the right-hand side of the case and above the other controls is a dial which should be set at the temperature of the sample solution, within a degree or two. This knob adjusts the number of millivolts that will show as one unit on the pH scale; 59.15 millivolts at 25°C, 62.13 millivolts at 40°C, etc. (see page 20). Like most direct-reading pH meters the Zeromatic has provision for connecting an automatic temperature sensor that can be placed in the test solution, like a thermometer, and automatically keeps the instrument in adjustment when the sample temperature changes. If this control is used, press the button called "Auto" under "pH Temp. Control" at the bottom left of the case. If it is not used, press the button "Manual." If the automatic temperature control switch is turned on by mistake, it will be impossible to adjust or standardize the meter.

4. Millivolt scale. In the middle of the row of buttons (bottom, front) are two labeled "Millivolts," ±700 and 1400. These are used when readings in millivolts are desired. When one of these buttons is depressed the "temperature compensation" buttons are automatically released, for the temperature adjustment now has no meaning.

The millivolt scale may be used when silver or platinum or ion-selective electrodes are substituted for the glass, but there is no obligation to use it; one can stay on the pH scale if one wishes, remembering that this is a millivolt scale too, but with one large division = 59.15 millivolts (25°C) instead of 100 millivolts.

5. "Standardize." (In the older Zeromatic models this knob is labeled "Asymm.") With this control the needle of the meter can be moved back and forth over the scale to read anything one wants it to read. It is used in pH measurements to calibrate the scale, to make the meter read the correct pH of a standard buffer. The meter should always be checked against a standard buffer, like 0.05 molar potassium hydrogen phthalate, before the pH of an unknown solution is read, and it should be rechecked if there is any reason to think the "Standardize" control has been moved inadvertently.

In titrations with electrodes other than glass, such as the silver electrode, it is generally the changes in the meter reading, rather than the reading itself, that are of interest. The "Standardize" control may be set before the titration is begun, to make the needle come in a convenient position on the scale and to prevent its going off the scale as the titration proceeds. Once it is set, of course, it should not be moved during the titration.

XIV. THE CARE OF GLASS AND CALOMEL ELECTRODES

Glass electrodes are sealed and can be stored dry without special precautions, except to avoid scratching the sensitive glass bulb. A dry glass electrode should be soaked in water for at least half an hour before use.

Calomel electrodes are not sealed. They must be kept filled with saturated potassium chloride solution. When the electrodes are in use the level of solution inside the electrode vessel should be higher than the level of the test solution, and the upper rubber sleeve should be pushed up or down to leave the filling hole open, so that potassium chloride will slowly flow out through the fiber junction and keep it clean. Before putting an electrode into storage the upper hole should be closed and a rubber cap should be placed over the lower end. The electrode must not be allowed to dry out. Once the potassium chloride solution has dried, the electrode is spoiled and it is almost impossible to restore it.

EXPERIMENTS

Experiment 1-1

Potentiometric Titration of Sodium Carbonate; Standardization of Hydrochloric Acid and Sodium Hydroxide

Materials Needed. Sodium carbonate, anhydrous; concentrated hydrochloric acid; sodium hydroxide, solid or (preferably) a solution 50% by weight, stored in a polyethylene bottle.

Magnetic stirrer; pH meter; buret; beakers (125 - 250 ml); pipet (25 ml); graduated cylinders. Standard buffer solution or solid potassium hydrogen phthalate.

First prepare about 1 l. of hydrochloric acid of concentration about 0.2 M. Pour 17 ml concentrated (12 M) hydrochloric acid into 1 l. of distilled water in a bottle with a glass or rubber stopper, mix well.

Place some anhydrous sodium carbonate in a weighing bottle and dry in an oven at 180° - 200°C. If no suitable oven is available, the sodium carbonate may be dried by heating carefully over a flame in a large test tube, wiping with a filter paper the drops of water that condense in the upper part of the tube, and continuing to heat carefully until no more water appears. Cool, place in a bottle, and keep tightly closed. Sodium carbonate dried in this way is liable to contain 1 or 2% of sodium oxide, Na_2O, or (more likely) peroxide, Na_2O_2. This impurity will be detected in the potentiometric titration if it is present. The best temperature for drying sodium carbonate is 270° - 300°C.

Weigh out accurately about 200 mg of sodium carbonate and dissolve it in about 50 ml of distilled water. There should be enough water to cover the ends of the electrodes of the pH meter, but no more.

Before starting the titration, standardize the pH meter with a known buffer. This may be the buffer solution supplied by the manufacturer, diluted according to instructions, or a solution of potassium acid phthalate; weigh 0.50 g potassium acid phthalate and dissolve in 50 ml distilled water; this gives a solution 0.05 M whose pH is 4.01. In using the pH meter, always be sure to rinse the electrodes with distilled water and wipe them gently with soft tissue paper when changing from one solution to another; and, if the glass electrode is dry when the experiment is started, let it stand in water for about 15-30 min before making any measurements.

When all is ready and the pH meter is standardized, set the beaker of sodium carbonate solution on the magnetic stirrer, put the stirrer bar in the beaker, and set the beaker in place by the pH meter with the electrodes dipping in the solution. See that the stirrer bar does not strike the electrodes. Fill the buret with 0.2 M hydrochloric acid and set it in place over the beaker. Press the button "Read" on the pH meter. Leave this button depressed until the titration is finished.

Note the pH reading and the initial reading of the buret (which may or may not be zero). Add the acid, at first, about 2 ml at a time, reading the pH after each addition. When the pH starts to change more rapidly, around pH 9.5 and below, take readings closer together, every 0.2 ml, say. The idea is to have readings close together in the equivalence regions, more widely spaced in the intermediate or "buffer" regions; see Fig. 1-7. One almost always overshoots the equivalence point the first time one does a potentiometric titration. One should count on doing at least two duplicate titrations, the first exploratory, the second with the equivalence regions well defined. One should be able to locate the second equivalence point (formation of H_2CO_3) to a precision of 0.02 ml or better.

From your curves, calculate the ionization constants of carbonic acid and the normality of the hydrochloric acid. See whether or not the volume of acid required to reach the second equivalence point is exactly double that required to reach the first. If it is not, calculate as accurately as you can the proportion of the impurity, which may be Na_2O_2, Na_2O, or possibly $NaHCO_3$.

This is done as follows. Suppose the carbonate sample contains sodium bicarbonate as an impurity. Instead of looking like Fig. 1-7 the potentiometric titration curve will be like Fig. 1-14. It will take somewhat more hydrochloric acid to go from the first inflection to the second than it did to reach the first inflection point. The bicarbonate ions that were in the sample at the start will not combine with acid until the first equivalence point is passed.

The proportion of bicarbonate and the acid normality can be calculated in this way. Let the volume of acid to reach the first equivalence (see Fig. 1-14) be v ml and that to reach the second (2v + w) ml. The ratio (moles $NaHCO_3$:moles Na_2CO_3) = w/v; the weight ratio is 84w/106v, 84 and 106 being the molecular weights of $NaHCO_3$ and Na_2CO_3, respectively. The total weight of the sample, which is known, is (84w + 106v) x N mg, where N is the normality of the acid. From the sample weight, v, and w, one can calculate N and the proportion of $NaHCO_3$ by weight.

This calculation assumes that the distilled water taken to dissolve the sample contains no carbon dioxide and that no carbon dioxide is absorbed from the atmosphere. It also assumes that the sample was dry. The best way to standardize hydrochloric acid is of course to make sure the sodium carbonate is pure and dried at the proper temperature and to use only the second equivalence point in the calculation. Then it will not matter if some carbon dioxide is absorbed from the atmosphere during the titration.

If the potentiometric titration shows that more hydrochloric acid is used to reach the first inflection point than to go from the first to the second, that is, if w is negative, we can perform a similar calculation assuming the impurity to be sodium oxide, Na_2O. However, a simple calculation from a table of thermodynamic data shows that if we heat sodium carbonate in the presence of air, the product formed at 300° - 400^oC will be Na_2O_2, not Na_2O. We can easily tell from the titration curve whether the sodium carbonate has been overheated or not, but to make a quantitative calculation of the amount of impurity we need more information.

This discussion has been long but will serve to show that the analytical chemist must keep his wits about him and understand what his measurements mean.

Preparation and Standardization of 0.2 M Sodium Hydroxide. Measure about 12 ml of clear 50% sodium hydroxide solution (the white sediment is sodium carbonate and should be left behind) and add to it 1 l. of distilled water. Mix well, and store in a polyethylene bottle. One can use a glass bottle with a rubber stopper, but sodium hydroxide slowly attacks glass, and glass vessels are not good for long storage (more than a week or two).

Pipet 25.00 ml of your standardized hydrochloric acid into a beaker, dilute with distilled water if desired, and titrate it potentiometrically with your sodium hydroxide solution. It is not necessary to plot the entire curve; a few points close to the equivalence point will suffice. Repeat the titration at least once, and calculate the normality of the sodium hydroxide.

The sodium hydroxide will be used in the next experiment. It can also be standardized (and more accurately standardized) by potentiometric titration against potassium hydrogen phthalate.

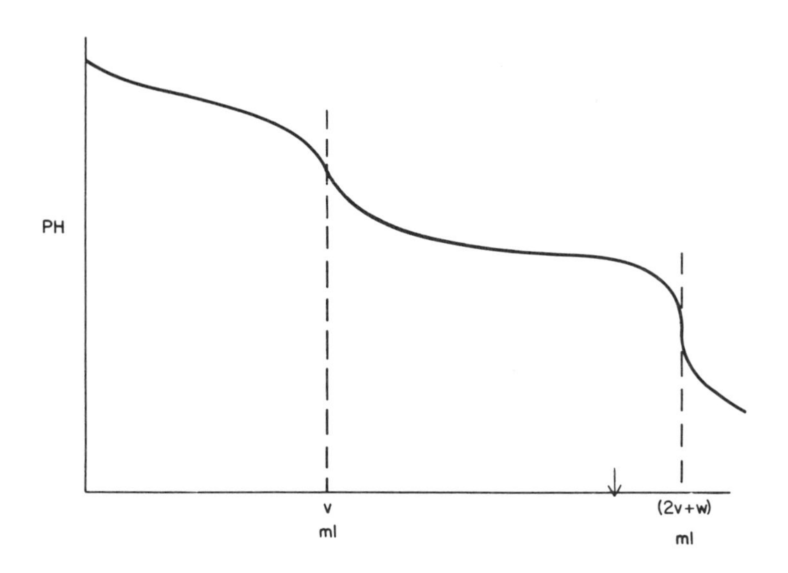

Fig. 1-14. Titration of a mixture of Na_2CO_3 and $NaHCO_3$.

Experiment 1-2

Determination of Phosphate in Baking Powder

Materials Needed. Standard sodium hydroxide solution, 0.2 M; hydrochloric acid, concentrated or 6 M; cation-exchange resin, Dowex-50W or equivalent, 50-100 mesh; tube for holding resin, see Fig. 1-15.

Magnetic stirrer; pH meter; buret; etc., as in Experiment 1-1.

Outline of the Problem. Baking powders contain sodium bicarbonate plus a solid acid, which in some products is an acid calcium phosphate, $CaHPO_4$. "Royal" baking powder is such a product. It also contains starch to keep the powder free-flowing. The phosphate cannot be titrated directly; it must first be separated from the accompanying calcium ions, and this is done by ion exchange.

Weigh accurately a sample of baking powder of about 0.8-1.0 g as it comes from the container. Place it in a small beaker, add about 10 ml water, then enough hydrochloric acid to convert all the bicarbonate to carbon dioxide and all the phosphate to phosphoric acid. The pH should be less

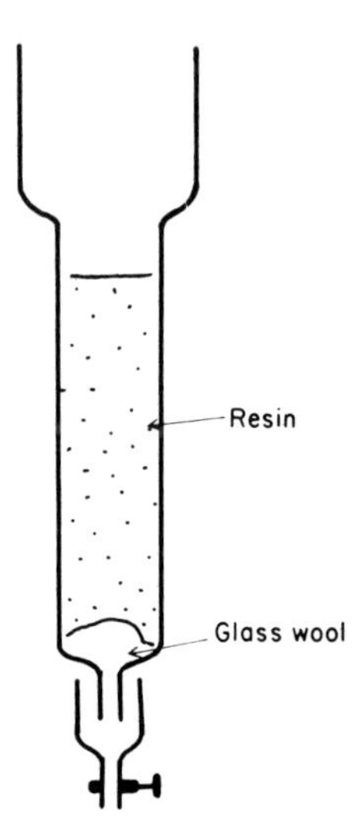

Fig. 1-15. Tube for holding resin.

than 2; confirm with test paper. Do not add more acid than necessary; 0.5 ml concentrated hydrochloric acid should be sufficient. Filter to remove the starch; wash the filter paper with a little water, but avoid diluting the solution unnecessarily. Boil the filtered solution for a minute or two to remove carbon dioxide.

Prepare a column of cation-exchange resin as shown in Fig. 1-15. The resin bed should be about 10 cm long and 1 cm wide, and it should be free from air bubbles. Wash it with 1-2 M hydrochloric acid to ensure its being in the hydrogen ion form, then wash with about 25 ml distilled water.

Pass the sample solution through the column at about 2 drops/sec or less, and collect the solution in a 250-ml beaker as it flows through the resin. Wash the resin with 25 - 50 ml water, adding a few milliliters at a time and letting the liquid settle down to the level of the top of the bed before passing more. When the water flowing out of the column is no longer acid to test paper (that is, pH is more than 4) the washing is complete.

Place the beaker containing the sample solution and washings on the magnetic stirrer and titrate potentiometrically with 0.2 M sodium hydroxide, locating carefully the inflections that come near pH 4.5 and pH 8.5. The volume of sodium hydroxide needed to reach the first inflection point is not important; this volume includes the base needed to neutralize the excess hydrochloric acid. The significant volume is that between the two inflection points, which gives the amount of phosphoric acid in the solution. Plot the curve between the two inflection points with care, and from it calculate (a) the proportion of phosphate, as PO_4, in the baking powder by weight; (b) the three ionization constants of phosphoric acid.

Do not discard the ion-exchange resin; the resin can be used over and over again. Wash the column with hydrochloric acid to displace the accumulated calcium ions, then with water. If air bubbles accumulate in the column or the column gets clogged with starch powder, pour the resin into a beaker, wash it by decantation, and pour it back into the column.

Instead of baking powder a phosphate fertilizer can be used as the "unknown" in this experiment. The fertilizer will probably contain iron as well as calcium, and iron, too, will be removed by the resin.

Experiment 1-3

Titration of an Amino Acid in Water and Glacial Acetic Acid

Materials Needed. Standard hydrochloric acid, 0.2 *M* or 0.1 *M*; perchloric acid, 72%; glacial acetic acid; potassium hydrogen phthalate; an amino acid, preferably beta-alanine; crystal violet or methyl violet, 0.5% solution in alcohol; pH meter; magnetic stirrer; buret; etc. A combination glass-calomel electrode is preferable to separate glass and calomel electrodes because of the smaller resistance offered by the solution.

Aims of the Experiment. Amino acids generally are weak acids and weak bases at the same time, and are so weak that it is impossible to titrate them in a solution in water (unless one blocks the amino groups by adding formaldehyde). They may, however, be titrated potentiometrically as bases in glacial acetic acid.

This experiment illustrates acid-base titrations in nonaqueous solvents, and it also shows that the ionization constant of a very weak acid or base can be found from a potentiometric titration even though there is no inflection to mark the equivalence point. The best amino acid to use in this experiment is beta-alanine, $H_2N \cdot CH_2CH_2COOH$, because this substance is freely soluble in water and most amino acids are only slightly soluble.

(a) Aqueous Titration. Weigh accurately about 2 mmol (2 x 89 = 178 mg) of beta-alanine and dissolve in a convenient volume of water, say 50 ml.

Titrate it potentiometrically with hydrochloric acid, carefully noting the initial pH and taking measurements at close intervals near the start of the titration. Continue until equivalence is passed, but remember that there is no inflection point at equivalence. Check the calibration of the pH meter before and after titration with a standard buffer such as 0.05 M potassium hydrogen phthalate (pH = 4.01 at 25°C.) Plot the curve, and from it determine the two ionization constants of the amino acid.

It is convenient to express the ionization constants as those of the protonated form of the amino acid, $HOOC \cdot CH_2CH_2NH_3^+$, considered as a dibasic acid. Writing the formula H_2A^+ for short, these constants are

$$K_1 = \frac{[H^+][HA]}{[H_2A^+]}, \quad K_2 = \frac{[H^+][A^-]}{[HA]}$$

When hydrochloric acid is added in our titration the reaction is $HA + H^+ = H_2A^+$. This reaction is reversible, but we can say approximately that (millimoles H_2A^+) = (millimoles HCl added) as long as $[H^+]$ is small compared with [HA] and $[H_2A^+]$. The number of millimoles of HCl is known from the volume and the normality; the total number of millimoles of HA and H_2A^+ is known from the weight of amino acid taken.

Construct a table such as the following:

Titration of Amino Acid

Amino acid taken = ... mg = X millimoles
Normality of HCl = ...

Ml HCl added	MMol HCl = Y	Ratio (X - Y)/Y	pH	$[H^+]$	$\frac{[H^+](X - Y)}{Y}$

The numbers in the last column are approximately the ionization constant, K_1, and will be very close to that constant between 0.1 and 0.5 mmol HCl per mole of amino acid. In your written account, explain the deviations that occur when Y is very small and when it is large.

To find K_2, use the best value of K_1 found from the table, together with the pH of the amino acid solution found originally, before any hydrochloric acid is added. This pH equals $\frac{1}{2}(pK_1 + pK_2)$.

(b) Titration in Glacial Acetic Acid. First prepare a 0.1 N solution of perchloric acid in glacial acetic acid by adding about 10 ml 72% perchloric acid to 1 l. of acetic acid. The small amount of water introduced with the perchloric acid does not affect the titration, but be careful not to get any more water into the solutions than you can help. Bottles and beakers should be dry; the electrodes of the pH meter should be wiped dry with tissue paper before placing them in the acetic acid.

Standardize the 0.1 N perchloric acid solution by weighing 1-2 mmol of potassium hydrogen phthalate and dissolving it in some 50 ml of glacial acetic acid. It dissolves faster if the acid is warmed, but the fumes of acetic acid are irritating, so heat the acid in the hood. Then cool the solution to about 30° - $40^{o}C$, and place the beaker on a magnetic stirrer and set the electrodes in it. Use a combination glass-calomel electrode if one is available. If the atmospheric humidity is high, cover the beaker with polyethylene film or paraffin-coated paper with holes cut to admit the electrodes and the buret tip. Before starting the titration, set the standardization control of the pH meter so that the needle reads high on the pH scale. The absolute reading is unimportant and has little or no relation to the aqueous pH scale; however, the pH reading falls considerably when the end point is reached.

In this titration potassium hydrogen phthalate acts as a base: $C_8H_5O_4^- + H^+ = C_8H_6O_4$. The reaction is helped by the fact that the other product, potassium perchlorate, is insoluble in acetic acid and precipitates out. The potential change at equivalence is extremely sharp, and it is easy to overrun the end point.

When the perchloric acid is standardized, titrate a weighed sample of the amino acid, plotting the curve carefully and comparing it with the titration curve in water. Observe the response of an indicator by adding a drop or two of a solution of crystal violet or methyl violet and noting how the color changes as the titration proceeds. Methyl violet shows two color changes.

The titration of other weak bases, such as pyridine or aniline, may be studied if desired.

Experiment 1-4

Titration of Chloride and Bromide Ions with the Silver Electrode

Materials Needed. Silver nitrate solution, about 0.1 M; dilute nitric acid; pure potassium chloride and potassium bromide; calcium or aluminum nitrate; pH meter; silver, glass, and calomel electrodes; stirrer.

Aims of Experiment. To use the pH meter for silver halide titrations, to measure solubility products, and to study the titration of mixtures.

Remove the glass electrode from the pH meter and replace it by an adapter that fits into the glass electrode socket and allows one to connect other electrodes. Connect a silver electrode to the adapter. If a commercial silver electrode is not available, use a short length of silver wire, completing the connection with a plain copper wire and an alligator clip. Clean the silver surface with fine sandpaper and rinse with water before using. Leave the calomel reference electrode in place, but before immersing this in the test solution, rinse it with water to wash it free from adhering potassium chloride.

Weigh accurately about 100 mg potassium chloride, or 150 mg potassium bromide, or the equivalent amounts of the sodium salts; dissolve in about 50 ml water and add 2 ml 5 M nitric acid. Set the electrodes in place, arrange a stirrer and buret, and titrate with 0.1 N silver nitrate. Before starting the titration set the pH meter to read in millivolts and adjust the "Standardization" knob so that the needle is in the upper part of the scale; the reading will drop as the titration proceeds. Perform the titration, taking points close together in the equivalence region, and use the titration curve to find (a) the normality of the silver nitrate, (b) the solubility product of the silver halide. Refer to page 27 for the method of calculating the solubility product. Note that the volume of solution at the equivalence point must be known to within some 5 - 10%.

Make a test to estimate the rate of leakage of potassium chloride from the calomel reference electrode. (The best way to do this is to titrate to a point close to the equivalence point, leave the solution and electrodes alone for half an hour, stir, and note the change in the EMF reading.) If the leakage is enough to affect the results, carry out another titration using the glass electrode as reference. Connect the glass electrode to the pH meter, take out the calomel and replace the calomel by the silver electrode. Add some 2 - 5 ml 5 M nitric acid to the halide solution before titrating, and note that the EMF reading will now go up, instead of down, as silver nitrate is added.

Note, also, that one may use the pH scale if one wishes, provided the temperature adjustment is properly set. With the Zeromatic pH meter, when returning to the pH scale after using the millivolt scale, be sure to depress the "Manual" temperature control button.

Titrating a Mixture of Chloride and Bromide. After titrating a single salt, titrate a mixture of chloride and bromide. Weigh known amounts of the two pure salts and dissolve them together in the same solution, add nitric acid, and titrate with your standardized silver nitrate, using either the calomel or the glass electrode for reference. Compare the volumes of silver nitrate used to reach each inflection point with those theoretically needed. You will find that the volume to reach the first inflection point is more than that which would be required if the silver bromide formed were pure. Now repeat the titration, using the same weights of chloride and bromide as before, and adding about 5 g of hydrated calcium nitrate or aluminum nitrate. Compare the curves obtained with and without the added nitrate; see Fig. 1-10 and page 29.

Experiment 1-5

The Fluoride Ion Electrode

Materials Needed. Sodium or potassium fluoride; sodium or potassium nitrate; calcium nitrate; pH meter; fluoride ion electrode (lanthanum fluoride single-crystal membrane electrode). The pH meter should preferably be one of high precision, like the Beckman Model G or GS, or an expanded-scale pH meter.

Object of Experiment. To investigate the response of the fluoride ion electrode over a range of concentrations, to note the effect of ionic strength, and to measure the solubility product of calcium fluoride.

Use a calomel reference electrode; remove the glass electrode from the pH meter and put the fluoride ion electrode in its place. No adapter is needed, for the fluoride ion electrode, being of high resistance, carries a shielded lead. Unlike the glass electrode, the fluoride ion electrode does not require soaking before use but may be placed straight into the test solution. Either the pH scale or the millivolt scale may be used; the pH scale is better because the divisions are larger, that is, they correspond to fewer millivolts. As the fluoride concentration falls the pH reading will also fall; set the standardization control accordingly. Remember that the actual position of the needle on the scale is not important, but the changes in reading must be measured as precisely as possible.

Prepare a stock 1.0 M potassium fluoride solution or an 0.50 M solution of sodium fluoride. (Sodium fluoride is less soluble than the potassium salt.) By careful quantitative dilution prepare 100-ml quantities of fluoride solutions, 0.10, 0.010, 0.0010, and 0.00010 M. Also prepare a solution that is 0.0010 M in fluoride and at the same time 1.0 M in potassium or sodium nitrate. The more dilute solutions are best prepared, not from the concentrated stock, but from one of the intermediate solutions; thus, the 0.0010 M fluoride is prepared by placing 10.0 ml 0.010 M or 1.00 ml 0.10 M fluoride in a 100-ml volumetric flask and making up to the mark with distilled water. These dilutions must be done carefully. If the solutions are to be kept for more than one laboratory period they should be stored in polyethylene bottles.

Place the most concentrated fluoride solution in a small beaker and immerse the electrodes in it. Set the standardization control to read pH 5 or more, and set the temperature adjustment. Read the meter as precisely as you can, and see that it is constant. Thereafter, do not touch the standardization control. Remove the fluoride solution and reserve it for future use. Rinse the electrodes and wipe them lightly with paper tissue. Now place the electrodes in the next most concentrated fluoride solution, that is, 0.10 M; take the reading on the pH scale; remove the solution, rinse and lightly dry the electrodes, place them in the 0.010 M solution, and so on. Read the 0.0010 M fluoride-1.0 M nitrate solution, and read any "unknown" solutions you may have, such as a fluoridated city water (these will contain about 1 mg F per liter, or 5×10^{-5} M).

Make a graph of the pH reading against the logarithm of the fluoride ion concentration. (Remember that the instrument is not now reading pH, but pF, and that the scale now goes in the opposite direction; high apparent pH means low pF.) Note the effects of ionic strength, and compare the observed effects with those expected theoretically; see page 31.

To measure the solubility product of calcium fluoride, prepare a solution 0.10 M in calcium nitrate or chloride, add just enough fluoride to produce a precipitate of calcium fluoride, place in a polyethylene bottle, and shake well. Leave overnight, if possible. Then read the "pF" with the fluoride ion electrode; if necessary, restandardize the meter with one or two known fluoride solutions. Calculate the solubility product from the observed fluoride ion activity.

If calcium chloride is used, remember that anhydrous calcium chloride often gives an acid reaction, and that acid affects the fluoride ion electrode (page 30).

Experiment 1-6

Oxidation-Reduction Titrations

Materials Needed. Ferrous sulfate or ferrous ammonium sulfate; ammonium metavanadate; granulated zinc; sulfuric acid, about 6 N; a solution of ceric sulfate, about 0.1 M in 1 N sulfuric acid; pH meter; magnetic stirrer; platinum electrode.

Object of the Experiment. To perform a typical oxidation-reduction titration.

Weigh about 1 mmol each of ferrous sulfate and of ammonium metavanadate, NH_4VO_3; dissolve them together in 25-50 ml of water with enough sulfuric acid to make the solution roughly 2 N in acid. Place in a small Erlenmeyer flask with about 5 g of granulated zinc (preferably amalgamated) and shake or stir (magnetic stirrer may be used) until the color is dark green, indicating reduction of vanadium to V^{3+}. Now transfer the solution to a beaker (150 - 250 ml), leaving the zinc behind; rinse the zinc and wash the rinsings into the beaker. Finally add 10 ml 6 N sulfuric acid, or enough to ensure that the acid concentration is at least 1 N.

Titrate the solution potentiometrically with ceric sulfate, using a platinum indicating electrode and calomel reference. The platinum electrode is connected to the glass electrode socket, using an adapter as described in Experiment 1-4. If a commercial platinum electrode is not available, one can easily be improvised from about 10 cm of platinum wire. Plot the curve, identify the inflection points with the help of a table of standard reduction potentials, and if the concentration of the ceric sulfate is known, calculate the amounts of iron and vanadium in the solution.

A variation on this experiment is to reduce the vanadium to V^{3+} as described, then, before titrating with ceric sulfate, raise the pH to about 2.0 (use test paper) by adding solid sodium sulfate. Acidity affects the reduction potential of vanadium(V) - vanadium(IV).

Experiment 1-7

Polarized Electrode Pair; the Karl Fischer Titration

Materials. Iodine; arsenious oxide; sodium bicarbonate; hydrochloric acid and sodium hydroxide solutions (about 6 N); anhydrous methanol; Karl Fischer reagent, ready made, or liquid sulfur dioxide (cylinder) and pyridine; volumetric flasks; buret; pipet; magnetic stirrer; pH meter and adapter; two platinum electrodes.

Aims of Experiment. To illustrate the polarized platinum electrode pair (this is also illustrated in the experiment on coulometric titration, Experiment 2-1, and an important analytical reagent, the Karl Fischer reagent for water.

Most pH meters have provision for applying a small polarizing current; see page 33. The way to connect the Zeromatic pH meter is shown in Fig. 1-16. Note that an adapter is needed, the same kind that is used for connecting silver or platinum electrodes, together with a "polarizing jumper" to make connection between the lead going to the glass electrode input and the socket labeled "Polarizing current." This connection can be made with a piece of wire and an alligator clip. Other pH meters have similar connections; consult the instruction manuals.

The platinum electrodes can be of the commercial type, or they can be platinum wires sealed into glass tubes. Two small pieces of platinum foil, about 1 cm^2 or less, welded to platinum wires are ideal. They should be mounted about 1 cm apart in the solution.

(a) The Iodine-Arsenite Titration. Prepare a standard sodium arsenite solution, 0.1 N, by weighing pure arsenious oxide (formula weight 197.4; equivalent weight one-fourth of this) and dissolving it in a few milliliters of 6 N sodium hydroxide, diluting, neutralizing with hydrochloric acid, and making up to the mark in a volumetric flask. A volume of 250 ml is convenient. Also prepare a solution of iodine, about 0.1 N; weigh the iodine roughly, put it in a beaker with a few grams of solid potassium iodide, add water, and when all the iodine has dissolved, dilute to the approximate volume needed; mix well.

Pipet 25 ml of standard arsenite into a beaker, arrange electrodes and stirrer, add enough water to cover the electrodes, and add a gram or two of solid sodium bicarbonate to bring the pH to about 8. Set the pH meter on the ±700 mV scale with the needle near the middle of the scale. Titrate with the iodine, and take readings of the potential around the equivalence point. Continue the titration until the potential no longer changes rapidly. The change at equivalence is very sharp. Observe the end point changes

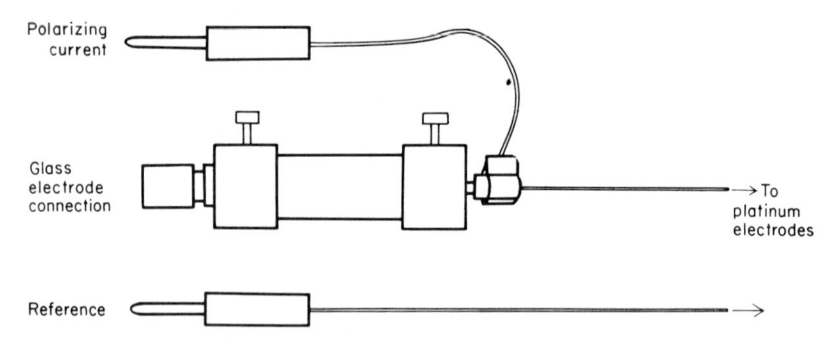

Fig. 1-16. Connection for dead-stop titrations; Beckman Zeromatic pH meter.

by adding arsenite and iodine in turn, going back and forth across the end point. You may note that the change is sharper in one direction than another, and that the readings are more stable on one side of the end point than the other. Describe fully in your report.

Once the iodine solution is standardized it may be used, if desired, to titrate arsenic or antimony in ores or other unknown samples, using the polarized electrode pair. A very useful titration that can be done with this equipment is the measurement of the "bromine number" of unsaturated fats and oils. The titrant can be a solution of bromine in glacial acetic acid.

(b) The Karl Fischer Titration; see page 33. Because water is being titrated it is important to protect the solutions from a humid atmosphere. For most purposes it is sufficient to cover the titration beaker with a plastic sheet or wax-coated paper, cutting holes to admit the electrodes and the buret. For fine work in humid atmospheres a more elaborate arrangement is necessary.

The reagent is a solution of iodine and sulfur dioxide in methanol and pyridine. It can be bought commercially, or prepared as follows: Dissolve 28 g of iodine in a mixture of 400 ml pyridine and 400 ml anhydrous methanol; then pass in sulfur dioxide gas from a small cylinder, taking care that it all dissolves, until 20 g have been added. This may be controlled by weighing the cylinder or by weighing the bottle of solution. It is important to have an excess of sulfur dioxide, rather than iodine. Finally add methanol to make about 1 l. The solution is then standardized against a standard solution of water in methanol or against solid sodium tartrate dihydrate, formula weight 230.1. One ml of the solution described is equivalent to about 2 mg of water.

A good way to proceed is this. Place in the titration vessel enough anhydrous methanol to cover the electrodes, and titrate with Karl Fischer reagent until the potential changes sharply. One now knows that there is no more water in the solvent or on the walls of the beaker. Note the buret reading, add a measured volume of standard water in methanol or a measured weight of solid sodium tartrate dihydrate, and continue titrating until the potential changes again.

For an "unknown" whose water content is to be determined, many common materials can be chosen, such as gasoline, lubricating oil, or flour. Methanol and benzene can be used as solvents, but their water content must be checked by titration. It is best to confirm the results of the Karl Fischer titration by determining the water content by an independent method. Infrared spectroscopy may be used, and a simple method, applicable to solids and liquids of low volatility whose water content is not too low, is

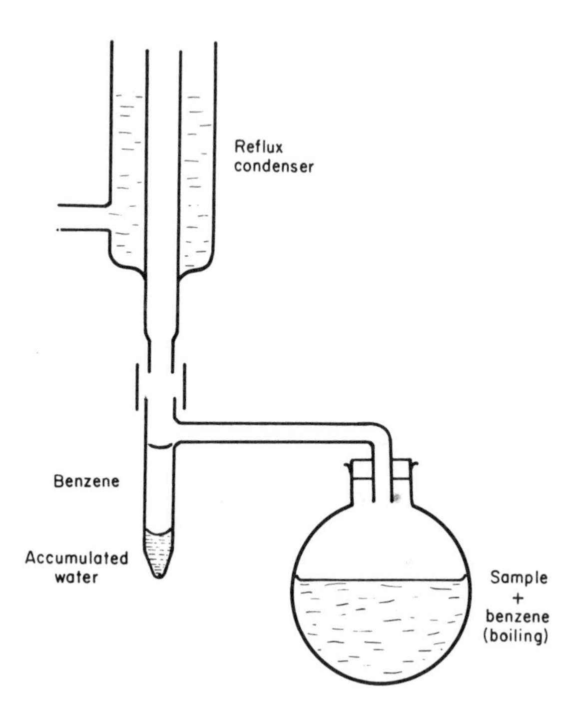

Fig. 1-17. Entrainment distillation.

entrainment distillation. The sample is refluxed with benzene in the apparatus shown in Fig. 1-17.

QUESTIONS

(Standard potentials, ionization constants, and other data may be found in the Appendix)

1. A solution of $SnCl_2$, 0.05 M, is titrated with $FeCl_3$, 0.10 M, using a platinum indicator electrode. What is the potential of this electrode at the equivalence point, (a) referred to a standard hydrogen electrode, (b) referred to a saturated calomel electrode?

2. The solubility product of $PbSO_4$ is 1.6×10^{-8}. What is the potential of a lead electrode in 0.0010 M sulfuric acid solution that is saturated with solid lead sulfate? (Consider that H_2SO_4 ionizes completely to H^+ and SO_4^{2-}.)

3. The solubility product of calomel, Hg_2Cl_2, at 25°C is 1.3×10^{-18} (see page 15, but use this value instead of the value calculated on page 16). Calculate the potential of a calomel electrode containing 0.10 M KCl.

4. The measured potential of the decinormal calomel electrode (that is, the electrode made with 0.10 M KCl) is +0.334 V referred to standard hydrogen. Comparing this value with the value calculated in Question 3, what is the activity coefficient of the chloride ion in 0.10 M KCl? On what assumption does this calculation of the activity coefficient depend?

5. The solubility product of AgCl is 1.8×10^{-10} at 25°C. 25.00 ml 0.100 M KCl is titrated with 0.100 M $AgNO_3$, using a silver indicating electrode. (a) By how much does the potential of this electrode change between 24.90 and 25.10 ml $AgNO_3$? (b) As $AgNO_3$ is added, does the silver electrode become more positive or more negative? (c) What per cent error would be made in the titration if the buret reading were taken when the potential of the silver electrode was 0.10 V greater or less than the true potential at the equivalence point? (d) Repeat this calculation for the addition of 0.0100 M $AgNO_3$ to 0.0100 M KCl. What general conclusion can you draw about the effect of concentration on the accuracy of titrations of this type?

6. Calculate the EMF of this cell, and indicate which electrode is positive:

Ag, AgCl /	a solution 0.10 M in NaCl and 0.20 M in NaF	/ LaF_3 membrane	/ NaF, 1.0×10^{-4} M	/ sat. calomel

7. From the ionization constants of H_2CO_3 given in the Appendix, calculate the pH of (a) 0.10 M $NaHCO_3$, (b) 0.10 M Na_2CO_3.

8. A solid salt contains Na_2CO_3 and $NaHCO_3$ together with water of hydration. A sample weighing 0.450 g is dissolved in water and titrated with 0.200 M hydrochloric acid. The first inflection point (pH 8.5) comes at 8.30 ml, the second (pH 4) at 22.0 ml. Calculate the weights of Na_2CO_3, $NaHCO_3$, and H_2O in the sample, and also the ratio of moles of H_2O to moles of Na_2CO_3.

9. Given the solid salt, tetramethylammonium sulfate, how would you prepare a standard solution of $N(CH_3)_4OH$ in anhydrous methanol?

10. A batch of methanol contains 0.30% of water by weight. The density of the liquid is 0.80 g/ml. What volume of a Karl Fischer reagent that is 0.150 M in I_2 and SO_2 will be used to titrate 25.0 ml of the "wet" methanol?

11. How would you measure the proportion of sodium bicarbonate in a baking powder like that described in Experiment 1-2?

12. A sample of phosphate rock weighing 0.300 g is dissolved in a slight excess of hydrochloric acid, and the solution is passed through a cation-exchange resin column as described in Experiment 1-2. All the phosphate is recovered as phosphoric acid, mixed with excess of hydrochloric acid. The solution is titrated with 0.200 M sodium hydroxide. It takes 28.0 ml to the first inflection (pH 4) and 48.0 ml (total) to the second inflection (pH 9). Calculate the per cent of phosphorus, as the element, P, in the sample.

13. Suppose you had a sample of pure hydroxyapatite, $Ca_5(PO_4)_3OH$, and dissolved 100.0 mg of this material in 50.0 ml 0.100 M HCl, then titrated the resulting solution with 0.100 M NaOH, stopping the titration at the first inflection in the pH curve, about pH 4. What volume of NaOH would be used?

14. A solution 0.20 M in potassium fluoride is titrated with 0.25 M calcium chloride, using a fluoride ion electrode. CaF_2 precipitates; its solubility product is 5.0×10^{-11}. By how many millivolts does the potential of the fluoride ion electrode change between the start of the titration and the equivalence point? Would you recommend this titration as an accurate way to determine Ca^{2+} or F^-?

Chapter 2

COULOMETRIC AND CONDUCTIMETRIC TITRATION

I. COULOMETRIC TITRATION

Coulometric titration is titration by a controlled electric current. An electrode takes the place of the buret, and a stream of electrons replaces the standard solution. The technique is applicable to oxidation-reduction reactions, for these are reactions of electron transfer. Ferrous ions are oxidized to ferric ions, for example, by removing electrons:

$$Fe^{2+} = Fe^{3+} + e$$

The electrons can be removed by a chemical oxidizing agent like ceric sulfate or potassium permanganate, or by a positively charged platinum electrode in an electrolytic cell. In the second case the quantity of ferrous ions oxidized is measured by the quantity of electricity passed, according to Faraday's law. The relations are as follows:

Quantity of electricity in coulombs = (current in amperes) x (time in seconds)

Charge on one mole of electrons (6.023×10^{23} electrons) = the Faraday constant = 96,487 coulombs

The amount of chemical reaction produced by electrolysis is found by integrating the current against time, and the easiest way to do this is to hold the current constant and measure the time taken to bring about a chemical reaction. For example:

If a current of 25.0 milliamperes (0.0250 ampere) must be passed for 320.0 seconds to oxidize all the ferrous ions in a solution, how much ferrous ions were present?

Quantity of electricity passed = 0.0250 x 320.0 coulombs = 8.00 coulombs = 8.00/96,500 faradays. This is the number of chemical equivalents of Fe^{2+}. Note that we rounded the value of the Faraday constant to three significant figures. If we want the weight of iron oxidized, we multiply the number of equivalents by the equivalent weight:

$$\text{Weight of } Fe^{2+} = \frac{8.00 \times 55.84}{96,500} \text{ grams, or more conveniently,}$$

$$\frac{8.00 \times 55.84}{96.5} \text{ milligrams} = \underline{4.63 \text{ milligrams}}$$

This calculation illustrates one of the advantages of coulometric titration, that it can accurately measure small quantities of substances. Other advantages are that one does not need to prepare standard solutions and that the operation lends itself well to automation.

For coulometric titration to succeed there must be no side reactions. The electron-transfer reaction must proceed quantitatively, or, as we say, with "100% current efficiency." It is not necessary that the substance we are determining be itself oxidized or reduced at the electrode surface, however. One can oxidize or reduce a second substance whose product then reacts with the material being titrated. The experiment described below provides an example of this procedure. We wish to determine arsenic by oxidizing arsenious acid, but arsenic(III) is not oxidized easily or quantitatively at a platinum electrode. We therefore add potassium bromide to the solution. Bromide ions are oxidized quantitatively at the platinum anode, producing bromine, which diffuses into the solution and there reacts rapidly and quantitatively with the arsenic; it is reduced back to bromide ions, which are oxidized to bromine again when they strike the electrode. The bromine acts as a carrier between the electrode and the substance being titrated. It is called an electrolytically generated titrant.

The use of electrolytically generated titrants is, in fact, obligatory if constant current is to be maintained. Near the end of the titration the substance being titrated becomes very dilute, and it cannot reach the electrode in time to keep pace with the flow of electrons. Bromine, chlorine, and iodine are the commonest electrolytically generated titrants used as oxidants; for reduction, ions of Fe(II) and Ti(III) are used. Silver ions are generated electrolytically from a silver metal anode; their commonest use in coulometric titration is to determine mercaptans, organic compounds containing the group -SH. This measurement is important in the petroleum industry. Electrolytically generated hydrogen and hydroxyl ions are used to titrate bases and acids, respectively, and as in most coulometric titrations, very great precision is possible.

End points are determined in a number of ways, the commonest being a polarized electron pair (see Chapter 1). One can add an indicator and watch the color change, just as one does in conventional titrations with a buret.

II. EXPERIMENTAL: CELL ARRANGEMENT

A simple cell design that is used in Experiment 2-1 is shown in Fig. 2-1. One electrode is enclosed in a glass tube 1-2 centimeters in diameter with a porous glass frit at one end. This electrode can be a small spiral of platinum wire. The other electrode, which is the working electrode, is placed just outside the porous frit, and may be a flat piece of platinum foil or preferably a piece bent in the form of a cylinder, as shown. Its area should be at least 1 centimeter square. It is necessary to keep the electrodes apart, so that the product of oxidation at one electrode shall not be reduced again at the other. The solution inside the glass tube should conduct electricity well; it is usually dilute sulfuric acid. The level of solution inside this tube should be higher than the level in the beaker.

Figure 2-1 shows a pair of "indicating electrodes" for use as a polarized pair to detect the end point, but obviously another detection system can be used, like a calomel and platinum electrode to measure the oxidation-reduction potential of the solution.

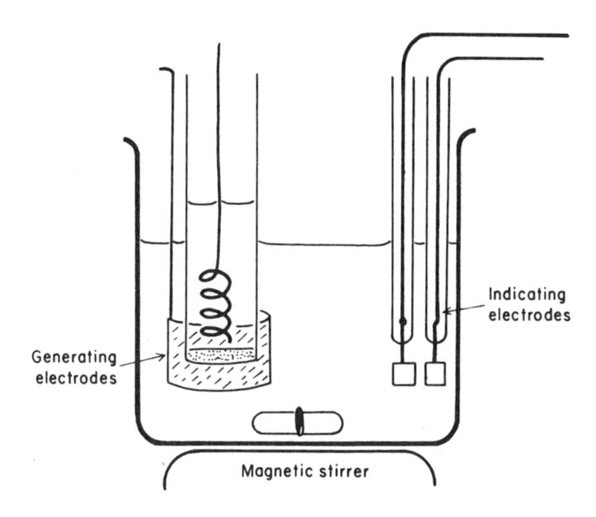

Fig. 2-1. Cell for coulometric titration.

III. SOURCE OF CONSTANT CURRENT

Commercial equipment is available that will supply a constant, accurately known current. The Leeds and Northrup coulometric titrator gives a choice of three currents, 0.9649, 9.649, and 96.49 milliamperes. The upper range, 96.49 milliamperes, supplies 10^{-6} faraday, or one microfaraday, per second, and one only has to read the number of seconds for which the current is passed to get the number of microfaradays of electricity, or microequivalents of chemical change. The instrument is also supplied with a timer that runs only when the electrolysis current is switched on.

If a commercial instrument is not available, one can build one's own. The simplest arrangement is a high-voltage battery, say 200 volts, with a high resistance in series; a resistance of 10,000 ohms would pass a current of 20 milliamperes. Changes in the resistance of the electrolysis cell make very little difference to the current, because the resistance of the cell is always much less than 10,000 ohms. This arrangement has several disadvantages, however. High-voltage batteries are not easily available commercially, the high voltage presents a certain hazard, and the voltage of the batteries changes slightly with time. The resistances change, too, as they warm up.

It is better to use a circuit with a solid-state regulating system built into it, like that shown in Fig. 2-2. In this circuit the primary regulating device is a "zener diode." This is a pair of silicon semiconductors joined together. On one side of the junction the silicon contains an impurity like arsenic, which supplies extra electrons. On the other side the silicon has an impurity like indium, whose atoms have one less valency electron than silicon. These atoms supply "holes" or places where an electron is missing. When this device is placed in an electric circuit, electrons flow easily

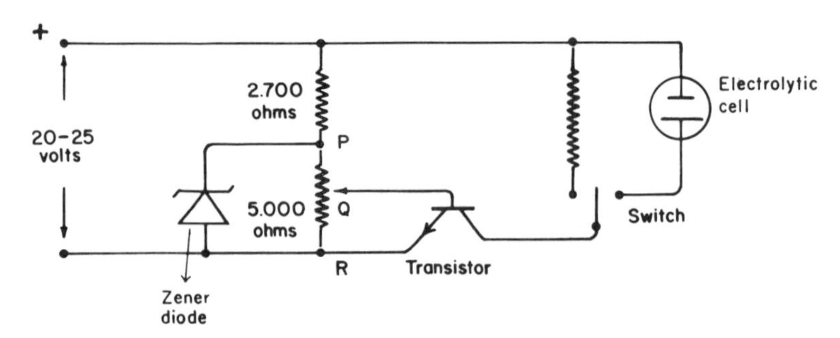

Fig. 2-2. Source of constant current (acc. Vincent and Ward).

from the electron-rich side (called an _n_-type semiconductor) to the electron-deficient side (called a _p_-type semiconductor) but they will not flow in the reverse direction unless a voltage is applied that is high enough to pull electrons out of the covalent silicon-silicon bonds. This critical voltage is very sharp. In a circuit like that of Fig. 2-2 it has the effect of maintaining a very nearly constant potential between points P and R, and hence between Q and R; the position of the sliding contact Q can be set to provide the particular voltage desired. The contact Q is connected to the "base" of a transistor which in its turn regulates the flow of current through the electrolysis cell. This simple circuit does not give the precision or stability of more complicated devices, but it will maintain currents of 10-20 milliamperes within one part in 1,000. An electric timer can be placed in the output circuit if desired. The circuit can be assembled easily and cheaply from standard parts.

The circuit of Fig. 2-2 is described by Vincent and Ward, Journal of Chemical Education, _46_, 613 (1969). Another circuit for coulometric titration, more complicated and more precise, is described in the same journal by Stock (_ibid._, p. 858).

EXPERIMENT ON COULOMETRIC TITRATION

Experiment 2-1

Titration of Arsenic(III) by Electrolytically Generated Bromine

Materials. Cell for coulometric titration, see Fig. 2-1; constant-current source, Fig. 2-2 or a commercial unit; polarizing current source, Chap. 1, Fig. 1-16, or the arrangement described below, Fig. 2-3; magnetic stirrer.

Arsenious oxide, pure; arsenious oxide, impure "unknown"; potassium bromide; sodium hydroxide; sulfuric acid; volumetric flask; pipet.

Set up the cell shown in Fig. 2-1. The electrode inside the tube with the fritted disk must be connected to the negative terminal of the constant-current source, and the working electrode, outside this tube, must be connected to the positive terminal. If the polarity of these terminals is not shown, test it by connecting wires to the terminals, holding the ends of these wires about 1 cm apart, and touching them to a piece of filter paper soaked in a concentrated potassium iodide solution. Iodine will appear at the positive wire, producing a brown stain.

The pair of polarized electrodes should be mounted on the opposite side of the beaker to the generating electrode to minimize stray currents. If a pH meter is not available to supply the polarizing current, the arrangement shown in Fig. 2-3 can be used. An ordinary flashlight battery can be used, and the resistances can be cheap ones used for radio sets. The principle is somewhat different from that described in Chapter 1, in that a constant potential is maintained across the pair of indicating electrodes, and changes in the current are noted.

If the constant-current coulometric source is home-made it is probably necessary to calibrate it. One should not rely on the reading of a milliammeter. Prepare a standard sodium arsenite solution by weighing accurately about 0.5 g of pure arsenious oxide, dissolving in a few milliliters of 6 N sodium hydroxide, diluting, transferring to a 250-ml volumetric flask, and making up to the mark, mixing well. Made as described, this solution is about 0.01 M in As_2O_3, or 0.04 N as a reducing agent. Pipet 10.0 ml of this solution into the titration vessel, acidify with hydrochloric or sulfuric acid, add about 5 g of solid potassium bromide, and dilute to cover the electrodes. Start the magnetic stirrer, turn on the electrolysis current and the timer, and measure the time needed to reach the end point. The end point is indicated by a sudden, permanent change in the polarizing potential or current. Just before the end point is reached the electrolysis current can be switched on and off, just as if one were adding titrant a drop at a time from a buret. The 10 ml of 0.04 N arsenic(III), specified above, equals 400 μequiv and will take 400 sec (6 min 40 sec) to oxidize at a current of 96.49 mA.

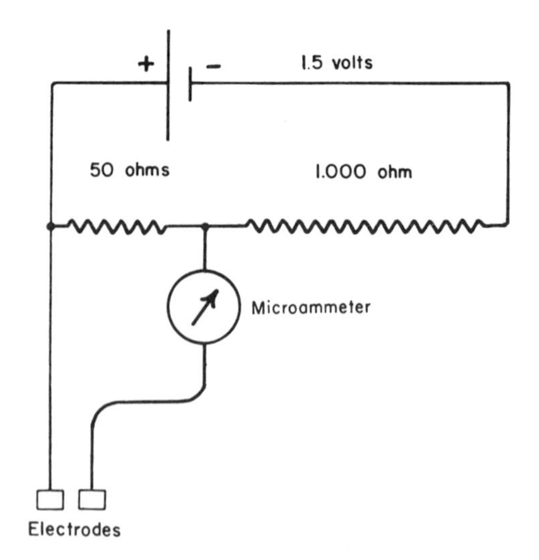

Fig. 2-3. Circuit for polarizing potential.

After standardizing the constant-current source, use it to titrate a solution prepared from an "unknown" impure arsenious oxide. Antimony-(III), from antimony trioxide, can be titrated in the same way, and so can many other substances, including hydrazine, nitrous acid, sulfur dioxide, and carbon-carbon double bonds in unsaturated fats and oils.

IV. CONDUCTIMETRIC TITRATION

Conductimetric titration uses the electrical conductivity of a solution to follow the progress of a titration. It can be used for reactions in which the number of ions changes or in which fast-moving ions are replaced by slow-moving ions or vice versa. Conductimetric titration is not often used in analytical practice because it is rather cumbersome, but there are certain tasks that it does conspicuously well. These relate to the analysis of mixtures of acids and bases and will be illustrated in the experiments that follow.

The basis of conductimetric titration is Kohlrausch's law of independent ionic mobilities. This says that in a dilute solution, the conductance of the solution is the sum of the conductances of the individual ions. For a salt AB,

equivalent conductance, $\lambda_{AB} = \lambda_A + \lambda_B$

Given a table of equivalent ionic conductances, therefore, we can predict the conductance of any solution that contains any combination of the ions listed in the table. To make an accurate prediction we must calculate the effects of the attractions between ions, which slow the ions down. Roughly, however, we can consider that these effects remain constant during a titration. A table of equivalent ionic conductances is given in the Appendix.

A few definitions will be helpful:

Specific resistance = resistance of a 1-centimeter cube of the solution, in ohms

Specific conductance = 1/(specific resistance); units, reciprocal ohms

$$\text{Equivalent conductance} = \frac{\text{specific conductance}}{\text{concentration in gram-equivalents per liter}}$$

What is measured experimentally is the resistance of the solution in a cell that contains two electrodes, and these do not necessarily have areas of 1 square centimeter, nor are they 1 centimeter apart. To convert the actual cell resistance to the specific resistance one multiplies by a "cell constant" that must be determined experimentally. In titrations, however, we are concerned only with changes in conductivity, and the value of the cell constant is of little interest.

Figure 2-4 shows how the course of a conductimetric titration, that of hydrochloric acid titrated with sodium hydroxide, can be predicted from the equivalent ionic conductances. This prediction supposes, however, that the solution is not diluted during the titration. This is never the case. A fair approximation to the curve shown is, however, obtained if the sodium hydroxide solution is at least ten times as concentrated as the hydrochloric acid. If the titrant solution is too dilute the branches of the conductance graph become curved instead of straight, and this is bad, as the precision of conductimetric titration depends on locating the intersection of straight lines.

If the straight-line portions of a conductimetric titration curve are well established it does not matter if the portions near the equivalence points are curved. The readings close to equivalence are not very important. One can therefore use conductivity to follow titrations where the reaction is appreciably reversible, where the equilibrium constant is not large. This is not true of potentiometric titrations. Another technique that gives linear titration graphs, and permits the use of reactions whose equilibrium constants are not large, is photometric titration.

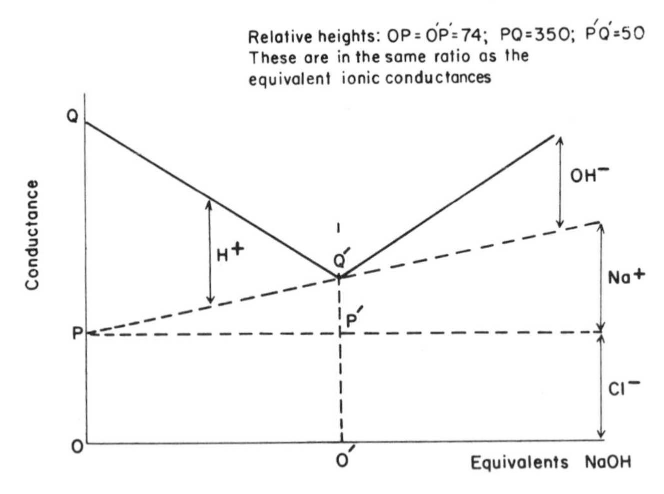

Fig. 2-4. Conductimetric titration of hydrochloric acid with NaOH.

A typical titration in which conductivity gives a good end point and potentiometry does not is the titration of an ammonium salt with sodium hydroxide. The curve is shown in Fig. 2-5. Conductimetric titration is also effective in determining the concentration of free acid in solutions of strongly hydrolyzed salts, for example free nitric acid in a solution of ferric nitrate or uranyl nitrate.

V. EXPERIMENTAL

A. Conductivity Cells

The most convenient arrangement for conductimetric titration is simply a pair of platinum foil electrodes placed side by side in a beaker; see Fig. 2-6. The electrodes can be about 1 centimeter square and 1-2 centimeters apart. They must be platinized, that is, covered with a black deposit of finely divided platinum. To "platinize" electrodes, first clean them well with concentrated hydrochloric acid and wash well with water, then place them in a solution which is 0.025 N in hydrochloric acid, 0.3% in platinic chloride, and 0.025% in lead acetate. Electroplate the platinum by passing a current of about 10 milliamperes per square centimeter, reversing the flow of current every 10 seconds and passing the current some 5 minutes in all.

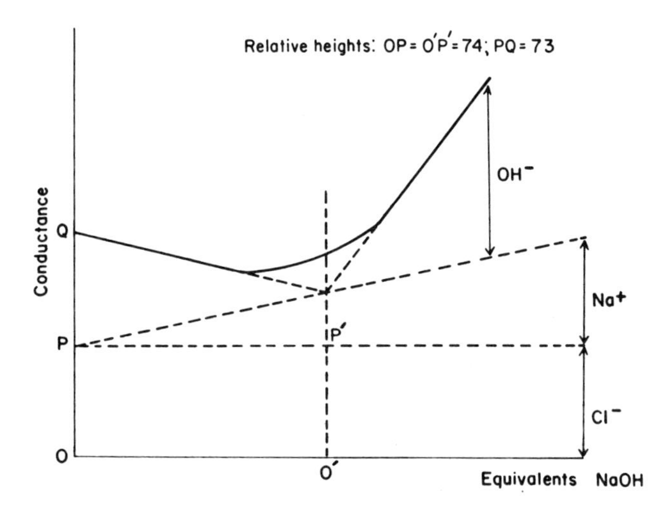

Fig. 2-5. Conductimetric titration of ammonium chloride with NaOH.

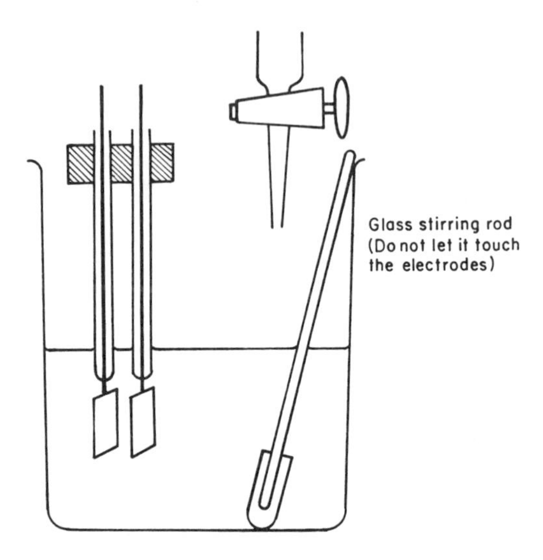

Fig. 2-6. Cell for conductimetric titration.

Dipping cells (Fig. 2-7) are convenient for measuring the conductivity of solutions but not convenient for titration. They may, however, be used in titrations by moving them up and down manually between additions of titrant, or by using the magnetic stirrer arrangement shown in Fig. 2-7.

The use of a magnetic stirrer in conductivity titration is undesirable because the motor of the stirrer heats the solution, raising its temperature as the titration proceeds, and distorting the shape of the graphs. Manual stirring is preferable.

B. Conductivity Bridges

The basic circuit used to measure electrical resistance, and hence conductance, is the Wheatstone bridge, illustrated in Fig. 2-8. The bridge has four resistances or "arms." One arm, AB, contains the cell. Along AC is a known, fixed resistance. The arms BD and DC are formed by a slide wire, and the ratio BD:DC can be adjusted at will by moving the sliding contact D. Between D and A is a source of alternating current with a frequency of about 1000 cycles; between B and C is a detecting device, which may be an oscilloscope or a "magic eye" tube. To make a measurement, D is moved until a position is found at which no current flows through

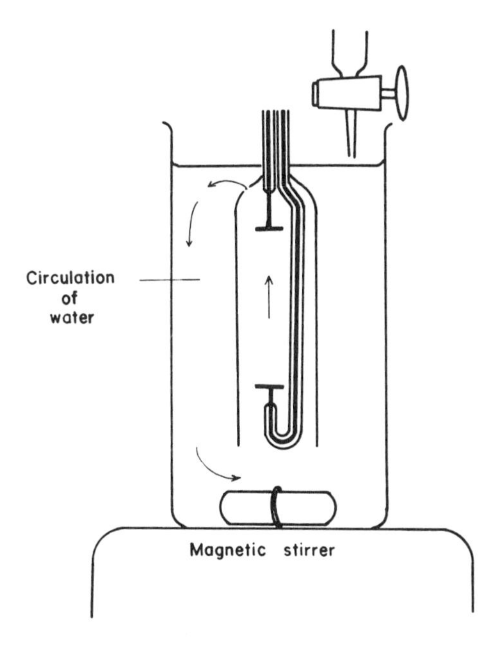

Fig. 2-7. Dipping electrode cell adapted to conductimetric titration.

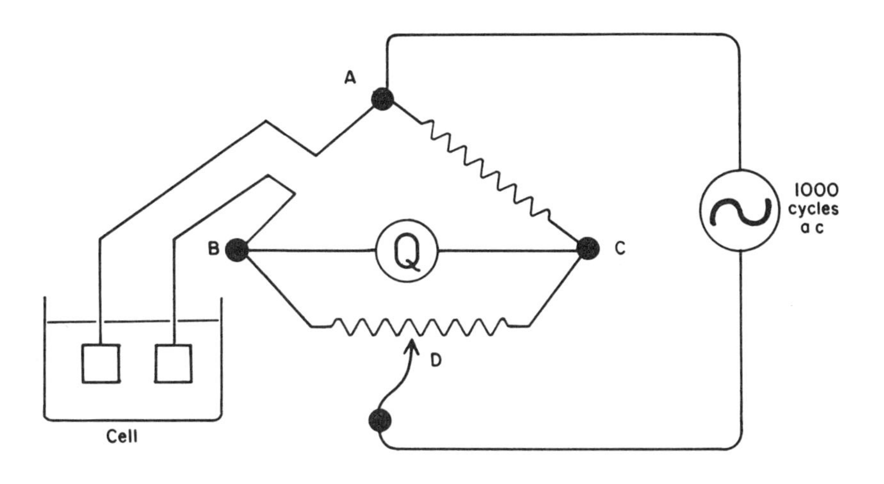

Fig. 2-8. Wheatstone bridge circuit. "Q" is a detecting device; see text.

the detecting device. The bridge is then in balance, and these resistance ratios are equal: AB/AC = BD/DC. Knowing the resistance AC and the ratio BD/DC the resistance of the cell, AB, is found.

Best precision is obtained if the contact D is somewhere near the middle of the slide wire, that is, if the cell resistance is not greatly different from the resistance AC. Since the cell resistance may vary greatly, a set of resistances with values differing by factors of 10 is used; one can place any one of these in the position AC by setting a switch. Commercial conductivity bridges have a "Range" or "Multiplier" switch that does this. They have a built-in 1000-cycle audio-frequency generator, a sensitivity control, and a "magic eye" null detector. The bridge is in balance when the dark sector of the "eye" is open to its maximum extent. Balancing is done by turning a knob which carries a dial graduated directly in ohms. Some bridges allow one to read directly in conductivity, which is a great convenience; otherwise one has to plot the reciprocals of the resistance readings.

A further refinement of some bridges is a small variable capacitance across the arm AC. This compensates for the electrostatic capacity of the conductivity cell and allows one to get sharper balance points. Alternating current is used to measure the resistance of solutions, because direct current would cause chemical decomposition.

EXPERIMENTS ON CONDUCTIMETRIC TITRATION

Experiment 2-2

Titration of a Mixture of Hydrochloric Acid and Ammonium Chloride

Materials. Conductivity cell and bridge; buret; pipet; solid ammonium chloride; standard 0.2 M hydrochloric acid; and sodium hydroxide.

In the titration beaker, Fig. 2-6, place a weighed amount of ammonium chloride, about 60 mg (1 mmol), and 5.0 ml 0.2 N hydrochloric acid; dilute with water to about 100 ml, or more if necessary to cover the electrodes. In the buret place 0.2 N sodium hydroxide. Use a rubber-tipped stirring rod, as indicated in Fig. 2-6, and keep it away from the electrodes. Read the conductivity of the original solution, and add sodium hydroxide 1 ml at a time, reading the conductivity after each addition. The conductivity first falls rapidly, then falls more slowly, then rises rapidly when the sodium hydroxide is in excess. Continue the titration until enough points have been taken to establish the slope of this last rising portion of the curve. Plot a graph of conductance against volume of 0.2 N sodium hydroxide, and from it calculate: (i) the number of milliequivalents of hydrochloric acid and ammonium chloride; compare these values with the known amounts that were added; (ii) the equivalent ionic conductance of the ammonium ion, taking the equivalent conductances of Na^+ and Cl^- at infinite dilution to be 51.0 and 75.5, respectively, at 25°C. It may also be possible to estimate the ionization constant of ammonia from your curve; do this if you can.

Other interesting titrations that can be done are: (i) a solution of ferric nitrate containing a small excess of nitric acid, to measure the concentration of free nitric acid; (ii) solutions of ammonium phosphates, to measure the concentration of NH_4^+ and total phosphate.

Experiment 2-3

Titration of Vanillin in a Vanilla Essence

Materials. As in Experiment 2-2, plus a separating funnel; ether; alcohol; sodium chloride (commercial grade); a good-quality vanilla extract; pure vanillin.

The flavoring material of vanilla is the substance vanillin, 4-hydroxy-3-methoxybenzaldehyde. This substance is a very weak acid. Its ionization in dilute aqueous solution is so small that the solution hardly conducts at all. When sodium hydroxide is added it forms the sodium salt of vanillin, which conducts moderately well, as it is completely ionized. The conductance rises linearly with the volume of sodium hydroxide added; then, when sodium hydroxide is added in excess, the conductance rises very rapidly owing to the high mobility of the hydroxyl ion. The conductimetric titration curve thus consists of two rising straight lines, the second rising much more steeply than the first. The equivalence point occurs at the intersection of these lines.

Before titrating the essence it is best to titrate a sample of pure vanillin, to become acquainted with the shape of the curve and the slopes of the two sections. Weigh accurately about 100 mg of vanillin (formula weight 152.15), place in a 150-ml beaker, dissolve in about 10 ml ethyl alcohol, and add water to about 75 ml. Titrate with 0.2 N sodium hydroxide and plot the curve.

Commercial vanilla essence contains about 1% of vanillin, though this concentration may vary widely. Coumarin is also likely to be present, and the cheaper flavoring essences may contain very little vanillin, only coumarin. It may be possible to titrate the essence directly, after diluting it with water, but commercial products contain so much other material that a preliminary separation is usually desirable. Measure 10 ml of the essence in a graduated cylinder, and mix it with 40 ml of water that has dissolved in it about 5 g of common salt (to diminish the miscibility of water and ether). Transfer to a separating funnel, add a few drops of 6 N hydrochloric acid to repress the ionization of vanillin; add 40-50 ml ether and shake. Let stand to separate the phases, and run off as much of the water phase (below) as it is convenient to do. Vanilla essences contain surface-active materials, and complete phase separation at this stage is difficult. Wash the ether layer with three or four successive small portions of distilled water; with each water wash, phase separation becomes easier. Finally decant the clean ether layer into a beaker and evaporate the ether on a steam bath.

Dissolve the oily residue in 10 ml ethyl alcohol, add water to about 75 ml, and titrate conductimetrically as before. Do the titration fairly fast, as excess sodium hydroxide slowly hydrolyzes the coumarin (which is a lactone) and forms the salt of a carboxylic acid.

In analyzing commercial products of complex and unknown composition, one should never take it for granted that one's analysis is correct. Repeat the determination using different sample quantities and different extraction procedures (for example, two ether extractions instead of one) until you can have confidence in your result. Report the number of grams of vanillin in 100 ml of the essence.

QUESTIONS

1. Ethyl mercaptan, C_2H_5SH, is titrated coulometrically at a silver anode, where it forms C_2H_5SAg. For how many seconds must a current of 25 mA be passed to titrate 10.0 mg of ethyl mercaptan?

2. The equivalent conductance of NaCl is 126 ohm^{-1} cm^2 equiv^{-1} at 25°C. Sea water contains 3.2 g of NaCl per 100 ml. What would be the resistance of a cell with plates 1 cm^2, placed 2 cm apart, and filled with sea water?

3. Acetic acid, 0.10 M, is titrated conductimetrically with 1.0 M sodium hydroxide. The equivalence point is reached when 10.0 ml NaOH has been added, and the conductance reading at this point is 250 units. Calculate the conductance readings at 5.0 ml and 15.0 ml NaOH, given these equivalent ionic conductances: H^+ 350, OH^- 198, Na^+ 50, CH_3COO^- 41. Also calculate the conductance of the acetic acid alone, given that the ionization constant is 1.8×10^{-5}. (Note that you do not need to know the dimensions of the cell, nor the size of the conductance units; relative values are sufficient.)

4. Suppose diammonium phosphate, $(NH_4)_2HPO_4$, were titrated conductimetrically with HCl. How would the curve look?

5. If conductimetric titrations are done with a magnetic stirrer (Fig. 2-7) the temperature of the solution rises during the titrations unless precautions are taken to avoid temperature rise. What effect will a steadily rising temperature have on the shape of the curve?

Chapter 3

ELECTROGRAVIMETRY

In electrogravimetry a metal or a compound is deposited by an electric current and then weighed. In this chapter we shall consider this process, and we shall see how the electric current can be used selectively to deposit one element and not another, and so serve as a tool for chemical separation. We shall also include a discussion of "electrographic analysis," a technique in which an electric current liberates ions from the surface of a solid metal or alloy and permits qualitative identification of its constituents.

These analytical techniques depend on electrolysis. Electrolysis has two aspects, the "reversible" aspect and the "irreversible" one. To understand electrolysis we must first understand reversible voltaic cells, but then we must go one step further and study the factors that determine the rate of the process that occurs at the electrode.

I. ELECTRODE POTENTIALS AND THE NERNST EQUATION

We introduced this topic in Chapter 1. The potential of a metal electrode, placed in a conducting, ionic solution, is given by the equation

$$E = E^{o} + \frac{RT}{nF} \ln Q$$

$$= E^{o} + \frac{2.303RT}{nF} \log Q$$

The quantity E^{o} is the standard potential which is characteristic of the particular electrode and the reaction that occurs at the electrode (see page 9); R, T, and F are, respectively, the gas constant, absolute temperature, and the Faraday constant; n is the number of electrons gained or lost in the electrode reaction; and Q is the quotient or ratio of the concentrations of reactants and products of the electrode reaction. In equation 3-1, Q is written with the oxidized substances in the numerator and the reduced substances in the denominator:

$$E = E^o + \frac{2.303RT}{nF} \log \frac{\text{(conc. of oxidized substances)}}{\text{(conc. of reduced substances)}} \quad (3\text{-}1a)$$

One may equally well write the logarithm term with a minus sign before it, and then invert the fraction; some chemists prefer to write it this way:

$$E = E^o - \frac{2.303RT}{nF} \log \frac{\text{(conc. of reduced substances)}}{\text{(conc. of oxidized substances)}} \quad (3\text{-}1b)$$

Anyone familiar with logarithms will see that these two equations mean exactly the same thing, for $\log x = -\log (1/x)$.

Consider a reaction that is important in electrolysis, the reduction of cupric ions to metallic copper:

$$Cu^{2+} + 2e = Cu \quad (3\text{-}2)$$

The Nernst equation, (3-1), is written thus:

$$E = E^o + \frac{2.303RT}{2F} \log[Cu^{2+}] \quad (3\text{-}3)$$

The logic of the equation is this. Raising the concentration of cupric ions drives reaction (3-2) to the right, and tends to pull electrons out of the copper metal. Taking electrons out of a metal makes the metal more positive. In equation (3-3), increasing $[Cu^{2+}]$ makes E more positive, which agrees with common sense.

Considering the temperature to be 25^oC, or 298^oK, and inserting the values of R and T, the factor $2.303RT/F$ comes out to be 0.05915. Generally it is accurate enough to round this number off to 0.059.

Values of the standard potential, E^o, referred to the hydrogen electrode as reference are tabulated in the Appendix. With these values we can calculate the electromotive force of cells. This subject was briefly discussed in Chapter 1. Let us now calculate the electromotive force of a cell important in electrolysis, namely:

$$Cu/Cu^{2+} (0.01\ \underline{M}) // Cl^- (2.0\ \underline{M}) / Cl_2 (1\ atm),\ Pt$$

The standard reduction potentials are

$$Cu^{2+} + 2e = Cu,\quad E^o = +0.337\ volt$$
$$Cl_2 + 2e = 2Cl^-,\quad E^o = +1.36\ volt$$

Thus the potential of the copper electrode is

$$E_{Cu} = +0.337 + \frac{0.059}{2} \log 0.01 = +0.337 - 0.059$$
$$= +0.278\ volt$$

and the potential of the chlorine electrode is

$$E_{Cl} = +1.36 + \frac{0.059}{2} \log 2.0^2$$

$$= +1.36 + 0.059 \log 2.0 = +1.36 + 0.018$$

$$= +1.38\ volt$$

The electromotive force of the cell is the potential difference between the two metal electrodes, which is

$$\text{Cell EMF} = 1.38 - 0.28 = \underline{1.10\ volt},$$

with the chlorine electrode positive. (Note that we have rounded off the potentials on both sides of the cell to two decimal places.) There is so much confusion over the signs of cell potentials, and what is meant by positive and negative, that we shall try to explain just what we mean by saying "the chlorine electrode is positive." We mean that the platinum metal that makes the contact between chlorine gas and the chloride ions, and that conducts the electricity in and out of the cell, is positive with respect to the copper metal electrode on the other side. And by this statement we mean that if we join the copper and the platinum by a wire, electrons will flow of their own accord from the copper (more negative) to the platinum (more positive). As we noted in Chapter 1, it is the electrons, the negatively charged particles, that actually move when an electric current flows through a metal wire, not the positive charges.

The double line, //, indicated in the representation of the cell means a liquid junction or a "salt bridge" (Chapter 1). One must prevent the dissolved chlorine from coming into contact with the copper metal, otherwise chlorine and copper will react directly and the cell will be effectively short-circuited.

Now let us consider another example: the electromotive force of a cell formed by metallic zinc in contact with zinc ions, and oxygen gas in contact with water. Hydrogen ions take part in the cell reaction and their concentration must therefore be specified. Let it be 1.0×10^{-3} molar. Let the zinc ion concentration be 1.0×10^{-1}, and consider that the oxygen gas is at 1 atmosphere pressure. We diagram the cell as follows:

$$Zn \;/\; Zn^{2+}(0.10\ \underline{M})\quad H_2O,\quad H^+(0.0010\ \underline{M}) \;/\; O_2(1\ \text{atm}),\ Pt$$

We can omit the salt bridge if we want. Oxygen gas is only slightly soluble in water, and by proper cell design we can prevent the oxygen gas from coming into contact with the zinc. (If we want precise measurements we must carefully prevent even small concentrations of dissolved oxygen from touching the zinc, but let us overlook this point for the present.) One might think it necessary to keep the hydrogen ions (0.001 normal acid) from coming in contact with the zinc, but, owing to an effect called hydrogen overvoltage, which will be discussed later, the direct chemical reaction between zinc metal and 0.001 molar hydrogen ions is not enough to be appreciable.

We are now ready to calculate the electromotive force of our cell. It is simplest to consider one electrode at a time, as we did in the first example. The potential of the zinc metal is

$$E_{Zn} = E^o + \frac{0.059}{2} \log [Zn^{2+}] = -0.763 + \frac{0.059}{2}(-1)$$

$$= -0.763 - 0.0295 = -0.766 \text{ volt}$$

That of the oxygen electrode, that is, of the platinum or other inert conductor in contact with oxygen and the solution, is

$$E_{O_2} = E^o + \frac{0.059}{4} \log [O_2][H^+]^4 \qquad (3\text{-}4)$$

The reaction at the oxygen electrode is

$$O_2 + 4H^+ + 4e = 2H_2O$$

The oxygen is at 1 atmosphere pressure and in its standard state, so $[O_2]$ in the Nernst equation is considered to be unity. Equation (3-4) thus reads:

$$E_{O_2} = +1.23 + 0.059 \log [H^+] = +1.23 - 0.18$$

$$= +1.05 \text{ volt}$$

The electromotive force of the cell is therefore + 1.05 - (- 0.77) = 1.82 volt, and the oxygen electrode is positive.

If, instead of pure oxygen at 1 atmosphere pressure, we had air at 1 atmosphere pressure in contact with the electrode, the concentration of O_2 would be one-fifth of an atmosphere (remember that the molar concentrations of gases are proportional to their partial pressures at constant temperature), and substituting into equation (3-4) we would have

$$E_{O_2} = +1.23 + \frac{0.059}{4} \log 0.2 \times 0.0010^4$$

$$= +1.23 - 0.188 = +1.04 \text{ volt (to two decimal places)}$$

Because there are four electrons reacting to every molecule of oxygen, a change in the oxygen concentration does not make much difference to the electrode potential.

II. INERT ELECTRODES

The oxygen and the chlorine electrodes in the last two examples had platinum shown as an "inert metal" to conduct the electrons in and out of the cell and to catalyze the transfer of electrons. Platinum is often used for experimental purposes, but other materials can be used. Gold is sometimes used as an inert electrode in electrochemical experiments. In large-scale industrial electrolysis electrodes of carbon are used, either amorphous (microcrystalline) or crystalline graphite. Another inert electrode material is magnetite, Fe_3O_4. The requirements in electrolysis are, as we shall see, somewhat different from those for measuring electromotive force in the laboratory, but one requirement, at least, must be met; the material must conduct electricity, and it must be an electronic conductor.

III. ELECTRODES OF THE SECOND AND THIRD KIND

In Chapter 1 we discussed the "calomel electrode" and the silver-silver chloride electrode. These are known as "electrodes of the second kind." The electrode proper, the metal that conducts the electrons in and out of the cell, is mercury in the first case and silver in the second. The potential of the metal is determined by the concentration of the ions of that metal. In the silver - silver chloride electrode, for example, the potential of the silver metal is fixed by the concentration of silver ions in contact with the metal, in accordance with the Nernst equation. However, the silver ions are in equilibrium with solid silver chloride, and the concentration of silver ions is fixed by the concentration of chloride ions in the surrounding solution.

Thus the silver - silver chloride electrode responds to the concentration (properly speaking, the activity; see page 9, Chapter 1) of chloride ions. It does so by two reactions that are coupled together:

$$Cl^- + Ag^+ = AgCl \text{ (solid)}$$
$$Ag^+ + e = Ag \text{ (metal)}$$

Another electrode of the second kind is the lead - lead sulfate electrode, where lead metal is in contact with solid lead sulfate. The potential of the lead metal is determined by the concentration of sulfate ions in the surrounding solution.

As an example of an electrode of the third kind let us consider liquid mercury in contact with a solution containing a fixed concentration of the mercury complex of EDTA (ethylene diamine tetraacetic acid) and a variable concentration of another metal ion which we shall call M^{n+}. (This can be zinc, cadmium, cobalt, nickel, manganese, calcium, and many other ions. The only requirement is that the metal ion shall form a complex ion with EDTA that is significantly less stable than the Hg(II)-EDTA complex.) The potential of the mercury metal responds to the concentration of M^{n+} through the following sequence of reactions (the anion of EDTA is represented by the symbol Y):

$$M^{n+} + HgY^{2-} \rightleftharpoons MY^{(2-n)-} + Hg^{2+}$$

$$Hg^{2+} + 2e = Hg \text{ (metal)}$$

The concentration of the ions of M fixes the concentration of the anions Y^{4-}, which then determine the concentration of the cations Hg^{2+}, which in their turn determine the electrode potential.

This kind of electrode has uses in potentiometric titrations (Chapter 1) but little or no importance in electrolysis. Electrodes of the second kind, however, are important in electrolysis.

IV. REVERSIBLE AND IRREVERSIBLE PROCESSES: ELECTROLYSIS

A student of thermodynamics soon becomes used to the idea of a "reversible process." The whole of classical thermodynamics is based on the concept of reversible processes and equilibrium. A "reversible process" is one that is held in check, or balanced, by an external force. When this force is relaxed ever so slightly, the process proceeds; when the external force is increased slightly the process is driven in reverse or "retrocedes." The electrochemical cell that is balanced by a potentiometer is an excellent example of a reversible system.

Once we use an electrochemical cell to supply electrical energy it is no longer reversible. The familiar lead storage battery, while it is supplying current, is not behaving reversibly in the thermodynamic sense. In another sense this cell <u>is</u> reversible, for it can be discharged and recharged many times, driving the reactions at each electrode back and forth each time. However, there is always a loss of energy, or rather a degradation of energy in which electrical energy is converted irreversibly to heat.

This chapter is about <u>electrolysis</u> and its use in chemical analysis. A process much used in chemical analysis is the electrolysis of an acidic solution of copper sulfate between platinum electrodes. Copper metal forms at one electrode in the form of an adherent deposit that is removed and weighed, while oxygen gas forms at the other electrode. We could represent the cell in the following way:

$$Cu / Cu^{2+}, \; H^+, \; H_2O / O_2, \; Pt$$

We can calculate the reversible electromotive force of this cell by using the Nernst equation and a table of standard potentials. If the cupric ions are 0.010 molar (for example) and the hydrogen ions 1.0 molar, and the oxygen is at 1 atmosphere pressure, the cell electromotive force is calculated to be 0.95 volt, with the oxygen electrode positive. However, one must apply a larger potential than this to the cell if one wishes to deposit copper at an appreciable rate. There are at least three reasons for this:

A. The Resistance of the Solution

To drive an electric current through a conductor - any conductor, be it a metal or a solution - one must overcome the electrical resistance of that conductor. The potential drop is directly proportional to the current in accordance with Ohm's law:

$$E = IR$$

where E is the electromotive force, or the drop in potential across the conductor, I is the current, and R is the resistance. If E and I are in volts and amperes, R is expressed in ohms.

In the cell described in Experiment 3-1 the resistance is of the order of 1 ohm, more or less. To pass a current of 1 ampere one must therefore apply a potential of 1 volt. This is over and above the 0.95 volt that we calculated as the electromotive force of the reversible cell.

B. Concentration Polarization at the Copper Electrode

At the negative electrode, where electrons are entering the cell (we call this electrode the cathode) copper ions are depositing as copper metal, and the solution in immediate contact with the electrode contains less copper ions per unit volume than does the bulk of the solution. This effect is called concentration polarization. Because the copper ion concentration next to the electrode is smaller than the bulk concentration, the potential of the electrode is less positive than it would have been in a reversible cell. Instead of $Cu^{2+} = 0.010$, as we assumed in the calculation on page 72, the true concentration may be several powers of ten smaller. This has the effect of making the cell electromotive force greater than the 0.95 volt that we calculated.

Concentration polarization is greater, the greater the current. It can be reduced, but not eliminated, by efficient stirring.

C. Chemical Polarization, or Overvoltage, at the Oxygen Electrode

The word "polarization," applied to an electrochemical cell, means any process or effect that obstructs the passage of current at or near the electrode surface. It does not include the Ohm's law potential drop, nor does it include reversible potentials, but it does include concentration changes near the electrode, like the one we have just discussed, and it includes any chemical process that obstructs or hinders the transfer of electrons across the surface of the electrode.

The removal of electrons from water to yield oxygen gas is such a process. It requires a certain "activation energy," just as does any slow chemical reaction. This activation energy can be overcome by raising the electrode potential, by making the platinum electrode more positive. The faster we want oxygen to be produced, the more additional potential, or overvoltage, we must provide.

This effect adds an additional 1 volt, or so, to the potential drop that must be placed across the cell. To electrolyze an acidic solution of a copper salt, depositing copper at one electrode and liberating oxygen gas at the other, therefore requires a considerably larger potential than the calculated reversible potential of 0.95 volt. A voltage of at least 3 volts, and probably more, is needed to pass a current of 0.5 - 1 ampere.

V. THE TERMS "ANODE" AND "CATHODE"

The words electrode, anode, and cathode were introduced in the early nineteenth century by Michael Faraday to describe the phenomena of electrolysis; "electrode" means a path for electricity, "anode" and "cathode" literally mean a "way upward" and a "way downward," respectively. Today we understand the terms as follows:

The cathode is the electrode through which electrons from the external circuit enter the cell. The cathode supplies electrons to the solution; therefore, reduction occurs at the cathode.

The anode is the electrode through which electrons leave the cell. In the solution inside the cell, therefore, oxidation occurs at the anode.

The words "anode" and "cathode" may be used to describe electrolysis, and they may also be used to describe a voltaic cell that is supplying current. Thus, in an ordinary flashlight battery with zinc and carbon electrodes, the zinc is the "anode" because it becomes oxidized while current is flowing, and the carbon is the "cathode." Where the use of these words does not make any sense is in describing a reversible cell at equilibrium. In such a cell, current may flow one way or the other, and a very small change in the externally applied electromotive force will cause the direction of current flow to reverse. Yet some authors do use the terms "anode" and "cathode" in connection with reversible cells. It is no wonder that their students get confused!

VI. OVERVOLTAGE

As we have seen, the production of oxygen at a platinum anode is accompanied by an overvoltage, a potential that must be applied in addition to the reversible potential calculated by the Nernst equation. Overvoltage is also observed when hydrogen is liberated at a cathode. Overvoltage depends on current, increasing with increasing current according to the equation

$$V = V_o + \frac{RT}{\alpha F} \ln I \qquad (3\text{-}5)$$

V is the overvoltage, V_o a number that depends on the electrode material and the temperature, α a number that is between 0 and 1 and is generally about 0.5, and I is the current density, the current divided by the area of the electrode.

The exact value of the overvoltage depends on factors that are difficult to control, such as the roughness of the electrode surface and the presence of minute films of impurities. Thus we cannot quote precise figures for overvoltages, but we can make some general statements that are a very useful guide. The overvoltage of hydrogen is greatest on mercury; here it is about 1.0 volt. It is nearly as high on cadmium and lead, somewhat smaller on tin and zinc. On copper and silver it is about 0.5 volt. On nickel and platinum it is smaller, and on platinized platinum, platinum covered with a layer of platinum black, it is almost zero. One can correlate the overvoltage of hydrogen on a metal with the effectiveness of that metal as a catalyst for hydrogenations; the better the catalyst, the lower is its hydrogen overvoltage.

It is because of hydrogen overvoltage that one can deposit metals from acid solutions that are more active than hydrogen. The electrolysis of zinc sulfate solutions is a very important industrial process; a large proportion of the zinc that is produced is made in this way. The standard reduction potential of zinc ions is -0.763 volt, and thermodynamically, the electrolysis of a zinc sulfate solution should produce hydrogen at the cathode and not zinc. Yet zinc is in fact produced, provided the acid concentration is not too high, and this is a result of the high overvoltage of hydrogen on zinc.

VII. THE MERCURY CATHODE

The overvoltage of hydrogen on mercury is high; furthermore, many metals dissolve in liquid mercury to form amalgams. Therefore, many metals, including metals more active than hydrogen, can be deposited in a cathode of liquid mercury and plated out of solution without interference from the liberation of hydrogen. Figure 3-1 shows the kind of cell that is used. Provision is made for good stirring, and a stopcock is provided so that the mercury, with the metals dissolved in it or sticking to it, can be separated from the solution. The electrical potential is maintained to prevent the metals from going back into the solution.

The following metals can be quantitatively deposited in a mercury cathode from dilute acid solutions: Zn, Cd, Fe, Co, Ni, Sn, Pb, Bi, Cu, Ag, Au. Metals not deposited include Al, U, Ti, V, the alkali and alkaline-earth metals. Manganese and molybdenum can be deposited from slightly alkaline solutions.

The mercury cathode is used in practice to "clean up" solutions, to remove unwanted elements before determining others. Suppose one wants to determine small amounts of aluminum, titanium, or uranium in a solution containing much iron; one removes the iron from the solution by electrolysis with a mercury cathode, then determines the element in the solution that is left. The quantity of iron is generally large; if so, the time needed for the electrolysis will also be large unless a high current is used, say 10 amperes or more. A high current means a rapid production of heat. For this reason most mercury cathode cells have cooling jackets; see Fig. 3-1.

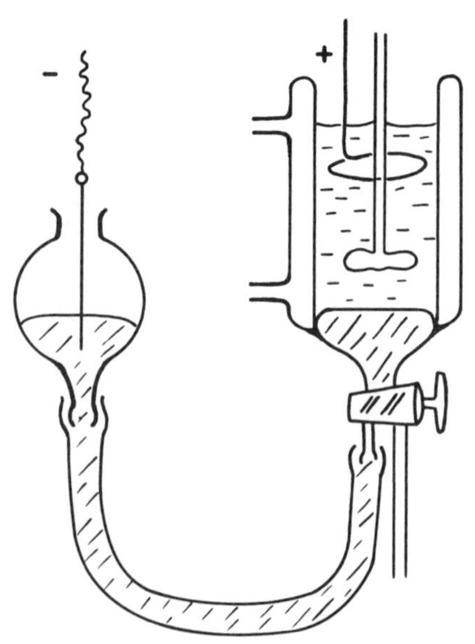

Fig. 3-1. Electrolytic cell with mercury cathode.

VIII. CONSTANT-CURRENT ELECTROLYSIS

The commonest arrangement used in electrolytic analysis is shown in Fig. 3-2. The electrodes are of platinum gauze mounted on frames of stout platinum wire. The inner electrode is cylindrical and is attached, while in use, to the chuck of an electric motor. This rotates the electrode and provides efficient stirring. The outer electrode may be cylindrical too, larger than the inner one and concentric with it, but it is more economical in platinum to use a flat rectangular frame of gauze as shown in Fig. 3-2. Either electrode can be made the cathode. It is customary to make the central, rotating electrode the cathode, but commercial instruments have a switch that allows one to reverse the polarity and make either electrode the anode or the cathode.

Experiments 3-1 and 3-2 describe the use of this equipment. Usually no special effort is made to hold the current or the voltage constant; one adjusts the rheostat to give the approximate current one wants, usually 0.5 - 1 ampere, and passes this current until one is sure that all the metal has been deposited. After the metal has been removed from the solution, hydrogen is deposited instead. This method of operation is simple and quite satisfactory is only one metal is to be deposited.

The time that the electrolysis will take may be estimated from Faraday's

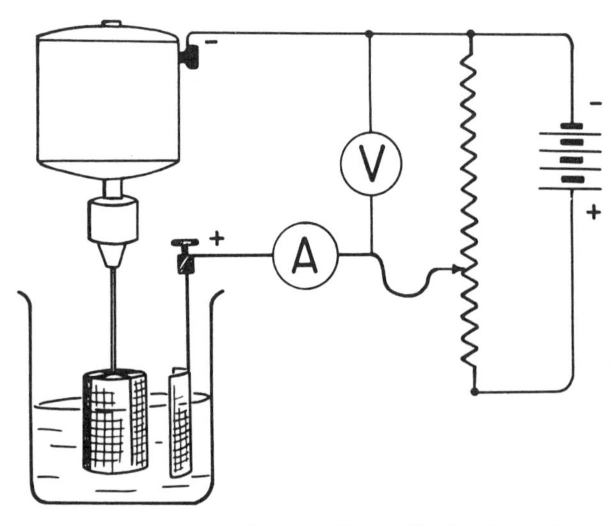

Fig. 3-2. Electrolysis with rotating cathode at constant current. (The direction of the current can be reversed if desired.)

law, which states that a certain quantity of electricity, 96,500 coulombs, is required to deposit one gram-equivalent of any chemical element; see Chapter 2. Suppose that a solution contains 0.100 gram of copper(II). How long will it take a current of 0.50 ampere to deposit this copper?

The cupric ions carry two charges; the atomic weight of copper is 63.6. Thus 1 millimole of copper(II), 63.6 milligrams, is 2 milliequivalents and needs $2 \times 96,500/1,000$ or 2×96.5 coulombs to reduce it to copper metal. Therefore, 100 milligrams Cu(II) needs $(2 \times 96.5 \times 100)/63.6 = 303$ coulombs of electricity. One coulomb of electricity is passed by 1 ampere in 1 second. Thus, 303 coulombs requires 1.0 ampere for 303 seconds, or 0.50 ampere for 606 seconds = 10 minutes, 6 seconds.

A calculation like this gives a minimum estimate of the time needed. The electrolysis always takes longer than this, because in its last stages, hydrogen is being liberated along with the metal, and this consumes current. One tests for complete deposition, either by a chemical test, or by raising the level of the solution, passing more current, and looking to see whether any new deposit of metal appears on the newly immersed surface.

IX. ELECTROLYSIS WITH CONTROLLED POTENTIAL

By controlling the potential of the cathode one can separate metals having different reduction potentials. An example is the separation of silver from copper. The reduction potentials of their ions are

$$Ag^{+} + e = Ag, \quad E^{o} = +0.799 \text{ volt}$$

$$Cu^{2+} + 2e = Cu, \quad E^{o} = +0.344 \text{ volt}$$

By maintaining the cathode potential at a value intermediate between these two standard potentials, say at +0.4 volt with respect to the standard hydrogen electrode, it should be possible to deposit silver from a solution without depositing any copper.

One can judge the effectiveness of such a separation by making a simple calculation from the Nernst equation.

Problem. Suppose the potential of a cathode is maintained at +0.400 V on the standard hydrogen scale. What is the minimum concentration of silver ions that would be left in the solution after electrolysis?

Answer. The potential

$$E = 0.400\ V = E^o + 0.059 \log Ag^+$$

$$E^o = +0.799, \quad \log Ag^+ = -\frac{0.799 - 0.400}{0.059} = -6.75$$

Ag^+ = about 2×10^{-7} M.

This is a very small value, undetectable by most techniques. We must emphasize that the calculated concentration is a minimum; the actual concentration that remains will be more than this, because of the delay in approaching equilibrium. Nevertheless the separation of silver from accompanying copper ions is virtually complete.

A way to control the cathode potential is shown in Fig. 3-3. The electrodes and the current supply for electrolysis are the same as shown in Fig. 3-2, but the outer electrode is now the cathode. Dipping into the solution and almost touching the surface of the cathode is the tip of a narrow tube, filled with potassium chloride or nitrate solution and forming a salt bridge to a saturated calomel reference electrode. The difference in potential between the cathode and this reference electrode is measured by the potentiometer circuit shown. This circuit is independent of the main electrolysis current. The procedure is to adjust the electrolysis current by the sliding contact shown in Fig. 3-2 so that the cathode potential, measured by the potentiometer shown in Fig. 3-3, remains within the limits desired. This adjustment may be made manually or automatically.

One may ask: Would it not be sufficient to watch the voltmeter shown in Fig. 3-2 and adjust the sliding contact so as to keep this voltage constant? The answer is no, for the voltage across the whole cell, from anode to cathode, depends on other things in addition to the cathode potential. It depends on the resistance of the solution between the electrodes and on the processes occurring at the anode. The reference electrode arrangement shown in Fig. 3-3 measures only the potential difference between the cathode and the solution next to the cathode.

By electrolysis at controlled potential one may deposit silver in the presence of copper, tin in the presence of cadmium, cadmium in the presence of zinc, and so on; complex-forming reagents may be added to the solution to increase selectivity. Controlled-potential electrolysis is not used much today, however, because other and more convenient methods of separation have become available.

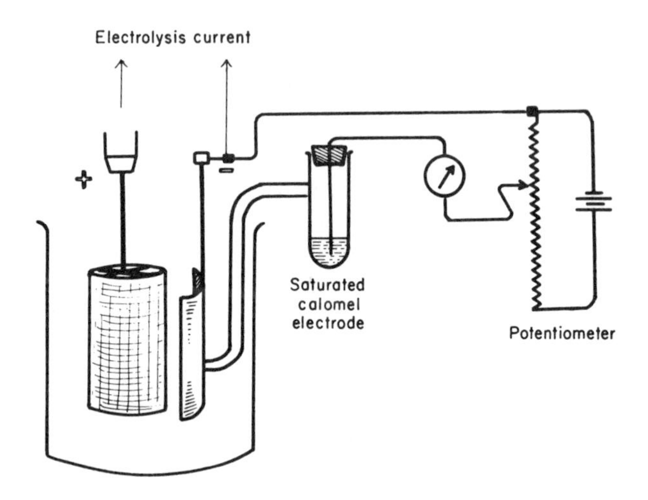

Fig. 3-3. Electrolysis at controlled cathode potential.

X. INTERNAL ELECTROLYSIS

This is a method of electrolysis that does not use an external source of current. It is also a method in which the potential is controlled. It is used to separate and measure small amounts of less active metals in the presence of large amounts of a more active metal.

Suppose one wants to determine copper, silver, and bismuth in "pig lead." This is impure lead that contains small amounts of less active metals as impurities. One may use the apparatus shown in Fig. 3-4 (see also Experiment 3-4). The sample of lead to be analyzed is dissolved in nitric acid and the solution, after dilution, is placed in the beaker. Mounted inside the beaker is a small porous pot or cup, usually of "alundum" or aluminum oxide, in which is a strip of pure lead foil or a spiral of pure lead wire surrounded by a dilute solution of lead nitrate. Outside this cup, and dipping into the sample solution, is a platinum gauze electrode which has been weighed beforehand. This platinum electrode is connected to the lead foil by means of a wire. Once this connection is made, copper and other metals more "noble" than lead start to deposit on the platinum electrode.

We have put together a cell in which the lead foil is the anode. It loses electrons and goes into solution as lead ions: $Pb = Pb^{2+} + 2e$

The electrons travel along the wire to the platinum electrode (the cathode) where they reduce the ions of any metals present that are less active than lead:

$$Cu^{2+} + 2e = Cu$$

$$Bi^{3+} + 3e = Bi, \text{ etc.}$$

When deposition has finished the cathode is removed, washed, and weighed. Deposition is hastened by stirring the solution.

The increase in weight of the cathode gives the total of all the metals whose ions are more easily reduced that Pb^{2+}. If one wishes to carry the analysis further and distinguish between the different metals (Cu, Bi, Ag, etc.) in the deposit, one may dissolve the deposit in nitric acid and apply further tests.

XI. ELECTROGRAPHY

This is a method of qualitative analysis, applicable to metals, alloys, and certain minerals and solid compounds that conduct electricity. The sample, which should have a smooth surface, is made the anode in an electrolytic cell that consists of a sheet of filter paper, wet with a salt solution, and pressed between the sample and a sheet of a pure metal like platinum, aluminum, or stainless steel. This metal is made the cathode.

A current is passed briefly between the two metal surfaces. Metal ions are drawn out of the anode surface (the sample) and into the filter paper, where they are detected by sensitive chemical tests. Very little of the sample is consumed, and the technique is applicable to many elements.

It is customary to use two sheets of paper, a sheet of close-textured or hardened filter paper next to the sample, and a sheet of thick, open-textured paper next to the cathode. Both sheets are wet with the same solution, a general-purpose electrolyte solution, which is a mixture of 3 volumes of 0.5 molar sodium carbonate to 1 volume of 0.5 molar sodium nitrate. Commonly a current of 20 milliamperes per square centimeter is passed for 10 - 60 seconds. If the sample surface is heterogeneous, with different metals predominating in different areas, the electrographic "print" produced in the hardened filter paper will show this fact.

Experiment 3-5 gives more information about this technique.

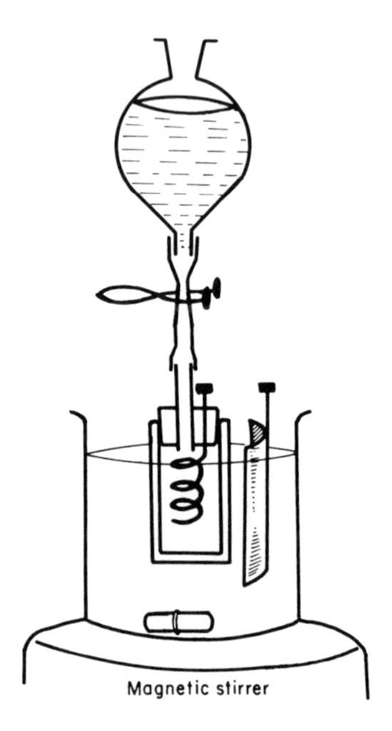

Fig. 3-4. Internal electrolysis.

EXPERIMENTS

Experiment 3-1

Determination of Copper and Nickel in a Solution

Equipment and Materials. Apparatus for electroanalysis as shown in Fig. 3-2. This may be home-made or a commercial model like the Eberbach or Sargent-Slomin analyzers. It must have provision for rotating the central electrode. A pair, or preferably two pairs, of platinum gauze electrodes are needed, as is an analytical balance.

Prepare a solution for analysis by weighing out accurately about 5 g each of hydrated copper sulfate, $CuSO_4 \cdot 5H_2O$, and nickel sulfate, $NiSO_4 \cdot 7H_2O$; dissolve these salts together in water in a 250-ml beaker, add (slowly and with constant stirring) about 25 ml concentrated sulfuric acid or an equivalent volume of diluted sulfuric acid, then transfer quantitatively to a 1-l. volumetric flask and make up to the mark. (Tap water can, of course, be used, since it contains no copper or nickel.)

You will also need beakers of a convenient size for the electrolysis (250 ml is suitable), a 25-ml pipet, and concentrated nitric acid and ammonia.

First, clean the platinum electrodes by rinsing in nitric acid of about 5 M concentration (this is made by mixing concentrated nitric acid with about twice its volume of water) and then in distilled water. Dry the cathode (this is the cylindrical electrode that is rotated during electrolysis) by holding it by the end of the central stem and waving it over a small burner flame or hot plate until it is just dry. Do not heat unnecessarily, and above all, do not touch the gauze with your fingers. Set the electrode down on a watch glass or clean surface to cool to room temperature, then weigh it to a tenth of a milligram.

Place the two electrodes in position on the electrolysis apparatus. Rotate the cathode once or twice by hand to make sure that the two electrodes do not touch one another.

Take a 250-ml beaker and add to it, using a pipet, 25.00 ml of the copper-nickel solution for analysis. Then add water to fill the beaker about halfway, and add about 1 ml of concentrated nitric acid. This improves the adhesion of the copper deposit. If the nitric acid is markedly yellow or brown, however, this means that it contains nitrous acid, which prevents complete deposition of the copper. If your nitric acid is brown or yellow, it is advisable to add about 1 g of urea to destroy the nitrous acid.

Raise the beaker or lower the electrodes, depending on how your apparatus is arranged, until the solution covers about two-thirds of the platinum cathode, leaving about one-third of it exposed. Now start the electrode rotating, and turn on the electrolysis current. Adjust the current to 0.5 A. A reddish deposit of copper appears immediately. It should appear on the central electrode, the one you have weighed. If it forms on the outer electrode instead, simply reverse the polarity; most instruments have a switch to do this. No harm is done; the copper dissolves from one electrode and deposits on the other.

Continue the electrolysis for a sufficient time (calculated as explained on page 82) that you expect all the copper to be deposited. One indication will be a rapid production of hydrogen gas at the cathode. Now test for complete deposition of copper as follows: Lower the cathode a little further into the solution, or raise the liquid level by adding water. If copper is still depositing, you will see more copper forming on the newly immersed platinum surface. If this happens, pass current for 10 min more and repeat the test.

When all the copper has deposited, raise the electrodes out of the solution while keeping the electrolysis current turned on. This is necessary to prevent copper from going back into solution. Rinse the electrodes with water from the wash bottle, then turn off the electrolysis power and carefully remove the cathode from its holder. Rinse the cathode (with the copper deposit) with some more distilled water, then dry it.

Do not dry the electrode in the oven, or you will cause the copper deposit to oxidize. It is porous and very reactive. A quick way to dry it is to rinse it in acetone, then let the acetone evaporate in the air.

Be especially careful not to touch the deposit of copper, for it is soft and easily brushed off. Place the dry electrode on the balance and weigh again to a tenth of a milligram. Leave the electrode on the balance pan for a minute or two and re-weigh; if it has lost weight, this shows that it was not completely dry.

After weighing, dissolve the copper coating by dipping the electrode in a freshly made mixture of nitric acid and water (about 1 volume concentrated acid to 2 of water), then wash with water. There is no need to weigh it again, except as a check.

Replace the cathode in the electrolysis apparatus. To the solution in the beaker, which contains the nickel, add concentrated ammonia, slowly and with stirring. First the ammonia neutralizes the excess acid, then it precipitates green nickel hydroxide, then, as more is added, it dissolves the nickel hydroxide, giving a deep blue-violet solution of nickel-ammonia complex ions. You are now ready to continue the electrolysis. Immerse the electrodes in the solution as you did before, start the stirring, and pass current as before. Nickel deposits on the cathode as a black coating. When the solution has lost its blue color, this is a good indication that the nickel has all deposited, but a test should be made in the same way as was done with the copper. (You can also remove 1 ml of the solution and test it for nickel ions by adding dimethylglyoxime; nickel forms a bright red precipitate.)

Remove the cathode, wash, dry, and weigh it as before. Report the concentrations of copper and nickel in the test solution, in grams per liter.

Experiment 3-1a

Determination of Copper in Copper Ore

This experiment can be done instead of Experiment 3-1. The "unknown" is a copper sulfide or copper oxide ore, finely ground. The main difficulties in this determination are: (1) obtaining a properly representative sample, for the powdered ore tends to segregate, the dense copper mineral settling to the bottom of the bottle that contains it, while the lighter impurities, like silica, tend to move toward the top; (2) getting the ore into solution.

If the ore sample is already ground, mix it thoroughly before taking a portion for weighing. First dry it in the oven at 110°C. Then weigh two samples, each of sufficient size to contain 50 - 150 mg of copper. Place them in separate 250-ml beakers, and add 10-15 ml concentrated nitric acid. Cover with a watch glass and heat gently over the hot plate until all dark material has dissolved. If nitric acid alone does not break up the ore, add a few milliliters of concentrated sulfuric acid and heat until dense white fumes of sulfuric acid vapor appear. Then cool thoroughly and add water. The solution should be blue, and any solid residue should be white. If the residue is still black, pour (decant) the clear solution into another beaker and heat the residue with some more nitric and sulfuric acids.

Filter the solution before electrolysis to remove any solid residue, then electrolyze to deposit the copper, as described in Experiment 3-1. If nitric acid is present, add 1-2 g of urea. As was noted above, the function of the urea is to decompose the nitrous acid; the nitric acid itself does no harm.

Experiment 3-2

Simultaneous Determination of Copper and Lead in Brass

Brass is an alloy of copper and zinc, containing between 50% and 80% of copper, with up to 4% of lead and up to 1.5% of tin, together with other minor constituents. On treatment with nitric acid the copper, zinc, and lead dissolve, while tin forms a white insoluble oxide that can be filtered off. When the solution is electrolyzed, copper deposits at the cathode while lead dioxide deposits at the anode. Both deposits are weighed.

If much tin is present the precipitated tin dioxide may absorb significant quantities of copper and lead, and a special step is necessary to recover the absorbed copper and lead.

Equipment and Materials. The equipment is the same as that in Experiment 3-1. Materials needed are: brass sample; nitric and sulfuric acids, concentrated; sodium hydroxide and sodium sulfide, solid; ammonium hydroxide, concentrated; hydrogen peroxide, 3%; urea; acetone; alcohol; filter paper; and filter paper pulp.

If the brass contains more than 0.05% of tin, proceed as follows. Weigh accurately 2 g of brass, place in a 250-ml beaker, and add 25 ml of 1:1 nitric acid (1 volume of concentrated acid to 1 volume of water). Cover the beaker with a watch glass. Warm gently until all the brass has dissolved; boil until most of the brown fumes have gone. (Note: this must be done in the hood.) Now add 50 ml of water, heat to boiling; if there is no white precipitate, no tin is present, and one can start the electrolysis immediately. But if there is more than a trace of white cloudy precipitate, add some filter paper pulp and stir until this has all dispersed, then filter the hot solution through close-textured filter paper into a second 250-ml beaker. Wash the filter paper with several small portions of hot dilute nitric acid (1 volume concentrated acid to 100 volumes of water) until no blue color is seen on the paper. Reserve the filtrate and washings, which contain most of the copper, and treat the precipitate to recover the absorbed copper and lead, as follows:

Take the filter paper out of the funnel and place it in the beaker that originally contained the sample. Add to it 25 ml concentrated nitric acid and 15 ml concentrated sulfuric acid. Heat (in the hood) until the color of the solution, which at first is black, turns to transparent red or brown. Transfer the solution to a larger beaker, 400 or 600 ml, and dilute with water to 250 ml. Then, carefully and slowly, stir in a solution of sodium hydroxide, about 50 g sodium hydroxide to 100 ml water, until the solution is alkaline. The tin is now in solution. Add 5 g of sodium sulfide dissolved in 20 ml of water and stir. Heat gently. Copper and lead, if present, form a black precipitate. Cool; filter the black precipitated sulfides on close-textured filter paper; wash the precipitate and paper with a solution of 2 g sodium sulfide in 100 ml water.

Dissolve the precipitated sulfides in a minimum volume of 1:1 nitric acid, and add this solution to the "reserved" solution (the filtrate from the tin precipitation) that contains most of the copper and lead. Add 2 g of urea to the solution.

You are now ready to start the electrolysis. First, clean both electrodes, the anode as well as the cathode; dry and weigh each electrode. Read the instructions for Experiment 3-1; be careful not to touch the electrodes except by the wire or shaft that is not immersed in the electrolyte. Set the electrodes in place on the electrolysis apparatus, check to see that they do not touch when the center electrode is rotated, then start the stirring motor. Start the electrolysis current, making the center electrode the cathode; pass a current of 0.5 A, increasing this to 1 A after a few minutes.

Calculate the minimum time needed to plate out all the copper (see page 82) and electrolyze for twice this time. Then test for complete deposition in the way described in Experiment 3-1.

When deposition of both copper and lead is complete, raise the electrodes out of the solution without turning off the electrolysis current, as described in Experiment 3-1, and wash them with water from the wash bottle. Then turn off the electrolysis current and carefully disconnect the electrodes. Dry them by washing with alcohol and then acetone, and leaving in the air until the acetone evaporates. Weigh them immediately. When you have verified that the weight does not decrease on the balance pan, and that they are therefore dry (see Experiment 3-1), clean the metal coatings from the electrodes by dissolving them in 1:1 nitric acid. If the lead dioxide is slow to dissolve, add to the nitric acid a few milliliters of hydrogen peroxide (3%). Clean the electrodes thoroughly.

If the brass contains less than 0.05% of tin, it is not necessary to redissolve the precipitated tin dioxide. Simply filter any slight precipitate that may have formed when the alloy was first dissolved, using filter paper pulp if necessary, wash the filter, and electrolyze the combined filtrate and washings, after adding 2 g of urea to destroy nitrous acid.

To calculate the percentage of lead in the brass, consider that the fraction of lead in the lead dioxide deposit is 0.866. This is the theoretical fraction in the pure compound PbO_2. Actually, the proportion of lead is somewhat less than this. Lead dioxide is what is called a "nonstoichiometric compound" whose composition does not correspond precisely to a simple chemical formula, because of irregularities in the crystal structure. The difference is small, less than 0.5%, and since it is not reproducible it is best to use the theoretical factor of 0.866. When lead must be determined with great accuracy, another weighing form must be used, such as lead sulfate. One may also use titration with EDTA.

Experiment 3-3

Aluminum in a Zinc Alloy

In this experiment, electrolysis at a mercury cathode is used to remove zinc and metals less active than zinc, leaving aluminum in solution. The "unknown" can be a zinc-base die-casting alloy containing up to 5% of aluminum. The alloy may also contain magnesium; this will not interfere with the determination.

Equipment and Materials. A cell is needed like that shown in Fig. 3-1. These may be purchased commercially. Also a platinum crucible of capacity 15-20 ml, with cover. Reagents include hydrofluoric acid, potassium bisulfate (solid), potassium ferrocyanide (5% solution), and methyl red indicator.

Weigh accurately 1 g of the sample, place in a 400-ml beaker, add 25 ml of water and 5 ml concentrated sulfuric acid. Cover the beaker with a watch glass and warm gently until the alloy has all dissolved. There may be a residue of copper metal; do not try to dissolve this. Transfer the entire solution to the electrolysis cell shown in Fig. 3-1, adding water so that the platinum wire anode is covered. Electrolyze at about 5 A, with stirring.

Theoretically, if no other reaction occurred at the cathode but the deposition of zinc, it should take 10 min for a current of 5 A to deposit 1 g of zinc. (Check this calculation!) However, a good deal of hydrogen is produced at the cathode and more time is needed to remove the zinc from the solution. After 1 hr, test for the presence of zinc by removing about 1 ml of the solution, placing it in a test tube, and adding a few drops of 5% potassium ferrocyanide solution. A white precipitate indicates zinc. If no zinc is found, pass current for 10 min more to be on the safe side, then, without turning off the electrolysis current, lower the reservoir shown on the left in Fig. 3-1 until the mercury just drains out of the cell. Now turn the two-way stopcock through 180° and run the solution (the electrolyte) out into a 400-ml beaker. Rinse the inside of the cell and run the rinse solution into the beaker. Now the electrolysis power can be turned off. The aluminum should all be in the beaker. If any mercury or insoluble material is present, remove it by filtering or decanting.

Now heat the solution to boiling and add 5 ml concentrated hydrochloric acid, or 2 g of ammonium chloride (this is to help control the pH). Add a few drops of methyl red indicator. This should turn red, though if the solution is very strongly acid it will appear orange. While the solution continues to boil, add concentrated ammonia solution. As soon as the indicator starts to turn yellow, add the ammonia very slowly, drop by drop, and stop as soon as the color is a uniform yellow (pH 6). A cloudy white precipitate of aluminum oxide should have appeared. Add a little filter paper pulp, boil for 1 min, then filter through fast filter paper, and wash the precipitate on the filter with hot 1% ammonium chloride solution.

Precipitated aluminum oxide (or hydroxide) is colloidal and some of it may pass through the filter. Look carefully at the first filtrate, and if it is cloudy, pour it back through the filter paper.

If much magnesium is present, some of it may have coprecipitated with the hydrous aluminum oxide, and the aluminum oxide should be redissolved and reprecipitated. This only takes a few minutes. Lift the filter paper and precipitate out of the funnel, put it back in the beaker, pour on it 25 ml 6 M hydrochloric acid, and break up the filter paper with a glass rod. Heat to boiling, add 100 ml water, and boil again. Add a drop of methyl red, then ammonia, as before, until the indicator just turns yellow. Boil for 1 min longer, then filter, as before, through fast filter paper, washing with 1% ammonium chloride solution.

Transfer the filter paper and precipitate to a weighed platinum crucible, capacity 15-20 ml. Dry carefully over a low flame (do not let the water spurt out of the crucible) and when all the water has gone, raise the temperature to carbonize the paper, then heat strongly until all the paper has burned and white aluminum oxide, Al_2O_3, remains. This solid is hygroscopic, so place the crucible in a desiccator as soon as it is cool enough, cover the crucible with its cover, cool, and weigh.

Heat and cool once more, and repeat if necessary until the weight has become constant.

The precipitate may contain silica, SiO_2. If so, take the platinum crucible containing the ignited precipitate, and add 4-5 g of potassium hydrogen sulfate, $KHSO_4$, or potassium pyrosulfate, $K_2S_2O_7$. Carefully melt the sulfate over a flame and heat until a clear red-hot liquid is obtained. Cool, place the crucible in a beaker, extract the solid mass with water containing about 5% of sulfuric acid by volume, remove and rinse the crucible, and evaporate the solution until dense white fumes of sulfuric acid appear. Cool, add 100 ml water, boil, and filter the solution through close-textured filter paper. Silica will be held on the paper while aluminum sulfate passes through. Wash the filter paper with a little hot water, then transfer it to the weighed platinum crucible, dry and ignite. Weigh the crucible and contents. The residue in the crucible is silica, but the silica will still contain some absorbed aluminum. Therefore, cool the crucible, moisten the silica with a few drops of sulfuric acid (1:1 by volume, or about 10 M), and add several milliliters of hydrofluoric acid (concentrated, 48%). Evaporate in the hood, carefully; hydrofluoric acid tends to spurt when it is boiled. Heat the crucible to redness, cool, and weigh. The loss in weight caused by the hydrofluoric acid treatment gives the weight of silica.

You have made four weighings:
(a) Weight of empty crucible and cover;
(b) Weight of crucible, cover, and ignited aluminum oxide;
(c) Weight of crucible and ignited silica precipitate;
(d) Weight of crucible and residue after HF evaporation.
The weight of SiO_2 equals (c) - (d); the weight of Al_2O_3 is (b) - (a) - (c) + (d).

Cleaning the Mercury. The mercury left in the reservoir of the electrolytic cell contains dissolved zinc and probably small amounts of other metals. The same quantity of mercury can be used for several electrolyses before it is necessary to clean it. To clean mercury, put it in a thick-walled suction flask, cover it with 5 M nitric acid, and using a water pump or aspirator, draw air through the mercury and acid for several hours; see Fig. 3-5. Then pour off the acid (which contains the impurities as well as a little mercurous nitrate) and wash the mercury several times with water.

IMPORTANT NOTE: MERCURY IS POISONOUS. Mercury vapor, breathed over a long time, accumulates in the body and can have very serious effects. Therefore, be very careful not to spill mercury or let little

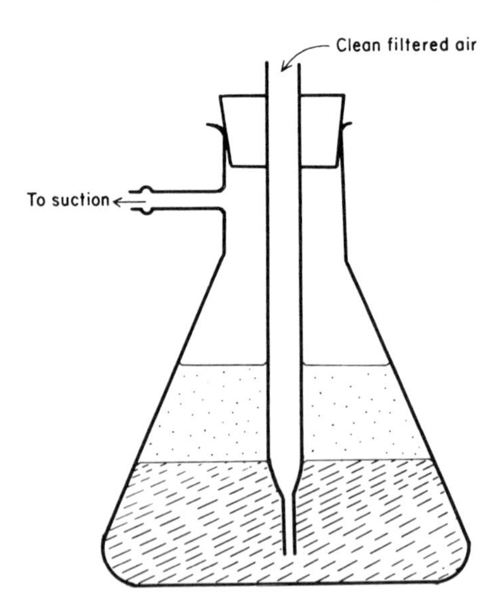

Fig. 3-5. Arrangement for cleaning mercury.

drops of mercury escape where their vapor may contaminate the room.

An Alternative Way to Precipitate Aluminum. The method we have described is the "standard" method and gives maximum accuracy in experienced hands. An easier and faster way to determine aluminum gravimetrically is by precipitation with 8-hydroxyquinoline. To the acid solution containing aluminum add an excess of a 5% solution of 8-hydroxyquinoline in 2 M acetic acid, heat to boiling, and add ammonia carefully. At about pH 4 a yellow precipitate of $Al(C_9H_6ON)_3$ appears. Filter through a weighed sintered glass crucible, wash with hot water, and dry at 130°C. The precipitate may contain absorbed silica, but it has less tendency to absorb silica than does hydrous aluminum oxide, and its molecular weight is much higher than that of aluminum oxide, which makes the error less.

Experiment 3-4

Internal Electrolysis: Bismuth, Copper, and Silver in "Pig Lead"

Equipment, Materials. See Fig. 3-3. The figure shows one porous cup and anode compartment, but two cups can be used with anodes in parallel, one on each side of the platinum gauze cathode. The anodes can be lead wires, strips of lead foil or sheet, or sections of lead pipe from a plumbing shop. Reagents needed include lead nitrate, tartaric acid, urea, a dilute solution of potassium permanganate.

The "unknown" can be "pig lead," impure lead that has not been refined and contains up to 0.5% of other metals, including bismuth, copper, and silver. It can also be galena, lead sulfide mineral.

Weigh accurately between 10 and 20 g of the sample, place it in a 400-ml beaker, and dissolve it in 100 ml of nitric acid, freshly made by mixing 25 ml concentrated acid with 75 ml water. Heat gently in the hood, and add 1 g of solid tartaric acid. Lead metal dissolves to a clear solution; galena will yield a mass of sulfur which liquifies when the solution is boiled, and the contents of the beaker must be kept stirred so that the molten sulfur does not prevent lead sulfide from dissolving. Keep boiling the solution gently until the brown nitrous fumes are gone. Add more nitric acid if needed to dissolve the sample.

If sulfur, silica, or other undissolved material is present, cool to let the sulfur solidify, and filter or decant the liquid into another beaker. If the solution is clear, this step is unnecessary. Dilute the clear solution with water at about 250 ml, cool to 40°C, and add dilute permanganate (about 1 g/100 ml) drop by drop until a pink color remains for 1 or 2 min. Heat and add about 0.1 g of urea. This solution is now ready to be placed in the electrolysis cell.

While the above solution (the catholyte) is being prepared, prepare the solution to be placed in the porous cup, the anolyte, by dissolving 5.0 g of lead nitrate in 100 ml water and adding 3-5 ml concentrated nitric acid. Also clean and weigh the platinum gauze cathode.

Assemble the apparatus shown in Fig. 3-3. Fill the porous cup and the reservoir above it with the anolyte, and pour the catholyte into the beaker. Adjust the level of the solution so that about two-thirds of the platinum gauze cathode is immersed. Connect the lead anode and the platinum cathode with a wire. A dark metallic deposit should start to form on the cathode. Keep the solution stirred, either by a magnetic stirrer as shown, or by a glass paddle stirrer introduced from above. Heat the solution to about 70°C if possible; this will hasten the deposition. Use a combination magnetic stirrer - hot plate if one is available. At intervals during the electrolysis, allow some of the anolyte solution to flow out of the reservoir through the porous cup. This will serve the double purpose of keeping catholyte from diffusing into the porous cup and immersing new areas of the cathode. When no dark deposit appears on a newly immersed part of the cathode, electrolysis is complete.

When you are satisfied that no more metal is being deposited, lower the beaker or raise the electrodes until the cathode is out of the solution. Disconnect the cathode, wash it well with distilled water from a wash bottle, then wash it with acetone and let it dry. Warm it, if you like, to not over 80°C to make it dry faster. As soon as the cathode is dry, cool and weigh it. Report the percentage of inactive metals in the sample.

Remove the deposit from the cathode by dissolving it in freshly mixed nitric acid, 1:3. If you wish, you can determine the copper content of the solution by titration with thiosulfate (after adding potassium iodide; directions will not be given here) or by another method of your choice.

Experiment 3-5

Electrographic Analysis

The apparatus is shown in Fig. 3-6. Equipment can be bought commer-

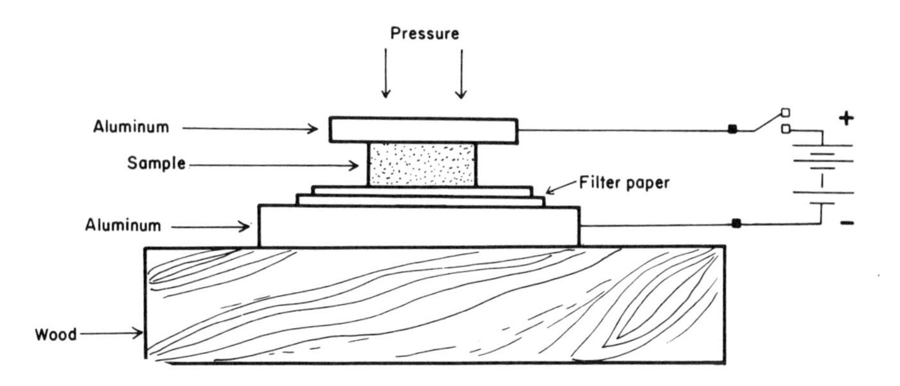

Fig. 3-6. Electrographic analysis.

cially or can be assembled in the laboratory. It is important to have sufficient pressure on the sample and to use two layers of filter paper, one (next to the sample) a hardened paper such as S&S No. 575, the other a soft paper like S&S No. 601.

Some typical tests are described here, but any metal can be identified if the proper "spot test" or color reaction is used.

A. Identification of a Pure Metal. Place the papers between the anode (the sample) and the cathode as shown, and wet them with a solution made by mixing 3 volumes of 0.5 M sodium carbonate and 1 volume of 0.5 M sodium nitrate. Pass a current of some 25 mA per square centimeter of the sample for about 10 sec. Remove the upper layer of paper, and test it as follows:

(1) Preliminary examination: Hold the paper in the vapor above concentrated ammonia solution. A strong yellow color indicates Cr; red-brown, Fe, pale green, Cu or Ni; gray (after exposure to light), Ag.

If Fe and Cr are absent, cut the paper into four segments and test each individually as follows:

(2) Tests for Ag, Cu, Ni: Place one segment of paper on a watch glass and place on it a few drops of photographic developer. A black stain indicates Ag. Confirm Ag by testing another part of the sample paper with potassium dichromate; brick-red silver chromate confirms Ag.

If the first test for silver was negative, add a 1% solution of potassium ethyl xanthate. A light yellow color indicates Cu; orange indicates Ni.

The reagent rubeanic acid (see Chapter 11) may also be used to test for Co, Ni, and Cu. The presence of Ni may be confirmed by dimethylglyoxime (red color).

(3) Tests for Cd, Sn, Pb: Place another segment of the paper on a watch glass and add a few drops of a solution of 1 g sodium sulfide and 2 g ammonium acetate in 100 ml water. Let the paper soak for 1 or 2 min, then wash it free from sulfide by adding 1% acetic acid and decanting, then washing with water by decantation. Check, by testing the washings with lead acetate, that all the sulfide has been washed out.

A yellow color remaining in the paper indicates cadmium; a dark brown or black, Sn or Pb. If Sn or Pb is indicated, wash the paper on the watch glass with a solution containing 5 g sodium hydroxide, 5 g sodium sulfide, and 1 g sulfur in 25 ml water. Wash several times with small portions of this solution. Tin sulfide dissolves and disappears; lead sulfide (black) remains.

To confirm Pb, take another segment of the paper and wet it with 5% potassium dichromate solution, then wash it with several portions of 5% (1 M) acetic acid. A yellow stain of lead chromate confirms Pb.

B. Identification of Zn, Sn, Pb, and Fe in a Copper Alloy. Place papers between the sample and the cathode and wet them with the same sodium carbonate - sodium nitrate solution as before. Electrolyze for 30 sec at about 20 mA/cm^2. Remove the papers, cut the hardened paper into two, and test one-half for Zn and the other for Pb, as follows:

(1) Test for Zn: Place the paper on a watch glass and cover it with a freshly prepared solution that is 1 M in KCN and 1 M in Na_2S. Stir for 2 min, and then wash with the following four reagents in the order given:

(i) A solution 0.25 M in KCN, 0.25 M in NH_4Cl, and 1 M in NH_3;
(ii) Sodium polysulfide solution (Na_2S plus added sulfur);
(iii) Water;
(iv) Acetic acid, 1 M.

If zinc is present, the paper will still contain precipitated zinc sulfide. This is white and not easily visible. To make it visible, moisten the paper with 5% lead acetate solution. Brown or black lead sulfide forms if zinc was originally present.

(2) Test for Pb: Place the second half of the paper on a watch glass and add equal volumes of 0.2 M potassium dichromate and 1 M acetic acid. Stir for 2 min; wash with 1 M acetic acid by decantation until the orange

dichromate color disappears. A yellow stain which remains is lead chromate and indicates Pb.

To test for iron, place another pair of papers between the sample and the cathode and wet the papers with the same solution as before. Electrolyze for 1-2 min at 20 mA/cm^2. Place the hardened paper on a watch glass and wash with a solution 0.5 M in ammonia and 0.5 M in potassium or ammonium nitrate until the blue color due to copper has disappeared; then wash with water. Add a few drops of 0.2 M potassium ferrocyanide and then a few drops of 1 M acetic acid. The dark blue color of "Prussian blue," $Fe_7(CN)_{18}$, indicates Fe.

Note that red-brown copper ferrocyanide will also form if the copper was not completely removed.

To test for tin, three layers of paper are used. The basis of this test is that stannous ions, Sn(II), reduce ammonium phosphomolybdate to molybdenum blue. Contact with any metal more active than silver will reduce this reagent also, therefore the reagent must not come into contact with the specimen.

Prepare a solution of 5 g precipitated ammonium phosphomolybdate (make this from ammonium molybdate and a soluble phosphate in the presence of dilute nitric acid) in 100 ml water plus enough ammonia to dissolve the yellow phosphomolybdate. Dip a sheet of the hardened filter paper in this solution in the absence of bright light - the solution is sensitive to light - and put it in the dark to dry; just before use, dip it in 1 M nitric acid.

Dip another piece of hardened paper (No. 575) and a piece of soft paper (No. 601) in 0.5 M sodium nitrate (omit the sodium carbonate this time), quickly place the three papers in position - hardened paper with $NaNO_3$ next to the sample, hardened paper with phosphomolybdate in the middle, soft absorbent paper next to the cathode - apply pressure and pass at 20 mA/cm^2 for 60 sec. A blue color indicates Sn.

A yellow background color can be removed by washing with 0.2 M potassium hydroxide solution and then water. The blue stain that is left is stable to light.

Report the elements that are present in your copper alloy. (A copper coin is a good "unknown" for this experiment.)

QUESTIONS

1. How many milligrams of copper are brought into solution at a copper anode when a current of 20 mA is passed for 30 sec?

2. Devise electrographic tests for manganese and molybdenum in steel.

3. In internal electrolysis with a lead anode in a solution 0.10 M in lead ions, what is the theoretical minimum concentration of cupric ions that would remain undeposited at the cathode?

4. A solution 0.20 M in $NiSO_4$ and 0.20 M in H_2SO_4 is electrolyzed between two platinum electrodes. The overvoltage of oxygen on platinum is 0.40 V; the overvoltage of hydrogen on nickel is 0.30 V. Assuming that the potential drop across the internal resistance of the solution is negligible, calculate (a) the voltage across the cell that will deposit nickel, (b) the voltage that would deposit hydrogen. (Remember that only the first stage of ionization of H_2SO_4 is appreciable in 0.2 M solution.)

   Note: Overvoltage is required to deposit nickel metal itself; this is unusual in the discharge of metals. Neglect this overvoltage in your calculation.

5. An electrode of copper metal, and another electrode of platinum coated with lead dioxide, dip into a solution which is 0.005 M in Cu^{2+}, 0.001 M in Pb^{2+}, and 3.0 M in H^+. Calculate the EMF of the cell thus formed. Which electrode is positive, that is, which metal has the lower affinity for electrons?

6. A mercury cathode is 2.5 cm in diameter; a current of 4.0 A is passed. What is the current density in amperes per square centimeter?

7. The overvoltage of hydrogen on mercury for the conditions of question 6 is 1:10 V. A cell is made with this mercury cathode and a platinum anode. The overvoltage of oxygen on the platinum anode is 0.45 V, and the internal resistance of the cell is 1.5 ohms. The current, given in question 6, is 4.0 A. Calculate the potential across the cell, if the electrolyte is 1.0 M sulfuric acid. Does the nature of the electrolyte make any difference in this problem? Does the concentration of hydrogen ions make any difference?

Chapter 4

POLAROGRAPHY

I. INTRODUCTION

Polarography is a method of analysis in which the solution to be analyzed is electrolyzed in such a way that the graph of current against voltage shows what is in the solution and how much is present. The method was developed before 1930 by the Czech chemist Jaroslav Heyrovsky, who won the Nobel Prize for his discovery.

The basic idea is to pass the current between two electrodes, one large in area, the other very small. Normally both electrodes are of mercury. The large electrode is a pool of mercury at the bottom of the cell. The small electrode is a drop of mercury coming out of a very fine capillary tube. Mercury flows out of this tube, and drops form and fall off every few seconds. The dropping complicates the current flow, but ensures that the mercury surface is always clean. The drop is generally the cathode, or negative pole of the cell. The solution contains one or more substances that can be reduced at the cathode. The substances may be organic compounds, but more usually are ions of "heavy" metals like copper, lead, or zinc. These ions gain electrons at the surface of the mercury drop and are reduced to metal atoms; the metal enters the mercury and forms an amalgam.

A simple cell and electrical circuit are shown in Fig. 4-1, and an idealized graph of current against applied potential (or voltage) is shown in Fig. 4-2. The circuit is that of a potentiometer; compare Fig. 1-13. The mercury drop is made increasingly more negative with respect to the anode (the mercury pool) by moving the sliding contact Q along the wire from P to R. As the voltage across the cell is increased, the current goes up in a series of steps (Fig. 4-2). The voltage, V_1 or V_2, at which the step occurs is characteristic of the ion being reduced. It is less negative for cad-

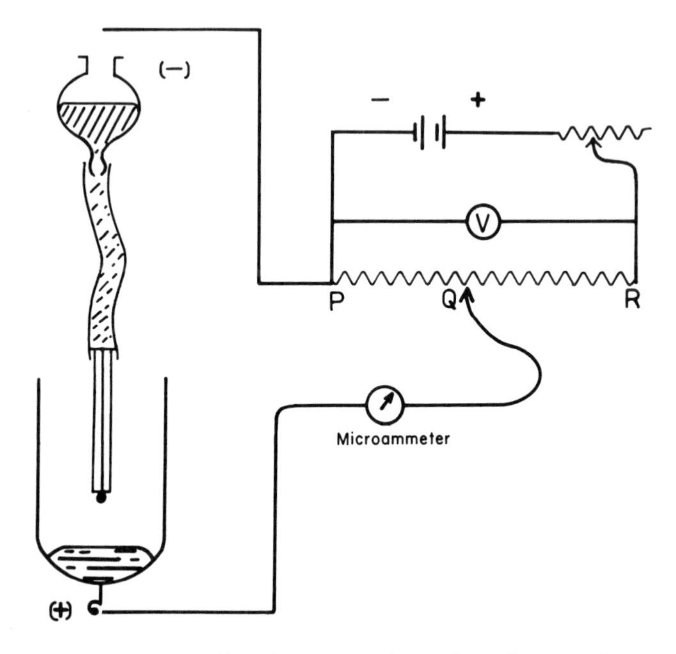

Fig. 4-1. Cell and circuit of simple polarograph.

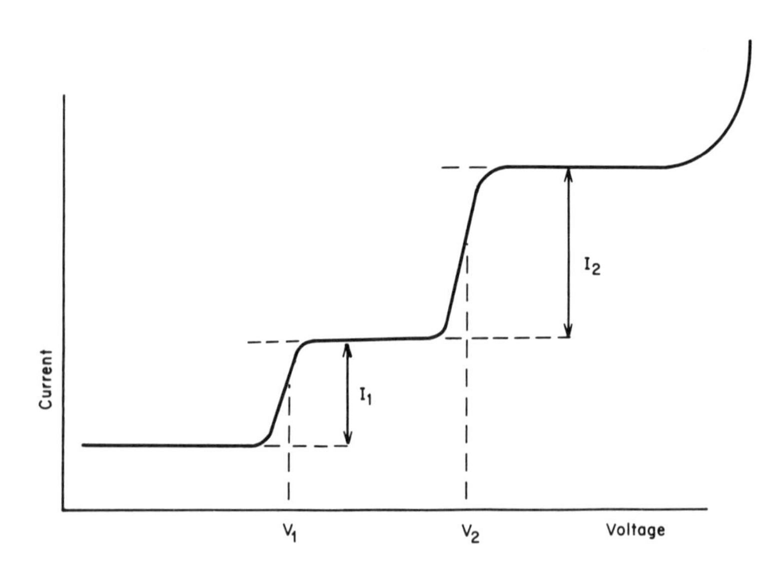

Fig. 4-2. Idealized current-voltage curve.

mium than for zinc, for example, because cadmium is lower in the electromotive series than zinc and its ions are easier to reduce. The voltage thus gives a qualitative indication of what ion is present. The height of the step gives a quantitative measure of the amount of this kind of ion. In favorable cases, four or five different steps or "waves" may be seen, permitting the measurement of four or five substances at once.

To get a curve like Fig. 4-2, however, special conditions are necessary. The reducible ions that cause the waves at V_1 and V_2 must be present in low concentrations, no more than 0.01 molar and preferably 0.001 molar or less. The solution must contain a great excess of other ions that are not easily reduced by the current. A salt (or acid or base) is added that is called the "supporting electrolyte." Frequently this is potassium chloride. The potassium ions do get reduced to potassium amalgam if the mercury drop is made very negative; this accounts for the steep rise in current at high applied voltage. At lower voltages the supporting electrolyte simply raises the conductivity of the solution. This has the important consequence that, by Ohm's law, the gradient of electrical potential between cathode and anode is very small, and the electrical force acting on the reducible metal ions is correspondingly small. For these ions to find their way to the surface of the mercury drop they must diffuse there. The electrical "pull" on the ions is almost zero. Now, the rate at which they can diffuse to the drop is limited by their concentration in the solution. The idea is that, surrounding the drop, there is a "diffusion layer," a well-defined layer of solution 0.01-0.05 millimeter thick, outside which the concentration of the diffusing species (the reducible ions) is uniform and the same as it is in the body of the solution, and within which the concentration falls in a linear manner as one approaches the surface of the mercury; see Fig. 4-3. At potential V_a, where there is no reduction, the concentration of reducible ions remains the same right up to the surface of the drop; at potential V_b, where the current has reached its maximum, the concentration of reducible ions at the drop surface is zero and the ions are diffusing as fast as they possible can; and at potential $E_{1/2}$, which is called the half-wave potential, the concentration of reducible ions at the drop surface is half its value in the bulk solution, and diffusion is proceeding at half of its maximum rate. This means that the current caused by the reduction of these ions is half of its maximum value.

The maximum current, shown as I_1 and I_2 in Fig. 4-2, is called the diffusion current. As one expects, the diffusion current is proportional to the concentration of the reducible ions. The current is given by the Ilkovic equation:

$$I_d = 607nC\, D^{1/2}\, m^{2/3}\, t^{1/6}$$

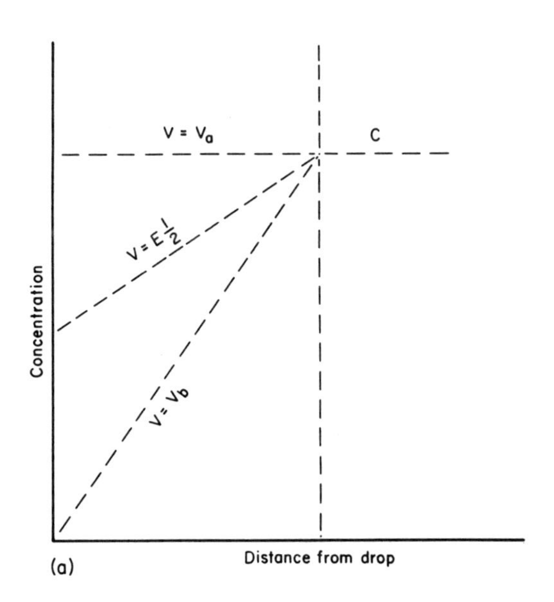

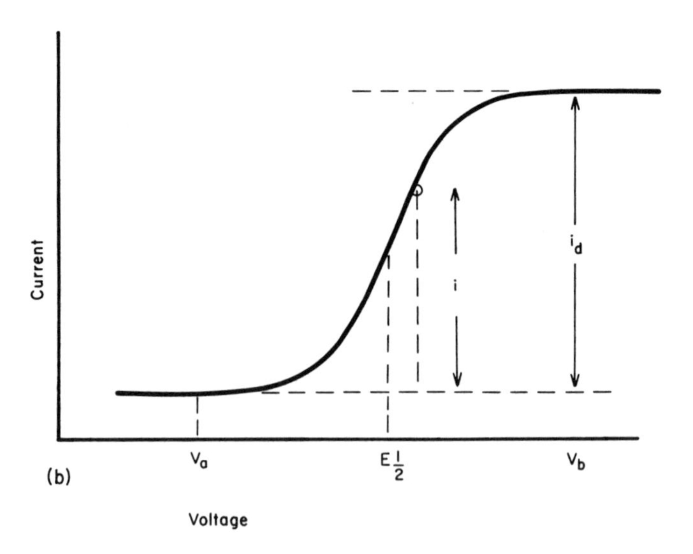

Fig. 4-3. (a) The diffusion layer at different potentials. (b) Currents corresponding to (a).

where: I_d = the diffusion current, averaged over the lifetime of the drop (in microamperes)

n = number of electrons transfered per ion

C = concentration in millimoles per liter

m = mass of mercury flowing per second, in milligrams

t = drop time in seconds

D = diffusion coefficient of the reducible ions, in centimeter square per second

One can calculate the concentration C from a measurement of I_d if one also measures m and t (this is easily done) and if one knows the diffusion coefficient D. The diffusion coefficient depends on the supporting electrolyte and also on the temperature. It is usually more practical to construct a calibration curve with solutions of known concentrations. Another way to obtain the concentration from the measured value of I_d is the "standard addition method." One plots the polarographic wave, then moves the sliding contact (Q, Fig. 4-1) back, adds to the cell a measured quantity of the ion that is being determined, and again plots the wave. It is higher the second time, and from the ratio of the wave heights the original concentration is easily calculated; see Fig. 4-4.

II. COMPLEXITIES: RESIDUAL CURRENT, OXYGEN, MAXIMA

Before going any deeper into the practical aspects of polarography we must mention some factors that affect the current-voltage curve.

A. Residual Current

The current is not zero when no reducible ions are present. As the mercury drop grows, ions from the supporting electrolyte gather around it. If the drop is negatively charged, these ions are positively charged; in potassium chloride solution, the potassium ions would be attracted to the drop. They are not reduced to potassium atoms (unless the negative potential is very high) but remain close to the mercury surface, one or two molecular diameters away, forming the "electrical double layer." The effect is like charging up a condenser. When the drop falls off, a new drop forms and a new condenser is charged up. This causes a continuous flow of electric current which increases linearly as the potential of the drop is in-

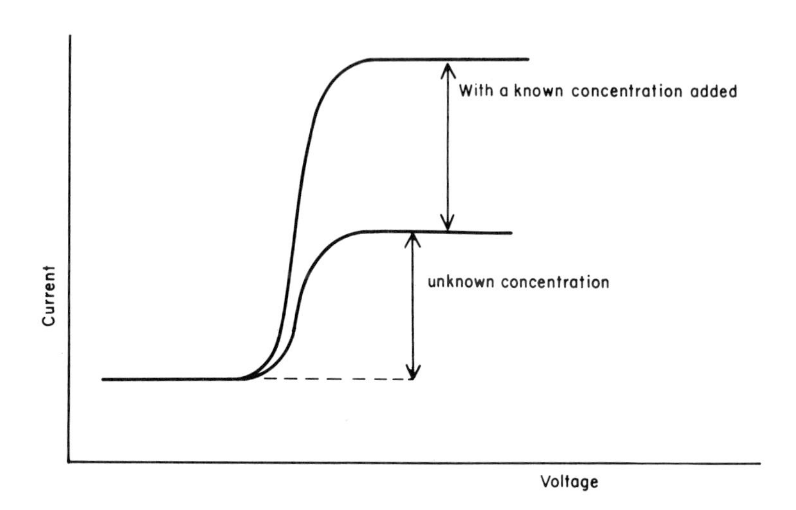

Fig. 4-4. Standard addition method.

creased. The indications are that this "charging current" is zero at the electrocapillary zero, that is, the point at which the surface tension of mercury is a maximum. This happens at about 0.52 volt more negative than the saturated calomel electrode. (As the drop is made more negative than this the surface tension falls; the graph of surface tension against potential is a parabola with its maximum at the electrocapillary zero. As the surface tension falls the drop time falls, and this affects the diffusion current through the Ilkovic equation.)

Thus, even when no reducible ions are present, the current rises in a linear manner as the potential rises. The graph of current against applied potential is not level, but is a straight line slanting upward. When the reducible ions are relatively concentrated (0.001 molar or more) this "slant" is negligible compared to the height of the polarographic waves, but when they are dilute, say 10^{-5} molar or less, charging current is quite large compared to the diffusion current, and it can by no means be neglected.

Small traces of reducible impurities can also cause a perceptible residual current. One impurity that is specially important is dissolved oxygen.

B. Dissolved Oxygen

Oxygen is reduced at the dropping mercury electrode and yields a long, trailing wave that covers most of the current-voltage graph. This effect is sometimes used to measure the concentration of dissolved oxygen, but more often it is a nuisance to be avoided. There are two ways to eliminate dissolved oxygen from solutions that are to be examined polarographically. One is to pass bubbles of nitrogen through the solution before the current-voltage readings are taken. The nitrogen normally obtained from cylinders contains enough oxygen to affect the readings and must be purified by passing it through a reducing solution before it enters the cell. The second way to remove oxygen is to add solid sodium sulfite to the solution in the cell. This method only works if the solution is alkaline; see Experiment 4-1.

C. Maxima

When a clean solution containing only the supporting electrolyte and the reducible ions is run in a clean cell, the wave generally does not have the neat form shown in Figs. 4-2 to 4-4. More usually, the current keeps on rising and then falls again as the voltage is increased, as shown in Fig. 4-5. The effect is caused by a flow of solution tangentially to the surface of the growing drop. It may be avoided by adding a small trace of a colloidal material that can be adsorbed at the surface of the mercury drop. Gelatin is frequently used, and so are nonionic detergents. The concentration must be very small, about 1 milligram in 100 milliliters. If it is too large the height of the wave is affected, as in curve (c), Fig. 4-5. Proper amounts suppress the maximum without changing the height of the wave; see curve (b).

D. Temperature

Diffusion coefficients rise rapidly with temperature; however, the diffusion current is proportional only to the square root of the diffusion coefficient. The current is so small that it produces virtually no heating in the cell, and for most analytical purposes there is no need to control the temperature. However, the cell should be placed in a constant temperature bath if great accuracy is needed.

E. Anodic Waves

Current can flow through the cell in both directions. The mercury drop need not be the cathode; it can be the anode, receiving electrons from the

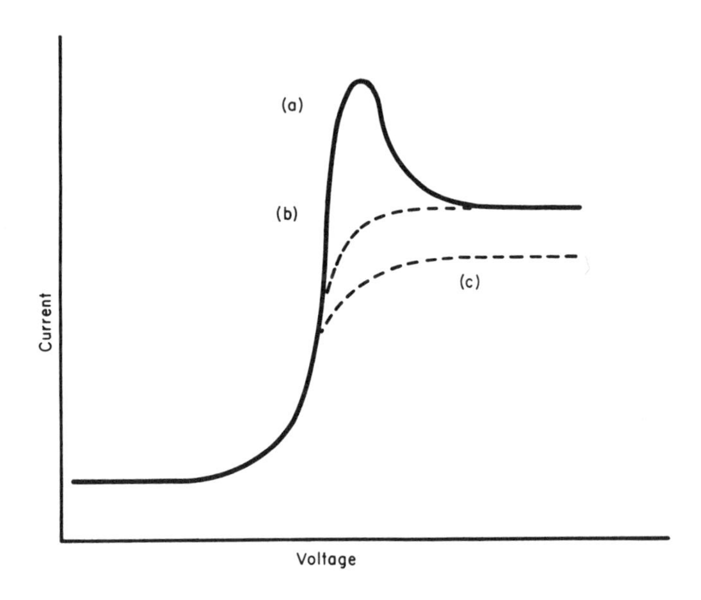

Fig. 4-5. Polarographic maxima.

solution instead of giving them, if its potential is on the positive side of the electrocapillary zero. Then we see an anodic wave. Alkaline solutions that contain sodium sulfite to remove oxygen show an anodic wave. As the negative potential on the drop is reduced to a point where it is only a tenth of a volt more negative than the saturated calomel electrode used for reference (see Experiment 4-1), current starts to flow backward through the cell, and this reverse (or anodic) current increases rapidly as the drop is made less negative. It corresponds to the oxidation of sulfite ions:

$$SO_3^{2-} + 2OH^- = SO_4^{2-} + H_2O + 2e$$

III. THE FORM OF THE WAVES

We can learn something of the process at the mercury drop, and get a clue as to what is being reduced, if we study the shape of the current-voltage graph in a polarographic wave. The reduction (metal ion + $\underline{n}$ electrons = metal atom) is only one of many processes that are possible. Processes like [Cu(II) + e = Cu(I)] give polarographic waves, and so do the reductions of many organic compounds.

If the transfer of electrons at the mercury surface is instantaneous and reversible in the thermodynamic sense, we can apply the Nernst equation (Chapter 1) to the equilibrium between oxidized and reduced species at the drop surface. The concentrations of these species are related to the diffusion current. Letting the symbols i and i_d have the significance shown in Fig. 4-3, and taking as an example the reduction of metal ions to metal,

$$M^{n+} + ne = M(\text{amalgam})$$

the concentration of M(amalgam) is proportional to the current i and the concentration of ions M^{n+} is proportional to $(i_d - i)$. By the Nernst equation the potential V is a constant plus $(0.059/n) \log [(i_d - i)/i]$. Or, since we are considering V to be larger the more negative it is,

$$V = \text{const.} + \frac{0.059}{n} \log \frac{i}{i_d - i} = E_{1/2} + \frac{0.059}{n} \log \frac{i}{i_d - i}$$

(The coefficient 0.059 is valid at 25°C, of course.) Plotting $\log [i/(i_d - i)]$ against V we get a straight line whose slope is $0.059/n$; the value of V for which $\log [i/(i_d - i)]$ equals zero is the half-wave potential, $E_{1/2}$. If the graph is not a straight line the process is not reversible.

IV. SUPPORTING ELECTROLYTES AND HALF-WAVE POTENTIALS

A supporting electrolyte must have a high electrical conductivity, and it must not contain ions that are easily reduced or oxidized. It must not form a precipitate with the ions that are being investigated. Thus, ferric ions can be studied only in an acid solution or a solution that forms a stable complex with ferric ions; in neutral solutions they hydrolyze to hydrous Fe_2O_3. Likewise lead ions cannot exist in an ammonium chloride - ammonia solution, for they form insoluble lead hydroxide. Acidic solutions have the disadvantage that the hydrogen ions are easily reduced to hydrogen gas, in spite of the high overvoltage of hydrogen on mercury. Supporting electrolytes that contain complexing agents are very useful in the analysis of certain mixtures. Suppose that small amounts of cadmium are to be measured in a solution that contains much copper; ordinarily, copper ions are reduced at a lower negative potential than cadmium and the cadmium would produce a small wave on top of a diffusion current that was already very large; however, if potassium cyanide solution is used as the supporting electrolyte, copper is not reduced at all, because its cyanide complex is so stable, and only the cadmium wave is seen.

Table 4-1 lists half-wave potentials for a limited number of metal ions in five supporting electrolytes. Much more extensive tables are available in reference books, such as Handbook of Analytical Chemistry, edited by L. Meites (McGraw-Hill, 1963) from which the data of Table 4-1 are taken.

Complex formation always shifts the half-wave potential to more negative values, and in favorable cases (reversible waves, only one or two complexes formed) the displacement of the half-wave potential can be used to calculate the formation constants of the complexes.

V. CELLS AND GLASSWARE

Figure 4-1 shows the general arrangement of a polarographic cell, but some refinements are necessary. Provision must be made to remove dissolved air. It should be easy to add and remove solutions for analysis. It is desirable to keep the anode separate from the solution in which the mercury is dropping, so that one can use a variety of supporting electrolytes. Furthermore, one should be able to collect the mercury drops that fall in a certain period of time, in order to weigh the mercury and determine m in the Ilkovic equation. Many different cell designs have been proposed, and cells of many types are sold commercially. A form of cell that is very convenient is shown in Fig. 4-6. The left-hand side of the cell contains saturated potassium chloride solution and a pool of mercury into which dips a stainless steel wire to make electrical contact. The horizontal tube that connects the two parts of the cell is closed at the right-hand end by a porous sintered glass disk. The disk by itself is not sufficient to prevent the solutions in the two sides of the cell from mixing, so it is backed by a plug of agar gel 2 or 3 centimeters long. This is formed by dissolving some 5 per cent of agar in hot, nearly saturated potassium chloride solution, pouring the hot solution into the connecting tube of the cell, and letting it set to a firm gel.

The capillary is mounted in a rubber stopper, as shown. A small groove is cut in the side of the stopper to let the nitrogen escape. The nitrogen inlet should be provided with a stopcock so that the gas stream can be turned off while readings are being taken.

TABLE 4-1

Half-Wave Potentials (against Saturated Calomel)[a]

Metal ion	1 M KCl	1 M HCl	1 M NH_3 1 M NH_4Cl	1 M KCN	Citrate, pH 6
As(III)	—	-0.43	-1.41	—	-1.46
Cd	-0.64	-0.64	-0.81	-1.2	-0.70
Co(II)	-1.20	—	-1.29	-1.3	None
Cr(VI)	—	—	-0.2 -1.6	—	-0.38
Cu(II)	+0.04(1) -0.22(1)	+0.04(1) -0.22(1)	-0.24(1) -0.51(1)	None	-0.17(2)
Fe(III)	—	+?	—	—	-0.23(1)
Ni	-1.1	—	-1.10	-1.36	None
Pb	-0.44	-0.44	—	-0.72	-0.54
Sb(III)	—	-0.15	—	-1.11	-0.77 -1.12
Sn(IV)	—	-0.47	—	None	-1.0
Tl(I)	-0.48	-0.48	-0.48	+?	-0.51
U(VI)	—	-0.2(1) -0.9(1)	-0.8(1) -1.4(2)	—	-0.45(2)
V(IV)	—	—	-0.94(1) -1.28(2)	—	-0.44 -1.49
Zn	-1.00	—	-1.35	None	-1.37

[a]The number in parentheses shows the number of electrons transferred, where this can be determined.

A very simple form of cell that needs no special glassware is shown in Fig. 4-7. The rubber stopper carries two tubes for nitrogen, one to bubble it through the solution before readings are taken, the other to pass

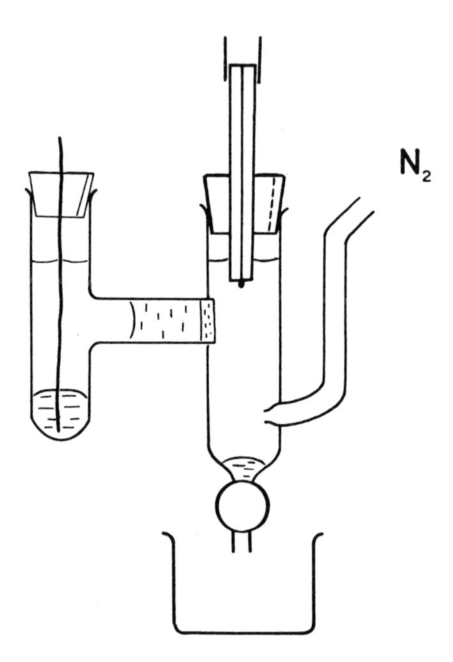

Fig. 4-6. Polarographic cell.

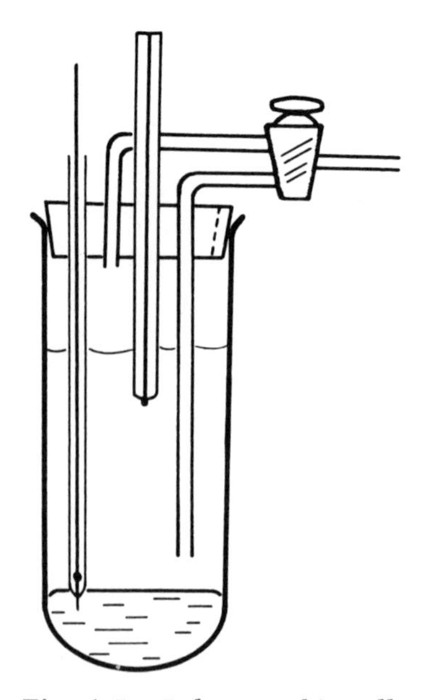

Fig. 4-7. Polarographic cell.

nitrogen over the surface and protect the solution from air while the current-voltage readings are taken. Again there is a groove cut in the side of the stopper to let the gas escape.

Nitrogen from a cylinder is usually not pure enough to use directly. To remove traces of oxygen, bubble the gas through a solution of chromous chloride. Make this solution by dissolving dichromate in water, about 10 grams in 150 milliliters, adding some 50 milliliters concentrated hydrochloric acid, then enough granulated zinc to reduce the dichromate to a blue solution of chromous chloride. Pour this into the gas bubbling bottle and add some granulated or mossy zinc to keep the solution reduced to blue Cr(II). The nitrogen is passed first through this solution, then through a second bottle filled with water to remove spray from the first bottle.

Essential to the polarograph is, of course, the capillary. Capillary tubes used in polarography are very fine, about 25 microns internal diameter. Special tubing is sold for the purpose. It comes in lengths of about 20 centimeters, and these are generally cut in two before use, with the freshly cut ends dipping into the solution. The mercury reservoir (Fig. 4-1) is adjusted so that the mercury falls at about one drop every 3 to 5 seconds when the potential is near the electrocapillary zero (- 0.52 volt versus saturated calomel; and note that the drops fall faster when the potential is raised above, or lowered below, the electrocapillary zero).

The end of the capillary must be kept clean; solution must not be allowed to enter it. Keep the mercury flowing continuously through a series of experiments, and when the experiments are over, lift the capillary out of the solution, rinse it with distilled water, dry it by touching with filter paper, and place the end under mercury in a tube kept for this purpose, as shown in Fig. 4-8. When the apparatus is dismantled, remove the capillary, blow out the mercury, boil the capillary in concentrated nitric acid, and wash with nitric acid and water. Then dry the tube and protect it from dust.

A final and very important note: <u>mercury is poisonous</u>. Drops of spilled mercury contaminate the air with mercury vapor, which is an insidious cumulative poison. Avoid spilling mercury, and as a safety precaution, place the stand that carries the cell and mercury reservoir in a tray.

Used mercury should be collected and cleaned by putting it in a bottle with 3-5 molar nitric acid and drawing air through it vigorously for a few hours, then washing it with water and drying; see page 94.

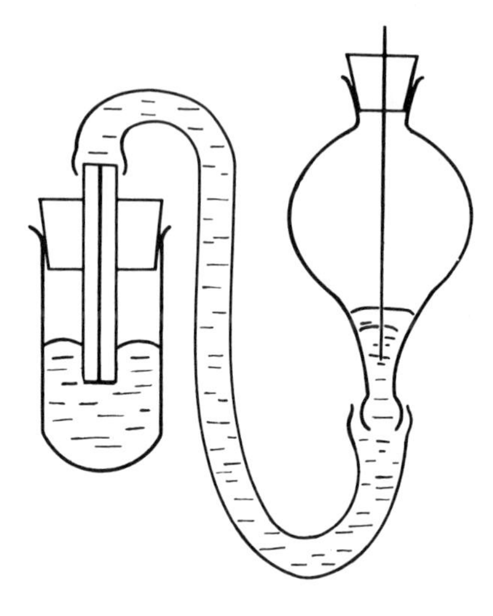

Fig. 4-8. How to protect the capillary between experiments.

VI. ELECTRICAL EQUIPMENT

A. Manual Polarographs

This equipment makes no provision for recording. It is therefore somewhat tedious to use, yet it is fully as precise as the recording equipment and probably better for research purposes, and it is cheaper. A good manual instrument is the Sargent Model III polarograph,* illustrated in Fig. 4-9. Along the top of the case is the scale of a sensitive, reflecting (lamp-and-scale) galvanometer. The image of the lamp filament appears on this scale as a vertical line. As the mercury drops form and fall the line moves to the right, then sharply moves back again. One notes the reading at maximum deflection, that is, furthest to the right. The inertia of the galvanometer coil is such that the back-and-forth swing is small.

*The word "polarograph" is the registered trade-mark of the Sargent Company. Like many trade-marks that describe successful products, the word has passed into general use as a generic term.

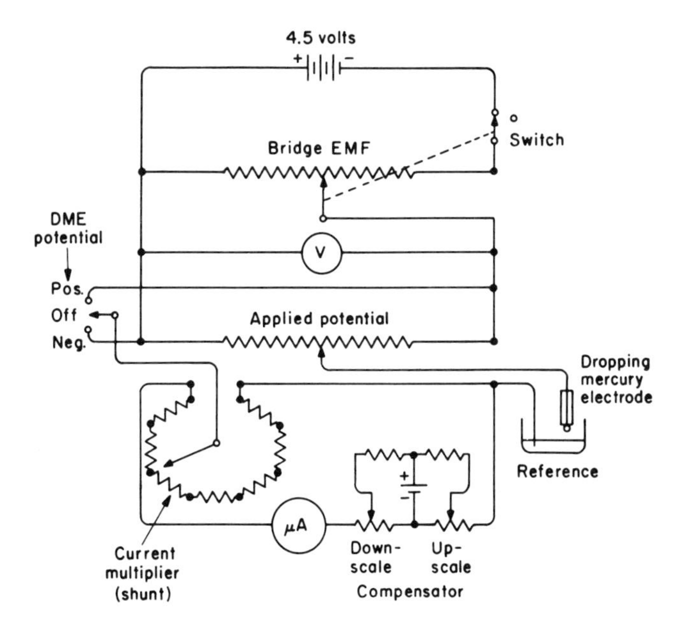

Fig. 4-9. Sargent Model III manual polarograph.

Controlled by knobs and switches are the following:

(i) Bridge electromotive force: sends current through the slide wire and adjusts the total potential drop across the slide wire; this potential drop is shown on the voltmeter in the center of the panel. A drop of 2 volts is adequate for most polarographic waves; it may be reduced to 1 volt to study reductions that proceed at low negative potentials, like those of cadmium and thallium ions.

(ii) Current multiplier: operates a shunt that controls the fraction of the drop current that actually passes through the galvanometer. That is, it controls the sensitivity of the galvanometer. Low multipliers mean high sensitivity and are used for small concentrations.

(iii) Applied potential: this control moves the sliding contact that regulates the potential across the cell. If the bridge electromotive force is exactly 1 volt, the potential is the same as the reading on the "Applied potential" dial; if the bridge electromotive force is 2 volts, the reading must be multiplied by 2.

(iv) Compensation, upscale and downscale: this sends a controlled current backward or forward through the galvanometer, if desired, to bring the galvanometer spot to any desired part of the scale. The function is like that of the "asymmetry" or "standardization" control on pH meters (Chapter 1). To illustrate its use, suppose one has a solution containing much cadmium and a little zinc. The wave for cadmium comes first and may move the galvanometer spot nearly to the end of the scale. When the zinc wave comes the spot may move off the scale. Then one can apply "downscale compensation" to bring the spot to the middle or lower part of the scale at the start of the zinc wave, so as to get all of the zinc wave on the scale. Also, one can increase the sensitivity (that is, decrease the current multiplier) before measuring the zinc wave.

The switches "D.M.E." (dropping mercury electrode) and "GALV" are normally in the positions (-) and (+), respectively. These positions are reversed for studying anodic waves.

The current for the slide wire and for the compensation comes from three dry cells that are placed within the instrument case. These are large-capacity 1.5-volt cells and have become difficult to obtain in some countries. If large-capacity cells are not available, use ordinary flashlight batteries, putting sets of two or three in parallel to give better current stability.

B. Recording Polarographs

These record current-voltage curves automatically. More correctly, they record current-time curves. The applied potential is increased at a steady controlled rate by means of a constant-speed motor, and simultaneously the chart paper is moved at a steady rate. The pen moves in accordance with the current passing through the cell. Because the pen is recording current against time, the graph that it traces moves up and down as the drops form and fall; see Fig. 4-10. The up-and-down movement that accompanies drop formation is a nuisance, and it may be minimized in two ways: one, by using a recorder with a long time constant, that is, slow pen travel (commercial polarographs use such recorders), and two, by placing a high-capacity condenser across the terminals of the recorder.

Recording polarographs have the same controls as manual instruments plus switches to adjust the damping, rate of chart paper movement, and the rate of "scanning" or change of applied potential.

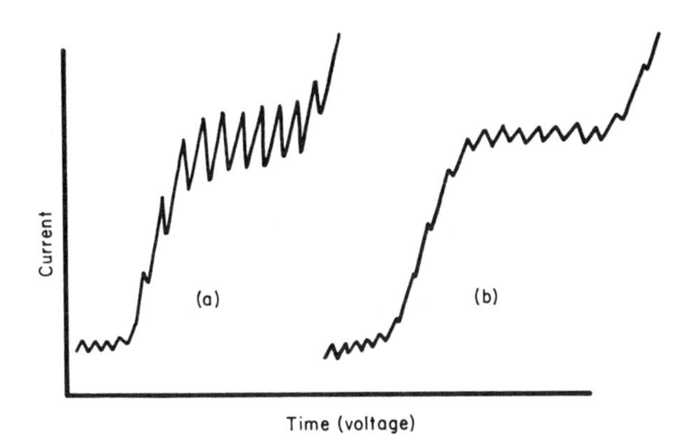

Fig. 4-10. Curves obtained with a recording polarograph: (a), little damping; (b), much damping.

EXPERIMENTS

Experiment 4-1

Plotting Current-Voltage Curves

This experiment will give experience in the use of the instrument, show the effect of the maximum suppressor, distinguish between reversible and nonreversible reductions, and show the effect of $\underline{n}$, the number of electrons transferred, on the height and shape of polarographic waves. For convenience, an alkaline supporting electrolyte will be used. This will permit the use of sodium sulfite to remove oxygen and avoid the need for passing nitrogen.

Materials. (a) Supporting electrolyte, 1 M NH_4Cl-NH_3. Weigh roughly 22 g of ammonium chloride, dissolve in water, add 35 ml concentrated (15 $\underline{M}$) ammonia, make to 500 ml with distilled water. (b) Solutions of salts of cadmium, zinc, nickel, copper(II), and thallium(I), each 0.010 $\underline{M}$ in metal ion. For thallium the nitrate is best as it is reasonably soluble; TlCl is sparingly soluble. (c) Gelatin, 100 mg dissolved in 25 ml hot water. This solution should be prepared fresh every two days. (d) Sodium sulfite, solid. Equipment: a two-compartment cell like that of Fig. 4-6 (however, the tube for passing nitrogen may be omitted); manual or recording polarograph.

Start with a clean cell; even a small trace of gelatin will affect the maximum that should be seen in the second part of the experiment. Fill the left-hand or reference part of the cell with saturated potassium chloride, and put enough mercury in the bottom to cover the end of the contact wire.

(i) Put supporting electrolyte in the right-hand part of the cell, filling it to about the same level as the solution in the reference compartment. The level must be high enough, obviously, to cover the end of the capillary. With a spatula, add about 100 mg of sodium sulfite and stir until most of this is dissolved. Set the capillary firmly in place, wait about 5 min to let the sulfite react with the dissolved oxygen, and then measure the current as a function of the applied potential, starting with the drop about 0.1 V negative with respect to the reference electrode. (That is, apply enough negative potential to prevent the anodic wave mentioned on page 107.) Use a fairly high sensitivity. The graph should be nearly level if oxygen has been eliminated and if the supporting electrolyte is free from metallic impurities.

(ii) Now add to the supporting electrolyte in the cell, using a pipet, 2.00 ml of each of the solutions of cadmium and zinc (0.010 M). Mix with a glass rod, and add a little more sodium sulfite. Again measure the current against applied potential, choosing a value of the sensitivity so that the waves stay nicely on the scale. You can raise or lower the potential as you wish, to check any part of the curve you wish to check. Plot the wave forms carefully. You will find that the zinc wave shows a maximum (Fig. 4-5) but the cadmium wave does not. The reason is that the cadmium wave, with $E_{1/2} = -0.81$ V in this supporting electrolyte, falls near the electrocapillary zero but the zinc wave does not.

(iii) Add 5 drops of gelatin solution to the solution in the cell and run the zinc wave again. The maximum should have disappeared, but the wave height should be unchanged. Compare the heights of the zinc and cadmium waves, and remembering that the concentrations of the two ions were made the same, account for the difference in the wave heights.

(iv) In a 50-ml volumetric flask place a measured volume, 5.0 or 10.0 ml, of 0.010 M thallium(I) nitrate; add 3-5 drops of gelatin and about 100 mg of sodium sulfite, and add supporting electrolyte to make up to the mark. Shake, let stand 5 min. Meanwhile, run the solution used in part (iii) out of the cell and rinse the cell with distilled water. Pour the solution of thallium nitrate into the cell (up to the same level, roughly, as the solution in the reference side of the cell), replace the capillary, and carefully plot the current-voltage curve. Include enough of the level portions before and after the wave to determine the maximum diffusion current i_d

with good accuracy. Analyze the data according to the equation on page 109, constructing a table with column headings as follows:

Applied potential, V	i	$i_d - i$	$\frac{i}{i_d - i}$	$\log \frac{i}{i_d - i}$

Note that the absolute value of the current need not be known; we only need ratios of currents. We can therefore express the current in scale divisions or chart-paper divisions. Plot the graph of $\log [i/(i_d - i)]$ against V, measure its slope, and calculate the number of electrons, $\underline{n}$, transferred when one thallium ion is reduced. Also find the half-wave potential $E_{1/2}$. Make the same calculations for the cadmium and zinc waves obtained in parts (ii) and (iii).

(v) In a 50-ml volumetric flask place 5.0 ml 0.010 $\underline{M}$ cupric sulfate [or other salt of Cu(II)] and 5.0 ml 0.010 $\underline{M}$ nickel salt. Add 3-5 drops of gelatin, then some 25 ml supporting electrolyte, then about 150 mg sodium sulfite; make up to the mark with supporting electrolyte, mix well, and let stand about 5 min before pouring into the cell. (First drain the other solution out, and rinse the cell.) Run the current-voltage curve, and interpret its form as completely as you can.

(vi) Take an unknown mixture containing any three of the metal ions studied (copper and thallium should not both be present, as their waves coincide), examine it polarographically, identify the ions, and estimate how much of each is present.

Experiment 4-2

The Ilkovic Equation: Diffusion Coefficients

The object of this experiment is to verify that the diffusion current is proportional to $m^{2/3} t^{1/6}$ and to measure, or at any rate compare, the diffusion coefficients of two or more ions. In this experiment the temperature of the cell must be constant within $\pm 0.5 - 1^{\circ}C$.

Measure the diffusion currents of two or three ions under carefully controlled conditions in the same supporting electrolyte that was used in Experiment 4-1; the ions Tl(I), Cd(II), and Ni(II) are suggested. Measure $\underline{m}$ and $\underline{t}$ as follows: first, empty any mercury out of the bottom of the polarographic cell by opening the stopcock briefly; next, allow exactly 20 drops to fall, and with a stopwatch, measure the time it takes for them to fall; then run the mercury that has accumulated from these 20 drops into a small beaker or crucible that has previously been weighed. Remove the solution that accompanies the mercury by absorbing it into filter paper; wash the mercury once or twice with distilled water, again absorbing the water with filter paper. Weigh the beaker with the clean dry mercury. From your observations, calculate $\underline{m}$, the mass of mercury flowing per second, and $\underline{t}$, the time for formation of one drop.

Change the values of $\underline{m}$ and $\underline{t}$ by raising or lowering the mercury reservoir, and again measure $\underline{m}$, $\underline{t}$, and the diffusion current. When comparing different ions, remember that $\underline{t}$ depends on the applied potential, but $\underline{m}$ depends only on the height of the mercury reservoir.

From the information you now have, which must include the concentrations of the metallic ions, determine the ratios of their diffusion coefficients. The diffusion coefficients themselves can be calculated from the Ilkovic equation if the absolute values of the current are known. To calibrate the galvanometer, connect a known resistance of 100,000 ohms in place of the polarograph cell and record the galvanometer reading for two or three different values of the applied voltage. Note that the current I_d in the Ilkovic equation is the average diffusion current; this is more easily read from a recording polarograph than from a manual instrument.

Experiment 4-3

Diffusion Current and Concentration: Quantitative Analysis

The object of this experiment is to plot a calibration curve of current against concentration and use it to determine the concentration of one or more unknown solutions. The standard addition method will also be illustrated. The same supporting electrolyte is used as in Experiment 4-1, with sulfite to remove the oxygen.

Take four 50-ml volumetric flasks, and add to them 0.50, 1.00, 2.00, and 5.00 ml 0.010 $\underline{M}$ cadmium chloride solution. (Higher concentrations can be used if desired.) Add supporting electrolyte (1 $\underline{M}$ NH_3-NH_4Cl, as used in Experiment 4-1), 4 drops gelatin and 100 mg sodium sulfite (as in Experiment 4-1), and more supporting electrolyte to make up to the mark. Mix well; run the current-voltage curves, taking care to rinse the cell between solutions; a good procedure is to start with the most dilute cadmium solution, run its current-voltage curve, empty out the solution, rinse the cell twice with small portions of the next most concentrated solution, then fill the cell with this solution and measure the new current-voltage curve. Be careful to keep the height of the mercury reservoir, and also the temperature, constant throughout this experiment. Plot a graph of diffusion current against concentration of cadmium. Measure the height of the mercury level in the reservoir, compared with the capillary tip, so that this height can be reproduced in later experiments.

If Experiment 4-4 is to be performed, make a calibration curve for zinc ions as well. This can be done at the same time as the cadmium calibration by mixing exactly equal volumes of 0.010 M cadmium and zinc salts beforehand, and taking known volumes of this mixture to prepare the standard solutions.

The "unknown solution" for analysis should be a cadmium salt solution, or a solution containing cadmium and zinc that is about 0.01-0.02 M in cadmium and zinc combined. Measure a known volume (of the order of 1-5 ml) into a 50-ml volumetric flask, add gelatin and sodium sulfite, and make up to the mark with the supporting electrolyte; run the current-voltage curve, and using the calibration curve, calculate the concentrations of cadmium and zinc in the unknown. For experience in using the compensation current and different current multipliers (galvanometer sensitivity), analyze an "unknown" that contains much more cadmium than zinc, say, a solution 0.020 M in cadmium ions and 0.0010 M in zinc.

After running the current-voltage curve for the "unknown" do not empty the solution out of the cell; add a small measured volume of standard solution, say 2.0 ml of 0.010 M cadmium chloride, and run the current-voltage curve again. Compare the diffusion currents before and after adding the 2.0 ml of standard. Now, to calculate the cadmium concentration it is necessary to know the volume of solution in the cell. Drain the solution out of the cell into a graduated cylinder, measure the volume, and calculate the cadmium concentration; first calculate the concentration in the polarograph cell before the standard was added, then calculate the concentration of the original "unknown" solution before it was mixed with supporting electrolyte.

A sample calculation by the standard addition method follows:

5.00 ml of an unknown cadmium solution is diluted to 50.0 ml with supporting electrolyte. Some of this solution is poured into the polarograph cell; it gives a diffusion current of 78 scale divisions. Now 2.00 ml 0.0100 M $CdCl_2$ is added to the cell; the current increases to 105 scale divisions. The total volume of solution in the cell is now 23.0 ml.

Adding the standard $CdCl_2$ solution increased the volume from 21 ml to 23 ml. The cadmium from the "unknown" therefore contributes (78 x 21)/23 = 71 scale divisions to the final current. We added 0.0200 mmol Cd; this raised the current to 105 scale divisions. The cadmium in the cell originally was therefore 0.0200 x 71/(105 - 71) = 0.0418 mmol. This was present in 21 ml; the cadmium in the original 5.00 ml of unknown was 0.0418 x 50/21 = 0.100 mmol, very nearly, and the concentration of the original unknown was 0.100/5.00 = 0.020 M.

Experiment 4-4

Ion-Exchange Separation of Cadmium and Zinc

This experiment illustrates the separation of metals by ion-exchange chromatography; more will be said about this technique in Chapter 10. Polarography is used to follow the course of the separation.

Materials. (a) Ion-exchange resin, strong-base anion exchanger such as Dowex-1x8, particle size 50-100 mesh. (b) Glass column; see Chapter 1, Fig. 1-15; 10-12 mm internal diameter, 25-30 cm long. (c) Supporting electrolyte, 1 M NH_3-NH_4Cl, as described in Experiment 4-1. (d) Gelatin solution, as in Experiment 4-1. (e) Sodium sulfite, solid powder. (f) Hydrochloric acid. (g) Zinc and cadmium chloride or sulfate solutions, each 0.010 or 0.020 M. (h) Two-compartment polarograph cell and polarograph.

The method depends on the different stabilities of the two complexes $CdCl_4^{2-}$ and $ZnCl_4^{2-}$. The cadmium complex is the more stable, and cadmium is absorbed by the anion-exchange resin under conditions when zinc is not absorbed. This kind of separation is discussed in Chapter 10.

Prepare a column of resin some 20 cm long, as shown in Fig. 1-15. The resin must be evenly packed and free from air bubbles. If the resin is new, and of commercial grade (that is, not specially purified), it must be washed very well before use. Washing with about 1 l. of 1 M hydrochloric acid, followed by some 200 ml of methanol to remove soluble organic impurities, should be sufficient; see Chapter 10; resin of purified grade, or resin that has been used before, will need much less washing. Finally pass about 25 ml 1 M hydrochloric acid to prepare the column for this experiment.

Measure 10.0 ml of each of the two solutions, cadmium and zinc; mix, and add 2 ml concentrated hydrochloric acid. Pour the mixed solution into the top of the column, and let it flow slowly into the column (5 ml/min is a suitable rate). Discard the solution that flows out of the bottom of the column. At the hydrochloric acid concentration used here, both zinc and cadmium are strongly absorbed and are held near the top of the resin column. Let the solution level fall until it reaches the top of the resin bed.

Now pour 25 ml 0.10 M hydrochloric acid into the column and let it flow no faster than 5 ml/min. Collect the solution flowing out of the column, using a small Erlenmeyer flask. When it has all passed, add 25 ml of supporting electrolyte and a few drops of gelatin, then 100 mg of sodium sulfite. Mix, pour some of the solution into the polarograph cell, and run the current-voltage curve. Probably the zinc wave will appear.

Pass another 25 ml 0.10 M hydrochloric acid, and treat the effluent (the solution flowing out of the column) in the same way. Keep pouring 25-ml portions until the zinc concentration falls to one-fifth its original value or less. Then, instead of 0.1 M hydrochloric acid, pass 25-ml portions of pure water. This will cause the cadmium to come out of the column. Plot the zinc and cadmium concentrations on a graph against the total volume passed. If the separation goes well, the zinc concentration will fall to nearly zero before the cadmium starts to come out.

If there is time, make a second and a third "run" using 0.20 M and 0.05 M hydrochloric acid for the first washings, followed by water. Make experiments to find the best conditions for separating zinc and cadmium by anion exchange.

QUESTIONS

1. Using the Ilkovic equation, calculate the diffusion current for a doubly charged ion, concentration 1.0×10^{-3} M, whose diffusion coefficient is 1.5×10^{-5} cm^2 sec^{-1}, if the mercury drops are exactly 1 mm in diameter and fall once every 4.0 sec. The density of mercury is 13.5 g/ml.

2. The diffusion coefficient of an individual ion can be calculated from its equivalent ionic conductance λ (see Chapter 2) using the equation

$$D = \frac{RT}{nF^2} \lambda$$

where D is the diffusion coefficient in cm^2 sec^{-1}, F is the faraday, n is the ionic charge or valence of the ion, and λ is in ohm^{-1} cm^2 $equiv^{-1}$.

Calculate D for a doubly charged ion whose λ is 60.

3. Suggest an experiment that will measure the diffusion coefficient of an individual ion.

4. The Nernst equation (Chapters 1 and 3) states that the reversible potential of a metal in a solution of its ions depends on the logarithm of the concentration of these ions. Why, then, is the half-wave potential in polarography independent of the ionic concentration?

5. In view of your answer to question 4, explain why the half-wave potential for the reduction of copper(II) is made more negative by adding ammonia, and explain how this effect may be used to measure the formation constants of complex ions.

Chapter 5

THE ABSORPTION OF RADIATION

I. LIGHT AND ELECTROMAGNETIC RADIATION

Light is radiation that affects our eyes, radiation that we can see. Light consists of an electric field that alternates back and forth some 10^{16} times a second and is propagated through space as a train of waves. Associated with the changing electric field is a changing magnetic field, at right angles to it but with the same wavelength; see Fig. 5-1. We call such waves <u>electromagnetic waves</u>, and the radiation <u>electromagnetic radiation</u>.

Visible light has waves that range in length, in a vacuum, from 400 to 700 nanometers (4 to 7 x 10^{-5} centimeters). The colors of light correspond to different wavelengths (see Table 5-1). Some persons can see beyond 700 nanometers, perhaps as far as 800 nanometers. Obviously the ranges given in Table 5-1 are approximate, for one color shades into another throughout the spectrum.

TABLE 5-1

Colors of Visible Light

Color	Wavelength, nm
Violet	400 - 480
Blue	480 - 500
Green	500 - 580
Yellow	580 - 600
Orange	600 - 620
Red	620 - 700 and over

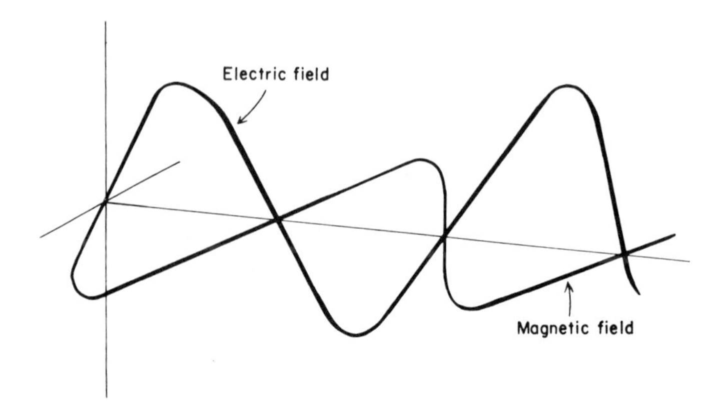

Fig. 5-1. Electromagnetic wave.

Radiation too short (in wavelength) to see is called ultraviolet; radiation whose wavelength is too long to see is called infrared. The visible spectrum is only a small part of a spectrum of electromagnetic waves that extends over a great range of wavelengths covering many factors of ten. To measure these wavelengths we need several different units of measurement. Those used in spectroscopy are the following:

One angstrom unit = 1 Å = 10^{-8} cm

One nanometer, or one millimicron = 1 nm or mμ = 10^{-7} cm = 10^{-9} m

One micron = 1μ = 10^{-3} mm, 10^{-4} cm or 10^{-6} m

The regions of the electromagnetic spectrum used in chemical analysis are given in Table 5-2. The divisions between the different spectral regions are a matter of convenience and are made partly to indicate the origin of the radiations and partly for experimental (instrumental) reasons. Gamma rays are emitted by disintegrating atomic nuclei; X-rays are emitted or absorbed by movements of electrons close to the nuclei of relatively heavy atoms. Ultraviolet and visible radiations come from movements of electrons in the outer parts of atoms and molecules. The distinction between "vacuum ultraviolet" and "ultraviolet" is made because air starts to absorb below 180 nanometers, and so does the quartz that is used to make cells and windows. This fact complicates enormously the task of observing and

TABLE 5-2

The Electromagnetic Spectrum

Gamma rays	0.01 - 0.1 Å
X-rays	1 - 10 Å
Vacuum ultraviolet	10 - 1800 Å
Ultraviolet	180 - 400 nm
Visible	400 - 700 nm
Infrared (near)	0.7 - 2.5 μ
Infrared	2.5 - 15 μ
Far infrared	15 - 200 μ
Microwaves	0.1 - 1 cm

measuring radiations below 180 nanometers in wavelength, and indeed this part of the spectrum is only just beginning to be used for chemical analysis.

The part of the spectrum called "infrared" is associated with changes in the vibration of molecules, and the distinctions between "near infrared," "infrared," and "far infrared" are mainly matters of instrumentation. The microwave region corresponds to changes in the rotation of molecules.

II. ENERGY AND THE ELECTROMAGNETIC SPECTRUM

According to the quantum theory, radiant energy is always produced or absorbed in multiples of a unit called the quantum. The size of the quantum of energy is related to the wavelength and the frequency of the radiation by the equation

$$\text{energy} = h\nu = \frac{hc}{\lambda}$$

where:

h = Planck's constant = 6.624×10^{-27} erg second
ν = frequency, or number of vibrations per second
λ = wavelength in centimeters
c = speed of light in centimeters per second

The quantities c and λ depend on the medium through which the radiation passes. In vacuum, $c = 2.998 \times 10^{10}$ centimeters per second.

For radiation to be produced, an atom or a molecule has to lose energy. When radiation is absorbed, an atom or molecule gains energy. The greater the energy gain or loss, the shorter is the wavelength of the corresponding radiation. We have already noted that radiations of different wavelength ranges correspond to different processes. The quanta with the largest energies come from changes within atomic nuclei and produce the radiations with the shortest wavelengths, the gamma rays. Next come radiations caused by, or causing, movements of electrons between different levels of energy in atoms and molecules. These radiations cover a very wide energy range, with wavelengths from 1 angstrom to 1 micron, or 10^4 angstroms. To get an idea of the wavelengths to be expected from electronic changes, consider the simplest atom of all, the hydrogen atom, with one proton and one electron. Its energy levels can be predicted accurately from the quantum theory and the Schrodinger equation. They are

$$E_n = -\frac{2\pi^2 m e^4}{h^2 n^2}$$

where:

E_n = energy of the level with quantum number n
m, e = mass and charge of the electron

An electron jumping from the first level to the second, that is, from $n = 1$ to $n = 2$, absorbs radiation of wavelength 1215 angstroms. The same transition in the singly charged helium atom, that is, an atom with one electron and a nucleus having double the charge of the hydrogen nucleus, absorbs at one-fourth of this wavelength, or 303.8 angstroms. These wavelengths come in the "vacuum ultraviolet," or what we may call the "hidden" region of the spectrum, "hidden" because it is so hard to study experimentally. The X-rays, with shorter wavelengths, are easier to study. These come from electronic transitions in heavier atoms, starting with the sodium atom, and correspond to electrons close to the nucleus, that is, in the "s" orbital, or "K" shell. Longer wavelengths, of 2000 angstroms (200 nanometers) and up, correspond to electron jumps between the outer orbitals or higher energy levels of atoms and molecules. These radiations, too, are easier to measure than those of the "hidden" region. It is convenient to consider the ultraviolet and visible regions from 180-200 nanometers to about 800 nanometers as one continuum. This part of the spectrum and the part of the infrared between 2.5 and 15 microns are the two spectral regions most used in chemical analysis. Ultraviolet-visible spectra are electronic spectra; they are used to detect and measure elements and certain types of chemical compounds, notably metal-coordination compounds and unsaturated organic compounds having conjugated double bonds. Infrared spectra are vibrational spectra; they are used for identification and, to some extent, for quantitative determination of all kinds of organic compounds and some inorganic compounds.

Both emission and absorption of radiation are used in chemical analysis, but absorption is used more often. To study absorption one selects a source of continuous radiation like white light. The source may be a heated solid, like the tungsten filament of an electric lamp, or an electrically excited gas under moderate pressure, like the "hydrogen lamp" used as an ultraviolet source. Radiation from the continuous source is passed through the sample and one observes what wavelengths are absorbed and how strongly. In this way one gets an "absorption spectrum" like that shown in Fig. 5-2. The ordinate in Fig. 5-2 is a quantity called the absorbance, which will be defined in a later section.

III. WAVELENGTH, FREQUENCY, WAVE NUMBER

It is customary to describe radiation by its wavelength, but frequency is a more significant quantity. There are two reasons for this. First, frequency is directly proportional to the energy of the quantum: $E = h\nu$.

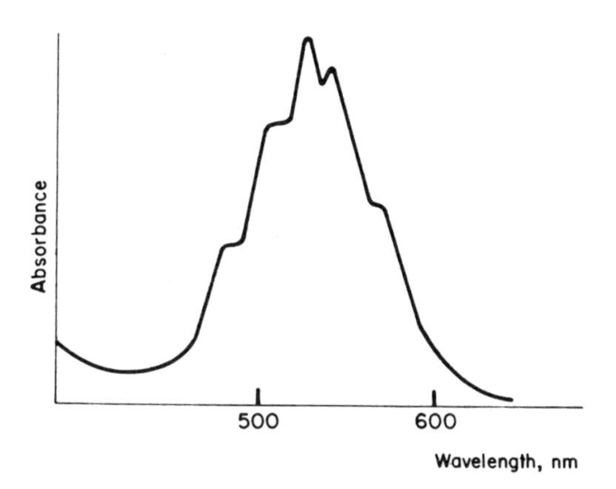

Fig. 5-2. Absorption spectrum of the permanganate ion (see also Fig. 6-15).

Second, the frequency depends only on the source of the radiation and not on the medium through which it travels. The wavelength does depend on the medium. Consider the orange-red line in the spectrum of krypton-86 that serves today as the international standard of length. Its frequency is 4.948866×10^{14} cycles per second. Its waves in a vacuum have length 6057.802 angstroms; 1,650,763.73 of these waves make 1 meter, and the meter is defined by international agreement as 1,650,763.73 times the wavelength of the orange-red krypton line in a vacuum. As soon as this light enters a material substance, however, its wavelength changes. The light is slowed down, and the wavelength becomes smaller in proportion: $\lambda = c/\nu$, where λ is the wavelength, ν the frequency, and c the speed of the light. The ratio of the speed of light in a vacuum to the speed of light in a given substance, or medium, is called the refractive index of that substance, and the refractive index depends not only on the substance, but also on the wavelength. The frequency, however, does not change as the light passes from one medium to another.

When we specify the wavelength of a certain radiation, therefore, we understand that we mean the wavelength measured in a vacuum.

Though the frequency is more fundamental than the wavelength, it is cumbersome, as it is usually a very large number. It is common practice, therefore, to express the frequency as the wave number $\tilde{\nu}$, or the number of waves per centimeter in a vacuum. The krypton line mentioned above has a wave number of 16,507 cm^{-1}, or, in common language, a "frequency" of 16,507 cm^{-1}. Its true frequency in cycles per second is this number multiplied by the velocity of light in a vacuum expressed in centimeters per second.

The wave number notation is used extensively in infrared spectroscopy; see Chapter 7.

IV. ABSORPTION OF LIGHT: THE BEER-LAMBERT LAW

The relation between the absorption of light and the concentration of a dilute solution or gas is called Beer's law. The relation between absorption and light path (or cell thickness) is called Lambert's law. It is convenient to consider both laws together.

In deriving the equation we postulate, first, that every quantum of light that enters the solution has an equal chance of being absorbed. This means that the light is monochromatic. Secondly, we postulate that every molecule of the light-absorbing substance has an equal chance of intercepting and absorbing a quantum of light, wherever in the cell it may be situated. This implies that the solution is sufficiently dilute that one molecule does not hide in the shadow of another.

Consider a beam of light of intensity I_0 entering a cell filled with a solution of concentration c; see Fig. 5-3. After the light has traveled a distance x within the cell, let its intensity be I. When it has traveled an additional (and infinitesimally small) distance dx, let the intensity be $I + dI$. (Obviously, the increment dI is negative; the light gets weaker as it passes through the cell.) Then

$$\frac{dI}{I} = -kc\,dx$$

The quantity $dI/I = d \ln I$, where ln I is the natural logarithm of I. This relation comes directly from the definition of logarithms and exponentials; we recall that, by definition, $d(e^x)/dx = e^x$ and if $y = e^x$, $x = \ln y$. Thus,

$$\frac{d \ln I}{dx} = -kc$$

and

$$\begin{aligned} \ln I &= kcx + \text{a constant} \\ &= -kcx + \ln I_0, \text{ since } I = I_0 \text{ when } x = 0. \end{aligned}$$

We recall, also, that to convert from natural logarithms to common logarithms, that is, logarithms to base 10, we divide by 2.303, which is the natural logarithm of 10. Rearranging, therefore, we get

$$\log \frac{I_0}{I} = \frac{kcx}{2.303} = \varepsilon cx, \qquad (5\text{-}1)$$

where ε is a new constant called the molar absorptivity. In using published values of ε it is understood that the concentration c is expressed in moles per liter, and the cell thickness x in centimeters.

The quantity $\log I_0/I$ is called the absorbance. In the older literature it is called optical density, and this term appears on the dials of older instruments.

The fraction of light transmitted, which is I/I_0, is called the transmittance. The per cent transmittance is simply $100I/I_0$. The dials of

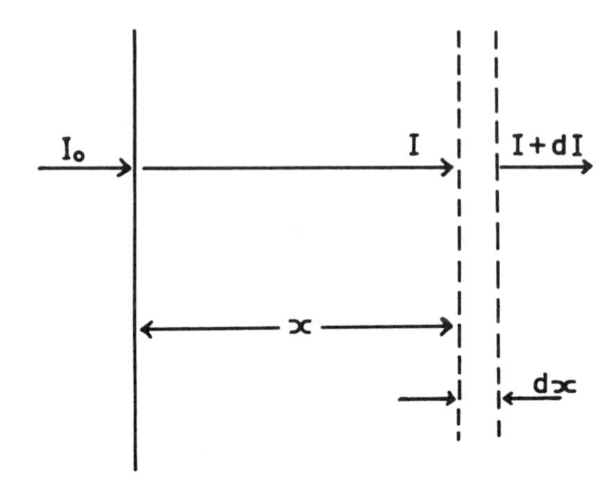

Fig. 5-3. Absorption of light: the Beer-Lambert law.

spectrophotometers generally carry two scales, one absorbance, the other per cent transmittance; see Fig. 5-4. In ordinary spectrophotometric analysis one reads absorbance, for absorbance is directly proportional to concentration if Beer's law holds. On the other hand, the transmittance scale is linear and it is easier to read by interpolation between the divisions, particularly when the transmittance is less than 10 per cent. In this range the absorbance scale is almost useless. It is best to read the transmittance as accurately as one can, then convert to absorbance by using a table of logarithms or the log scale on a slide rule.

A trend in modern instrument design is to have one scale, marked in equally spaced divisions from 0 to 100, then have a switch which makes this scale read in transmittance or absorbance, as desired. The conversion from direct to logarithmic reading is made by an operational amplifier circuit.

V. PROBLEMS USING THE BEER-LAMBERT RELATION

1. The per cent transmittance of a strongly colored solution is 8.4%. What is the absorbance?

   This is simply an exercise in using logarithms. The log of 8.4 is

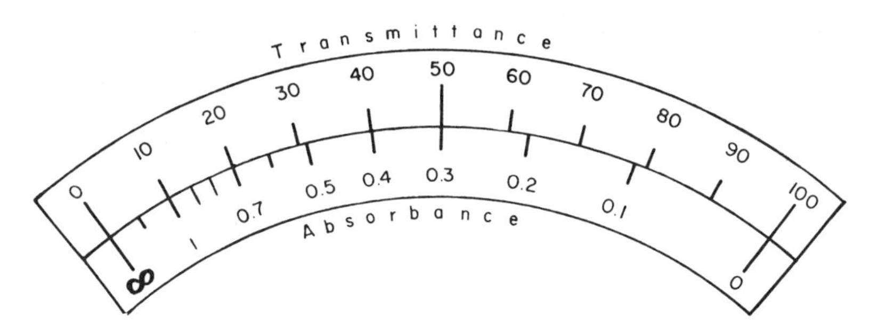

Fig. 5-4. Spectrophotometer scales.

0.925; absorbance = (log 100) - (log 8.4) = 2.000 - 0.925 = 1.075. Note that it is wrong to use more than three decimal places in the logarithm, even though four-figure tables may be available. Indeed, the accuracy of the measurements may only justify the use of two places in the log.

2. A solution placed in a 1-cm cell transmits 50% of the light of a certain wavelength. What percentage is transmitted by (a) a 4-cm cell, (b) a 0.5-cm cell?

   This problem is worked by "common sense"; there is no need to take logarithms or go through the motions of solving equation (5-1). If 1 cm transmits a fraction 0.5 of the incident light, 2 cm transmit $0.5 \times 0.5 = 0.5^2 = 0.25$, and 4 cm transmit a fraction $0.5^4 = 0.0625$, or 6.25%. Likewise a 0.5-cm cell transmits a fraction $0.5^{1/2} = 0.71$, or 71%. (The square root of a number is easily read from a slide rule.)

3. A solution of potassium chromate contains 3.0 g of the salt per liter and transmits 40% of the incident light of a certain wavelength in a 1-cm cell. What percentage of the light will a solution with 6.0 g of potassium chromate per liter transmit in the same cell?

   By doubling the concentration we square the fraction of light transmitted, just as if we doubled the cell length (problem 2). The answer is $(0.4)^2 = 0.16 = 16\%$.

4. The molar absorptivity of the permanganate ion, MnO_4^-, dissolved in water is 2500 cm^{-1} l. mol^{-1} at 525 nm. How many milligrams of manganese, present as MnO_4^-, in 1 l. of solution would give an absorbance of 0.50 in a 1-cm cell?

Substituting into equation (5-1): $A = 0.50 = \varepsilon cx = 2500c$, since the cell thickness x is 1 cm; thus, $c = 0.50/2500 = \underline{2.0 \times 10^{-4}\ M}$. The atomic weight of manganese is 55, and a 2.0×10^{-4} M permanganate solution contains $2.0 \times 10^{-4} \times 55 = 1.1 \times 10^{-2}$ g Mn per liter $= \underline{11\ mg/l}$. Note that we do not need to know the molecular weight of MnO_4^-.

VI. DEVIATIONS FROM BEER'S LAW

An analytical chemist should never assume that his solutions obey Beer's law. He should always measure three of four standard solutions of known concentration, and more if necessary, and plot the graph of absorbance against concentration. Ideally this should be a straight line passing through the origin ($A = 0$ when $c = 0$). If the reagents added to produce the color absorb light themselves or contain impurities, A may not be zero when the (added) concentration is zero. In addition the graph may not be a straight line. It may curve upward or downward from the ideal straight line projected from low concentrations (Fig. 5-5), giving positive or negative deviations. Various factors may cause deviations from Beer's law; the commonest are the following.

A. Use of Non-monochromatic Light

In deriving Beer's law [equation (5-1)] we postulated that every quantum of light had the same chance of being absorbed. This will generally be untrue if the light quanta entering the solution have different energies, that is, if the light is not monochromatic. Now, monochromatic light is an ideal which may be approached but never realized; the light entering a solution has a range of wavelengths, and this range may be narrow or it may be broad. If the range is broad there are bound to be some quanta that pass more easily than others. To take an extreme case, let us suppose that part of the light that passes through a solution is not absorbed at all. Then, no matter how concentrated the colored substance is, some light will always go through, and the measured absorbance, instead of rising continuously with concentration as Beer's law says it should, tends toward a constant upper limit, as shown in Fig. 5-6. The use of non-monochromatic light always causes negative deviations from Beer's law,

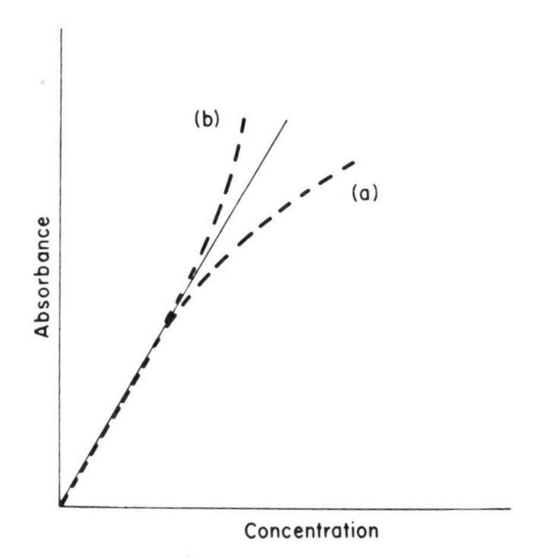

Fig. 5-5. Positive and negative deviations from Beer's law. Curve (a), negative deviations; curve (b), positive deviations.

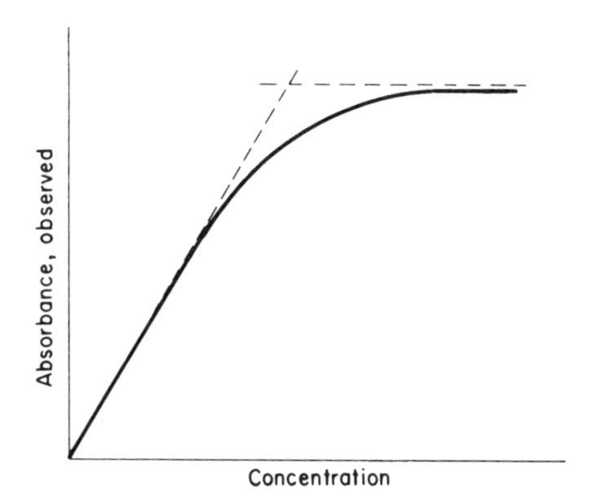

Fig. 5-6. Negative deviations and nonmonochromatic light. It is considered that two wavelengths pass through the solution, one absorbed and the other not absorbed. The solid curve shows the combined effect of the two rays.

and this effect may become quite serious when filters are used to select the wavelengths desired; see Chapter 6.

Errors due to non-monochromatic light are minimized by reading absorbances as near to the maximum as possible. If we are measuring permanganate ions with an instrument that gives a band of wavelengths

20 nanometers wide, for example, a glance at Fig. 5-2 tells us that we shall get better agreement with Beer's law in the range 510 to 530 nanometers than in the range 540 to 560 nanometers. Sometimes we are forced to read absorbances on the side of a steeply sloping absorption band. This happens especially with yellow solutions, that look yellow because they absorb violet light at the extreme lower end of the visible spectrum. An example is silicomolybdic acid, whose absorption spectrum is shown in Fig. 5-7. Below 400 nanometers the absorbance rises very rapidly as the wavelength falls, and it is impracticable to look for a maximum. One must choose a certain wavelength for one's analyses and stick to it as closely as possible. This matter will be discussed again in Chapter 6 when we consider instruments and their design.

B. Association or Dissociation of Light-Absorbing Species

It may happen that the concentration of the molecules that actually absorb the light is not the same as the total concentration of the material we want to measure. In such cases the concentration of the light-absorbing substance may not rise and fall in proportion to the total concentration. A

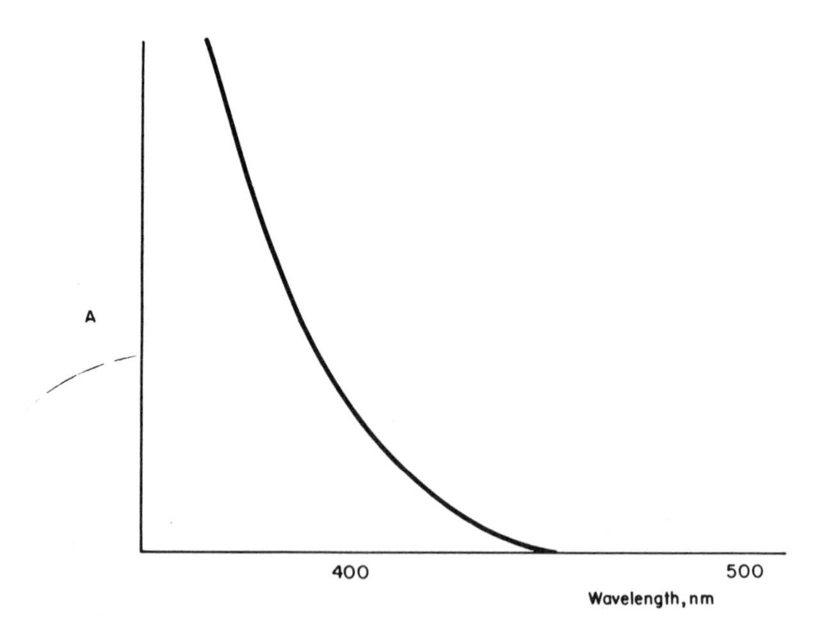

Fig. 5-7. Absorption spectrum of the silicomolybdate ion, $(SiO_4 \cdot 12MoO_3)^{4-}$.

simple example is the gas nitrogen dioxide, NO_2, a gas that is important in air pollution. It is deep brown in color, and small concentrations can be measured by light absorption. However, as the concentration of NO_2 rises so does the proportion that associates to form N_2O_4:

$$\underset{\text{(brown)}}{2NO_2} = \underset{\text{(colorless)}}{N_2O_4}$$

The dimer, N_2O_4, is colorless, so the light absorption does not rise as fast with concentration as would be expected from measurements at very low concentrations. Negative deviations from Beer's law are found.

Positive deviations from Beer's law are found if the associated species absorbs more strongly than the dissociated species. This happens with dichromate solutions at certain wavelengths, because of the reaction

$$Cr_2O_7^{2-} + H_2O = 2CrO_4^{2-} + 2H^+$$

C. Addition of Insufficient Color-Producing Reagent

Many chemical determinations are made by adding a reagent that forms a colored substance with the element that is to be determined. Thus, iron is determined by adding the reagent 1,10-phenanthroline to a solution containing iron as Fe(II). A complex is formed that is an intense red color. If one adds a sufficient excess of the reagent, over 99 per cent of the iron is converted to the red complex and the light absorption is proportional to the concentration of iron, in accordance with Beer's law. However, if only a small excess of reagent is added, the absorbance does not rise proportionately to concentration, and negative deviations are observed. This condition is easily corrected by adding enough reagent. It is merely a special case of B.

D. Fluorescence and Turbidity

Either of these conditions produces deviations from Beer's law. A slight turbidity in a sample to be analyzed may easily be overlooked, yet it increases the light absorption considerably, especially at low wavelengths.

VII. PHOTOMETRIC ERROR

Looking at the scale of a spectrophotometer (Fig. 5-4) and seeing how the absorbance divisions are crowded together at the low-transmittance end of the scale, one can easily see that a small error in reading the position of the needle in this range would cause a large error in the absorbance. Likewise, a small error in reading the position of the needle at high transmittances would cause a large percentage error in the absorbance. The most reliable readings are those taken somewhere in the middle of the transmittance scale.

Let us calculate mathematically the relative error in the concentration reading caused by a certain error in reading the position of the needle on a linear transmittance scale, like that shown in Fig. 5-4.

Let the fraction of light transmitted, I/I_0, be t. Let a small change dt in the transmittance correspond to a small concentration change dc. Then the relative change in the concentration caused by a change dt in the transmittance is

$$\frac{1}{c}\frac{dc}{dt}$$

By Beer's law, $c = -k \ln t$, where k is a proportionality constant that includes the molar absorptivity and the cell thickness. Differentiating,

$$\frac{dc}{dt} = -k \ln t, \quad \frac{1}{c}\frac{dc}{dt} = \frac{1}{t \ln t} \tag{5-2}$$

The graph of equation (5-2) appears in Fig. 5-8. To find the value of t for which the relative error $(1/c)\,dc/dt$ is a minimum, we differentiate this quantity with respect to t and find the value of t for which the derivative is zero:

$$\frac{d}{dt}\left(\frac{1}{t \ln t}\right) = -\frac{1}{t^2 \ln t} - \frac{1}{t\,(\ln t)^2} \cdot \frac{1}{t} = 0$$

or, $\ln t = -1;\ t = 1/e = 1/2.718 = \underline{0.368}$, or <u>36.8 per cent</u>.

Maximum accuracy (with respect to errors in the scale reading) is obtained at a transmittance reading of 36.8 per cent; however, the middle part of the curve of Fig. 5-8 is nearly flat, and any reading between about 10 per cent and 70 per cent transmittance is acceptable. If the reading

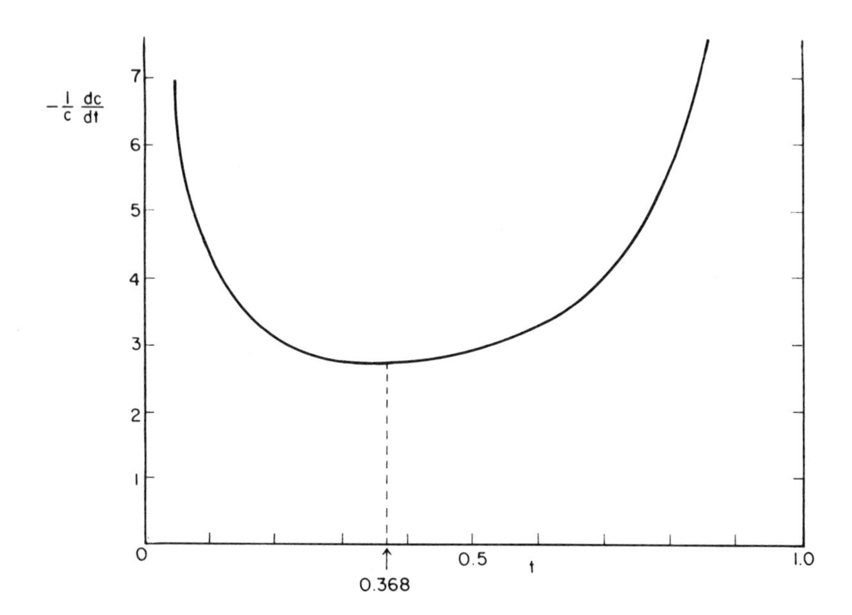

Fig. 5-8. Curve of photometric error.

obtained in an analysis falls outside this range, the chemist should consider preparing a new sample, with a concentration greater or smaller than the one first taken. If this is not possible, he may use a cell with a longer or shorter light path.

Problem. A waste water contains chromium at a concentration of 0.1 ppm (part per million). You will measure this concentration by forming the complex of Cr(VI) with diphenylcarbazide. The molar absorptivity of this complex is 41,700 at 540 nm. You have cells of thickness 2 mm, 1 cm, 5 cm, and 10 cm. Which cells should you use?

First, note that the calculation need only be rough. One significant figure, or two, is adequate.

The concentration of chromium is 0.1 mg Cr per liter, or $0.1/52 \times 10^{-3}$ M; the atomic weight of Cr is 52. Let us round off this value to 2×10^{-6} M. Multiplying this molar concentration by the molar absorptivity we have $2 \times 10^{-6} \times 4.2 \times 10^{4} = 0.084$. This is the absorbance in a 1-cm cell. In a 5-cm cell the absorbance would be 0.42, in a 10-cm cell it would be 0.84. The value that comes closest to $t = 0.368$ is $A = 0.42$ ($t = 38\%$). We therefore choose to work with 5-cm cells.

VIII. PRECISION SPECTROPHOTOMETRY: THE MEANING OF I_o/I

A way to increase the accuracy of photometric measurements, especially in cases where the solutions have a high absorbance, is to compare the intensity of the light passing through a solution with the intensity of a beam of light from the same lamp or light source that passes through a second solution whose concentration is accurately known and is somewhat smaller than that of the sample to be tested; see Fig. 5-9. This is easily done; one places the second, or reference, solution in the path of the beam and adjusts the meter to read 100 per cent transmittance (see Chapter 6). Then one puts the test solution in its place and reads the apparent transmittance and absorbance.

Let I_o be the intensity of the light before it passes through the cell (Fig. 5-9). Let I_1 be the intensity of the light emerging from the test solution, and I_2 the intensity of the light emerging from the reference solution. Let the concentrations of the two solutions be c_1 and c_2. Then,

$$\log I_1 = \log I_o - kc_1, \quad \log I_2 = \log I_o - kc_2$$

where k is the product of molar absorptivity and cell thickness and is the same constant for both solutions. Then it is clear that

$$\log \frac{I_2}{I_1} = k(c_2 - c_1)$$

The spectrophotometer now measures the difference between two concentrations, and an error in measuring this difference causes a relatively smaller error in measuring c_1 itself.

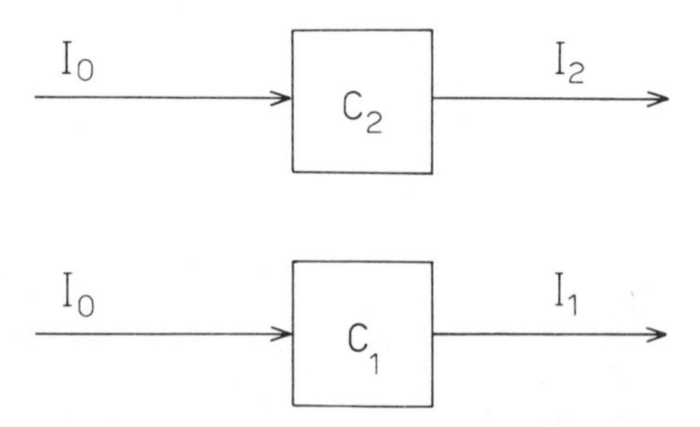

Fig. 5-9. Precision spectrophotometry at high absorbance.

This is a good moment to point out that the state of affairs shown in Fig. 5-3 is an idealization. In practice one never measures I_o/I in the sense that Fig. 5-3 implies. One places a cell with the pure solvent in the path of the beam, then one places a cell with the solution in the path of the beam, and one compares the intensities of the two emergent beams. The "I_o" that is used in applying Beer's law is actually the intensity of light transmitted by the cell that contains pure solvent. In this way one compensates for the light that is lost by reflection at the surfaces of the cell walls.

IX. SIMULTANEOUS DETERMINATION OF TWO OR MORE COMPONENTS

If two substances have different absorption spectra one can analyze mixtures of the two by measuring the absorbances at two wavelengths. Consider the spectra shown in Fig. 5-10. Substance I absorbs strongly at wavelength λ_1; substance II absorbs strongly at wavelength λ_2. By making measurements on known solutions of the pure substances one can evaluate four molar absorptivities:

Substance I: absorptivity = ε_{11} at λ_1, ε_{21} at λ_2
Substance II: absorptivity = ε_{12} at λ_1, ε_{22} at λ_2

If we now take a solution containing unknown concentrations of substances I and II, but no other substance that absorbs light at the two wavelengths chosen, and we measure the absorbances at λ_1 and λ_2, we can calculate the unknown concentrations by solving a pair of simultaneous equations. If the concentrations of I and II are c_1 and c_2, respectively, and if the absorbances at λ_1 and λ_2 are A_1 and A_2, then, using Beer's law,

$$A_1 = \varepsilon_{11}c_1 + \varepsilon_{12}c_2 \tag{5-3a}$$

$$A_2 = \varepsilon_{21}c_1 + \varepsilon_{22}c_2 \tag{5-3b}$$

An easy way to solve simultaneous equations is by determinants. For the pair of equations (5-3a) and (5-3b) the solutions are

$$c_1 = \frac{\begin{vmatrix} A_1 & \varepsilon_{12} \\ A_2 & \varepsilon_{22} \end{vmatrix}}{\begin{vmatrix} \varepsilon_{11} & \varepsilon_{12} \\ \varepsilon_{21} & \varepsilon_{22} \end{vmatrix}}, \quad c_2 = -\frac{\begin{vmatrix} A_1 & \varepsilon_{11} \\ A_2 & \varepsilon_{21} \end{vmatrix}}{\begin{vmatrix} \varepsilon_{11} & \varepsilon_{12} \\ \varepsilon_{21} & \varepsilon_{22} \end{vmatrix}}$$

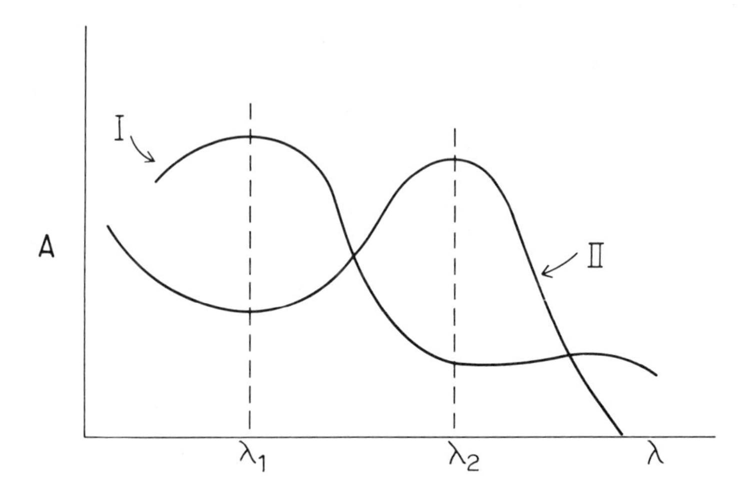

Fig. 5-10. Simultaneous determination of two substances.

To expand a determinant one multiplies crosswise and down and changes sign when one goes backward or up. Thus,

$$\begin{vmatrix} A_1 & \varepsilon_{12} \\ A_2 & \varepsilon_{22} \end{vmatrix}$$

is a shorthand notation for

$$A_1 \varepsilon_{22} - A_2 \varepsilon_{12}$$

The convenience of the determinant notation is seen when one goes to more complicated simultaneous equations, like those for a three-component mixture. Here, one must measure absorbances at three wavelengths and solve three simultaneous equations, as follows:

$$A_1 = \varepsilon_{11}c_1 + \varepsilon_{12}c_2 + \varepsilon_{13}c_3 \qquad (5\text{-}4a)$$

$$A_2 = \varepsilon_{21}c_1 + \varepsilon_{22}c_2 + \varepsilon_{23}c_3 \qquad (5\text{-}4b)$$

$$A_3 = \varepsilon_{31}c_1 + \varepsilon_{32}c_2 + \varepsilon_{33}c_3 \qquad (5\text{-}4c)$$

The solutions are

$$c_1 = \frac{\begin{vmatrix} A_1 & \varepsilon_{12} & \varepsilon_{13} \\ A_2 & \varepsilon_{22} & \varepsilon_{23} \\ A_3 & \varepsilon_{32} & \varepsilon_{33} \end{vmatrix}}{\begin{vmatrix} \varepsilon_{11} & \varepsilon_{12} & \varepsilon_{13} \\ \varepsilon_{21} & \varepsilon_{22} & \varepsilon_{23} \\ \varepsilon_{31} & \varepsilon_{32} & \varepsilon_{33} \end{vmatrix}}, \quad c_2 = -\frac{\begin{vmatrix} A_1 & \varepsilon_{11} & \varepsilon_{13} \\ A_2 & \varepsilon_{21} & \varepsilon_{23} \\ A_3 & \varepsilon_{31} & \varepsilon_{33} \end{vmatrix}}{\begin{vmatrix} \varepsilon_{11} & \varepsilon_{12} & \varepsilon_{13} \\ \varepsilon_{21} & \varepsilon_{22} & \varepsilon_{23} \\ \varepsilon_{31} & \varepsilon_{32} & \varepsilon_{33} \end{vmatrix}}, \quad c_3 = \frac{\begin{vmatrix} A_1 & \varepsilon_{11} & \varepsilon_{12} \\ A_2 & \varepsilon_{21} & \varepsilon_{22} \\ A_3 & \varepsilon_{31} & \varepsilon_{32} \end{vmatrix}}{\begin{vmatrix} \varepsilon_{11} & \varepsilon_{12} & \varepsilon_{13} \\ \varepsilon_{21} & \varepsilon_{22} & \varepsilon_{23} \\ \varepsilon_{31} & \varepsilon_{32} & \varepsilon_{33} \end{vmatrix}}$$

A determinant with three rows and three columns is evaluated as follows:

$$\begin{vmatrix} a & b & c \\ d & e & f \\ g & h & i \end{vmatrix} = a \begin{vmatrix} e & f \\ h & i \end{vmatrix} - b \begin{vmatrix} d & f \\ g & i \end{vmatrix} + c \begin{vmatrix} d & e \\ g & h \end{vmatrix}$$
$$= aei - afh - bdi + bfg + cdh - ceg$$

All these relationships can be proved by simple, if laborious, algebra. The quantity that appears in the denominator of each of the expressions for c_1, c_2, and c_3 is called the "secular determinant" and is useful in quantum mechanics.

QUESTIONS

1. Nitrogen dioxide is red-brown in color. What visible wavelengths, approximately, are absorbed most strongly?

2. The yellow sodium line has wavelength 589 nm in a vacuum. What is the wavelength in angstrom units? What is its wavelength in water, whose refractive index is 1.33? What is its frequency (a) in sec^{-1}, (b) in cm^{-1} in air?

3. The molar absorptivity of sodium benzoate in water is 560 cm^{-1} l. mol^{-1} at 268 nm. Its formula weight is 144. A solution contains 12.0 mg sodium benzoate in 25 ml; what is its (a) absorbance, (b) transmittance in a 1.0-cm cell at 268 nm?

4. A solution contains mercury dithizonate and excess dithizone (page 165); its absorbance in 1-cm cells is 0.75 at 500 nm, 0.35 at 600 nm. The molar absorptivities of mercury dithizonate and dithizone at these two wavelengths are:

   At 500 nm, $HgDz_2$ 6.0×10^4, HDz 1.0×10^4
   At 600 nm, $HgDz_2$ 1.0×10^3, HDz 4.0×10^4

   (These numbers are rounded off to make calculation easy.) Calculate the concentration of mercury in the solution in parts per million (milligrams Hg per liter).

5. The absorbance of a solution of sodium chromate containing 100.0 mg Cr per liter is 0.465 at 375 nm. This solution is placed in the "sample" beam of a spectrophotometer and the meter is adjusted to read exactly 100% transmittance. Then an unknown chromate solution is placed in the sample beam; the transmittance reads 92%. The cell thickness is 0.5 cm in every case. Calculate (a) the concentration of Cr in the unknown, (b) the molar absorptivity of sodium chromate.

Chapter 6

VISIBLE AND ULTRAVIOLET SPECTROSCOPY

I. INTRODUCTION

The ultraviolet-visible region of the spectrum from roughly 200 to 800 nanometers is the part of the spectrum most used in chemical analysis. The instruments used in the visible and ultraviolet are common and relatively simple and are well suited for quantitative analysis. As we have noted, absorption of light in this wavelength range causes electrons to move from one energy level to another in the upper, or outer, levels of atoms and molecules. We shall examine the relation of light absorption to molecular structure later in this chapter but shall note at this point that there are two broad classes of chemical compounds that absorb in the ultraviolet-visible region and can be quantitatively determined by measurements of this absorption. They are: (i) organic compounds having conjugated double bonds or aromatic rings; (ii) ions of transition metals, especially complex ions formed with organic reagents.

It will make our discussion more realistic if we begin with a description of the instruments. Essentially the same instrumentation is used throughout the range from 200 to 800 nanometers, but there is an important complication which enters when we go to wavelengths below about 340 nanometers. Below this wavelength glass becomes opaque, and it is necessary to use cells, windows, prisms, and lenses made of quartz. This fact immediately makes the instruments more expensive. It is also necessary to use a different source of light when we go below 340-350 nanometers.

II. INSTRUMENTS FOR MEASURING LIGHT ABSORPTION

A. General Features

All instruments designed for measuring the absorption of radiation have certain features in common. They are illustrated by Fig. 6-1.

First, there is a source of continuous radiation. For visible light this is a tungsten-filament lamp. This is the common "light bulb" used in homes, and the kind of bulb used for automobile headlights is ideal for a spectrophotometer. The way in which the energy emitted by this lamp varies with wavelength is shown in Fig. 6-2. The peak emission is actually in the near infrared, just beyond the end of the visible range. If the temperature of the filament is known the wavelength maximum can be calculated by the Wien displacement law:

$$\lambda_{max}T = \text{a constant} = 0.29 \text{ cm } ^{o}K$$

At $3000^{o}K$, the maximum emission comes at 960 nanometers. There is good emission throughout the visible and near infrared, but the intensity falls rapidly below 400 nanometers and is very weak at 300 nanometers. For work below 350-400 nanometers one uses a hydrogen gas discharge lamp. The light emitted by a glowing gas is normally a series of lines, but if the gas is under pressure the lines broaden and eventually overlap to give a continuum. Deuterium-filled lamps give much brighter emission than hydrogen-filled lamps, but cost more. The lamps have a window of quartz or fused silica to let the ultraviolet radiation pass, and they are good from 375 nanometers down to 180-200 nanometers, the wavelength range in which quartz starts to absorb. Hydrogen and deuterium lamps need a special power supply that provides a regulated current.

The light from the lamp is passed through a wavelength selector, or or monochromator, that selects a certain part of the continuous radiation, or in simple terms, light of a certain color, to be passed through the sample. In the simplest instruments this is simply a piece of colored glass. This transmits a band of wavelengths that may be quite broad. Typical transmission curves for glass filters are shown in Fig. 6-3.

More usually, however, the light is dispersed into a spectrum by means of a prism or a diffraction grating. The general scheme is shown in Fig. 6-4. Light from the lamp is focused on a narrow slit that serves as a line source; the light proceeding from this source is focused again to give a parallel beam, which then passes into the prism. The prism disperses

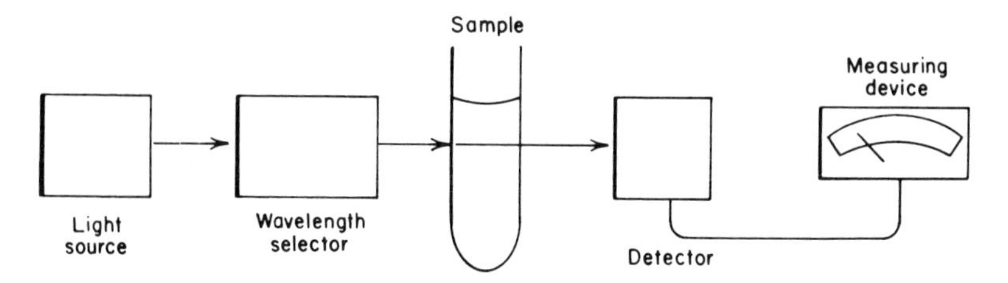

Fig. 6-1. Instrument for measuring light absorption (schematic).

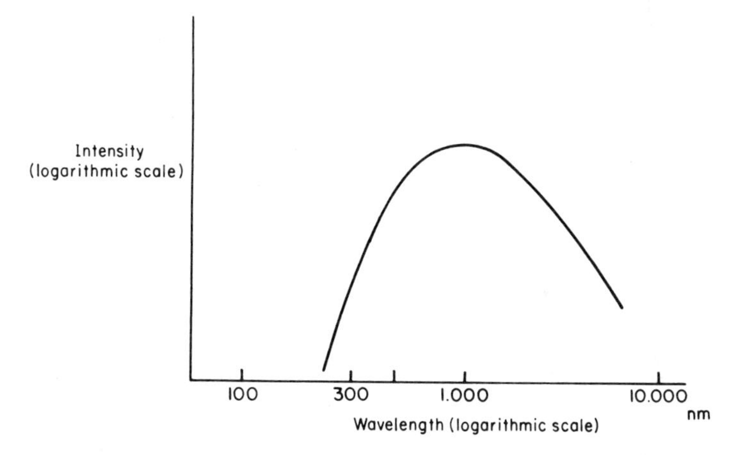

Fig. 6-2. Emission by a tungsten filament lamp (black body radiation at 3000°K).

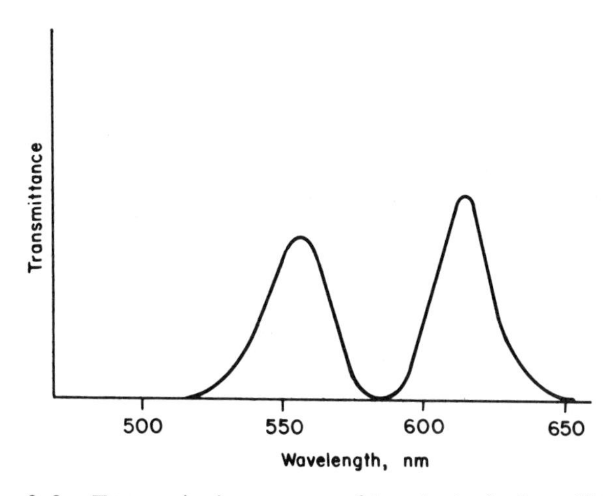

Fig. 6-3. Transmission curves of two typical glass filters.

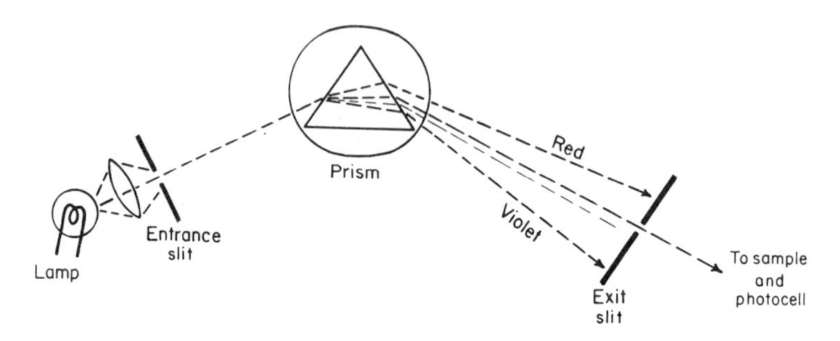

Fig. 6-4. A simple prism monochromator (schematic).

the beam of white light into a fan of wavelengths, the long wavelengths (red) being bent the least, the short wavelengths (violet) being bent the most. The dispersed light falls on a black surface in which is a slit. Only the part of the light that falls on the slit passes on to the sample. In most instruments the width of this slit can be adjusted to pass a wide band of wavelengths if one needs a lot of light, or a very narrow band of wavelengths if one wishes to isolate one small part of the spectrum. The wavelengths that pass out through this slit can be changed at will by rotating the prism. The prism is mounted on a table that can be turned by turning a knob on the outside of the instrument.

Instruments that use filters as wavelength selectors are called filter photometers; those that use prisms or diffraction gratings are called spectrophotometers. Spectrophotometers are more versatile and permit more precision; filter photometers are cheaper.

The cells used to contain the samples for photometric analysis may simply be test tubes. Very good results can be obtained with ordinary glass test tubes if care is taken to select a set that is matched, so that different tubes give the same transmission. Special test tubes can be bought for photometric analysis, and they are convenient because they are uniform and well-matched, but they are not necessary; with a little more trouble one can use ordinary test tubes. It must be remembered that one always needs at least two tubes, one to contain the pure solvent (which is usually water) and one to contain the solution.

For accurate work one uses special cells with optically flat sides. Commonly these cells are 1 centimeter wide internally, but cells of different widths are available. Glass cells are used for measurements down to 340-350 nanometers, cells of silica or fused quartz for measurements below this.

After the light passes through the cell, it strikes a detector that converts the light energy into an electric current. These detectors are of two types, the solid-state photocell and the vacuum phototube. The former are cheaper and more robust; the latter are more sensitive and are always preferred for precise work. Filter photometers usually use solid-state, "barrier-layer" photocells that have a layer of a semiconductor, like selenium, coated with a layer of silver that is so thin as to be transparent. When light strikes the interface between the silver and the selenium it makes the selenium give up electrons, and a small electric current passes, the electron stream flowing from the selenium to the silver. The current is passed without amplification through a sensitive microammeter, and provided the resistance of the microammeter and connections is not great, the current is proportional to the light intensity. No batteries are needed; the light itself provides the energy to deflect the microammeter.

If the light energy is very weak the barrier-layer cell does not respond. One must then use a vacuum photocell (or phototube). This is illustrated in Fig. 6-5. The light enters an evacuated bulb, where it strikes a metal surface coated with an alkali metal, preferably cesium or rubidium. The outermost electron in an atom of cesium is easy to dislodge, and when a light quantum is absorbed, an electron is emitted. It moves across the evacuated space toward a wire or rod that is charged positively with respect to the alkali metal surface. A potential difference of some 30 - 50 volts is required; it is supplied by batteries or a power supply. The current is proportional to the light intensity over a wide range. Since it originates in a circuit of high electrical resistance, it can be amplified and the amplified current is read on a meter.

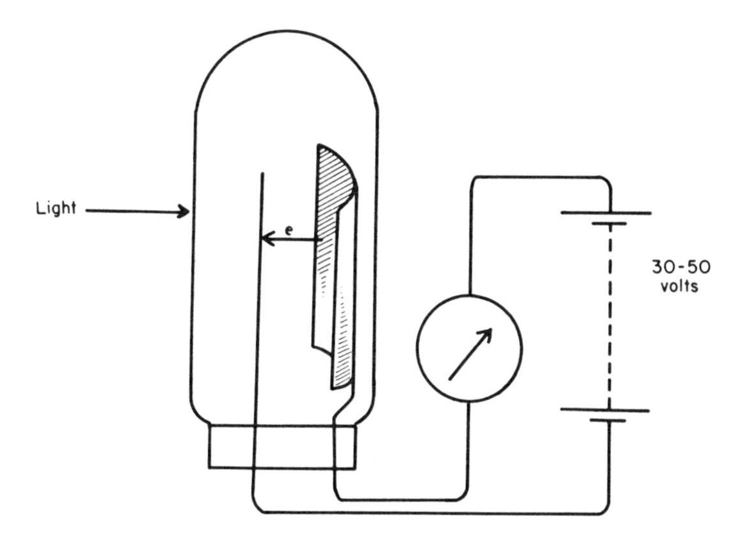

Fig. 6-5. Vacuum phototube.

The light-sensitive coating is not pure alkali metal, but contains the metal, cesium or rubidium or a mixture, mixed with the oxide of this metal and often other metals and their oxides. Different coatings give different responses. A phototube that responds well to wavelengths below 600 nanometers does not respond well to wavelengths above this; tubes that respond well above 600-800 nanometers do not work well for shorter wavelengths. It is thus usual for spectrophotometers to be equipped with two kinds of phototubes, a "red phototube" for readings above 650 nanometers, a "blue phototube" for readings at lower wavelengths. Sometimes both tubes are mounted permanently in the instrument and one switches from one to the other by moving a knob. More usually one has to unscrew the mounting, take one phototube out, and insert the other.

A special kind of phototube which is used for measuring very low light intensities is the photomultiplier tube. This is illustrated in Fig. 6-6. It works by a cascade effect; the electrons produced when the light strikes the sensitive surface do not pass directly to the positively charged collector but are attracted to another sensitive surface that is kept at a positive

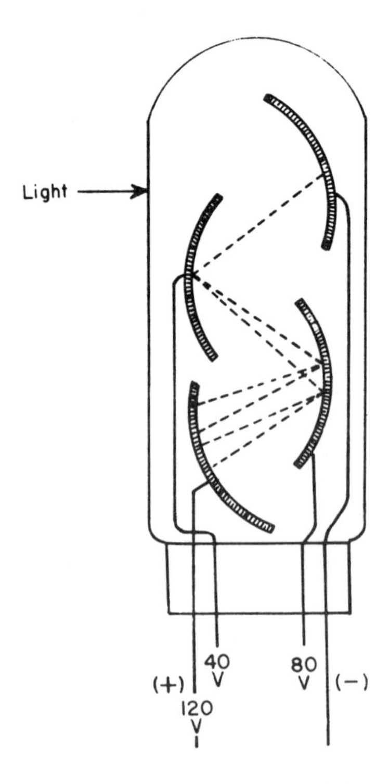

Fig. 6-6. Photomultiplier tube (schematic).

potential with respect to the original emitting surface. The extra energy that the electrons gain by falling through this potential difference causes them, when they arrive, to drive out more electrons; one electron striking the surface may cause the emission of three or four more electrons. These are attracted to yet another sensitive surface, and the process is repeated. There may be eight or ten of these surfaces altogether, giving a very large multiplication effect. For this effect to be useful it must be stable and reproducible. Special stabilized power supplies are therefore used with photomultiplier tubes.

B. Prisms and Diffraction Gratings

To disperse the light and form a spectrum, two devices are used. One is a prism of glass or quartz, such as is shown in Fig. 6-4. This looks simple, but has certain disadvantages. First, quartz prisms are expensive; moreover, they are doubly refracting. This problem is overcome by the "Littrow mounting" which will be discussed in Section III-C. A drawback to both glass and quartz prisms is that the dispersion changes greatly with wavelength; the short wavelengths are separated much more effectively than the longer ones. One will notice that the wavelength dial of a prism instrument has graduations that are far apart in the low-wavelength region but close together in the red end of the spectrum. The crowding together of the red wavelengths is more noticeable with quartz than it is with glass. On the other hand, the dispersion of quartz in the ultraviolet is excellent, and prisms are often preferred to gratings in the ultraviolet for this reason.

The diffraction grating works on the principle shown in Fig. 6-7. It consists of a reflecting surface that is ruled by many fine lines close together, of the order of 6,000 lines to the centimeter. When light strikes the surface of the grating, each one of these lines acts as a secondary light source, scattering light in all directions. There is one direction in which the light waves reinforce each other, where the waves scattered by one line (or ruling) are exactly in phase with the waves scattered by the ruling next to it. This direction depends on the length of the light waves. If the light from the source strikes the grating at right angles to its surface, light is scattered or "diffracted" at an angle α to the perpendicular, where

$$n\lambda = d \sin\alpha$$

An advantage of a grating over a prism is immediately apparent; the angle of diffraction, α, is simply related to the wavelength and the dispersion is the same throughout the spectrum. Wavelengths in the red are separated as effectively as wavelengths in the ultraviolet. Gratings do not

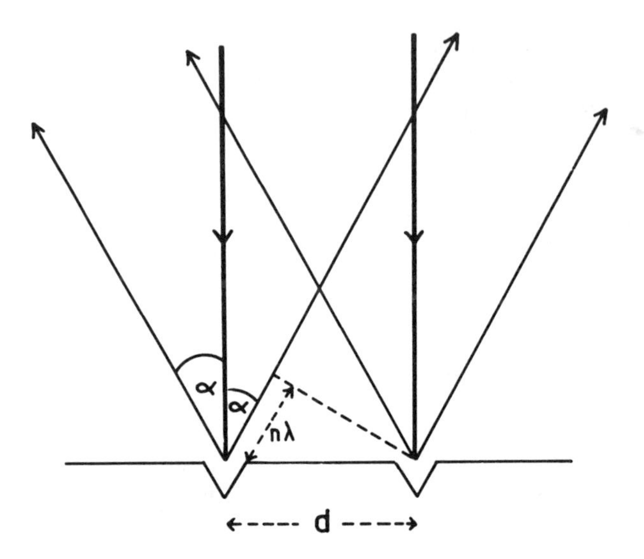

Fig. 6-7. Diffraction grating.

absorb light and, unlike glass prisms, can be used over the entire ultraviolet-visible wavelength range. However, a disadvantage of a grating is also evident from Fig. 6-7. It wastes a lot of light. Much of the energy is reflected back along the original light path, and the diffracted light is dissipated among various "orders," with n = 2, 3, etc. Gratings are made to use light more efficiently by "blazing" them, that is, by cutting the rulings at an angle (Fig. 6-8), but even so, light energy is lost, and it is impossible to prevent the appearance of spectra of second and third order. One pre-

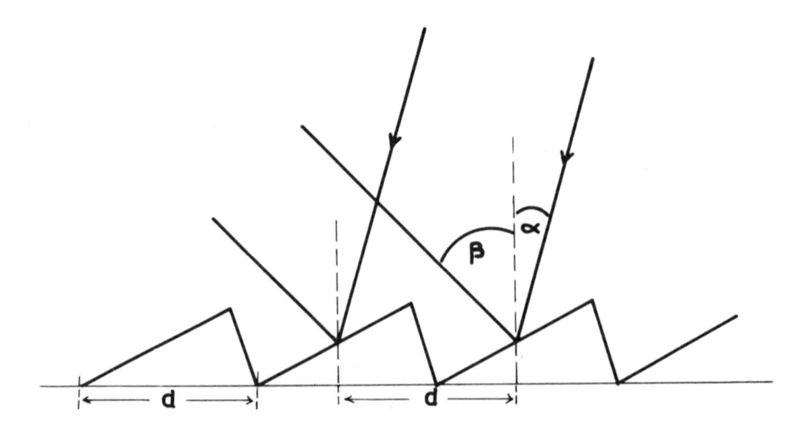

Fig. 6-8. Blazed diffraction grating.

vents higher-order spectra from interfering by placing a glass filter between the grating and the phototube. If one is measuring light of 800 nanometers, for example, one knows that it will be accompanied by second-order diffracted light of wavelength 400 nanometers, so one uses a red filter to block the unwanted light. In some instruments the filter must be inserted manually; in other and more expensive instruments, the filters are placed automatically as the wavelength selector dial is rotated.

C. Single-Beam and Double-Beam Instruments

In chemical analysis by light absorption one never measures the absolute intensity of a light beam. One always compares the intensities of two light beams, one passing through the pure solvent, the other through the test solution. The ratio of these two intensities is the ratio I_0/I to which one applies the Beer-Lambert law. There are two ways to find this intensity ratio. In the single-beam method, one first adjusts the galvanometer to read zero when no light is passing, then one adjusts it again to read 100 per cent transmission when a cell containing pure solvent is placed in the beam. Then one removes the cell containing solvent and replaces it by the cell that contains the solution. The galvanometer now reads the transmission of the solution. (One should now put the solvent cell back again to see if it still reads 100 per cent.)

Whenever the wavelength setting is changed the solvent cell must be replaced and the 100 per cent transmission reading reset. This is necessary because the phototube sensitivity and the power given out by the lamp both change with wavelength. It is difficult or impossible to adapt a single-beam instrument to automatic scanning; to plot an absorption spectrum like that shown in Fig. 5-2 one has to take one reading at a time and plot one point after another, which is tedious. On the other hand, very precise readings of absorbances can be obtained with single-beam instruments if the light source is sufficiently stable.

In double-beam instruments the beam of light leaving the exit slit of the spectrometer is split into two parts by a revolving or vibrating mirror. One part is passed through the cell containing the solvent while the other part is passed through the sample cell. Both cells are placed in the instrument together. A typical arrangement is shown in Fig. 6-9. After they have passed through the cells the two parts of the beam are brought together by mirrors and strike the photocell together. The photocell receives first one beam and then the other, alternating many times a second, giving a current that alternates in a step-like fashion. The current is

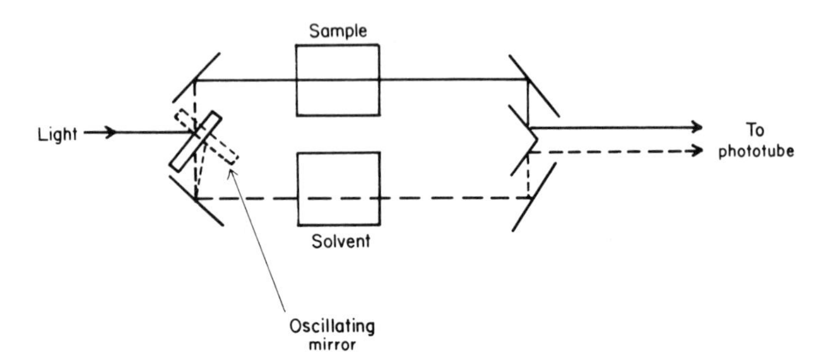

Fig. 6-9. Double-beam spectrophotometer.

amplified and the ratio of the maximum and minimum currents is registered on the galvanometer as the per cent transmittance.

Double-beam instruments are easier to use than single-beam instruments. Changes in the lamp intensity or in the photocell sensitivity affect the sample beam and the reference (or solvent) beam equally and do not affect the ratio which is read on the dial. Double-beam instruments are, therefore, suited to the automatic recording of absorption spectra.

III. COMMERCIAL INSTRUMENTS

There are many instruments on the market and it is impossible to describe them all. We shall describe a few representative instruments and indicate how to operate them. Once a person has become familiar with a few typical instruments, it is easy for him to learn to use other instruments of the same type. Before starting to work on an unfamiliar instrument, the operator should always consult the manual supplied by the manufacturer.

A. The Klett-Summerson Filter Photometer

This is an instrument that uses filters to select the light desired and barrier-layer photocells to measure the light. It is a double-beam instrument. The construction is shown in Fig. 6-10. The light source is a 100-watt bulb of the kind used in slide projectors. The light is collected and brought to a parallel beam by a large lens, then passed through a glass filter. Three or four filters are supplied with the instrument, and they are in holders so that they can be easily inserted and removed. After passing through the filter, part of the light goes through the sample, which is contained in a special test tube, accurately formed to specified dimensions. This light then falls on one of the two photocells. Another part of the filtered light beam passes through an iris stop of the kind used in cameras; its aperture can be varied continuously by turning a knob.

The two photocells are connected back-to-back so that the currents oppose one another. The cell that receives light through the sample has its current flowing through a resistance, and along this resistance moves a sliding contact, so that the circuit is like a potentiometer (see Chapter 1). The contact is moved by the large knob in front of the instrument. On this knob is mounted a scale which is graduated in absorbance units. Above the knob and scale is the pointer of a galvanometer. The galvanometer has only one mark in its center. It is a <u>null instrument</u>. The idea is to bring the pointer of the galvanometer back to the center mark by turning the scale, then take the reading from the scale.

To use the Klett-Summerson photometer, insert the filter desired, then turn on the light. Place a tube of pure water (or other solvent) in the sample well; put the galvanometer in circuit, if necessary, by opening the short-circuit switch on the side of the instrument. Set the absorbance

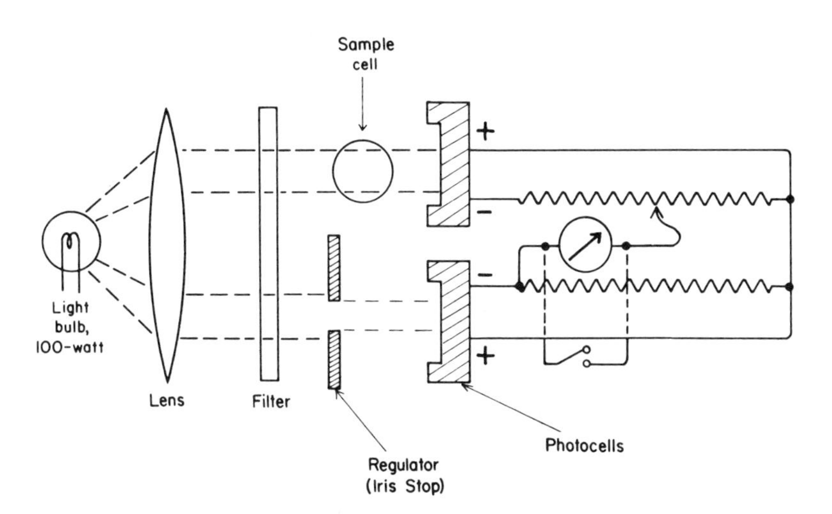

Fig. 6-10. Klett-Summerson filter photometer.

scale to read zero. Now adjust the iris stop using the knob next to the sample well, until the galvanometer balances with the needle at the center mark. Take out the tube of solvent, and replace it with the tube containing the test solution. The needle will swing away from the center mark. Bring it back by turning the large knob, and when the galvanometer is again in balance, read the absorbance from the scale. Readings that are very high or very low should be distrusted; very concentrated (highly absorbing) solutions should be diluted and read again.

The double-beam feature makes this photometer very stable and insensitive to fluctuations in the electric power supply. However, one should never assume that solutions obey Beer's law (for the reason discussed in Chapter 5) and one should always calibrate the instrument carefully. The calibration should be repeated whenever the light bulb is changed or a new filter is used.

An instrument very similar to the Klett-Summerson is the Fisher electrophotometer.

B. The Bausch and Lomb Spectronic-20

This is a spectrophotometer that uses a diffraction grating to disperse the light. It is a low-priced, mass-produced instrument that is very popular in laboratories all over the world. To keep the price low, the design has been kept simple. Thus the slit width is fixed, and passes a band of wavelengths 20 nanometers wide (hence the name, "Spectronic-20"). The instrument works in the glass-optics region, 340 - 900 nanometers. Above 600 nanometers one must replace the standard blue-sensitive phototube with a special red-sensitive tube, and at the same time insert a red glass filter to exclude second-order spectra (see page 152). The phototube can be reached by opening a door on the underside of the instrument.

The instrument is illustrated in Fig. 6-11. Another economy feature will be noted; the beam is focused by a compound lens and is converging, not parallel, when it strikes the grating. This causes a small loss in resolution.

There are only three controls: a wavelength selector, which turns the grating about its axis; a dark current control, which is electrical, and an adjustment for 100 per cent transmittance, which moves a V-shaped slit that controls the amount of light reaching the exit slit. To operate the instrument, proceed as follows: Select the wavelength desired; turn the

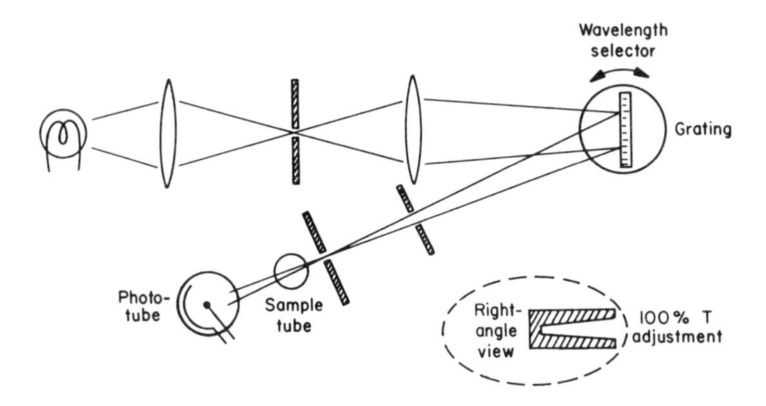

Fig. 6-11. Bausch and Lomb Spectronic-20.

instrument on by moving the dark current knob clockwise, and after warm-up, adjust this knob so that the galvanometer needle is at zero. Open the cap of the sample holder and place in the holder a tube containing pure water or other solvent; close the cap and adjust the right-hand knob on the front of the case to make the meter read 100 per cent transmittance. Then remove the water tube and insert the sample tube; read the position of the needle. The dial is graduated in absorbance and transmittance units.

The 100 per cent reading must be checked frequently. In the older models this reading drifts badly with changes in the lamp intensity; the drifting is reduced by using a special voltage-stabilizing transformer. The newer models have a second photocell, not shown in Fig. 6-11, which balances the principal, or working, photocell and compensates for small changes in the lamp intensity.

C. The Beckman Model DU Spectrophotometer

This was one of the first commercial spectrophotometers and is still one of the most precise. It covers the ultraviolet, visible, and near-infrared from 200 to 2000 nanometers. It is a single-beam instrument with a quartz prism as the dispersing element. The optical system is shown in Fig. 6-12. The light passes through the prism twice, being reflected at a silvered surface on one side of the prism, as shown. This arrangement is called a Littrow mounting. It compensates for the double refraction of quartz.

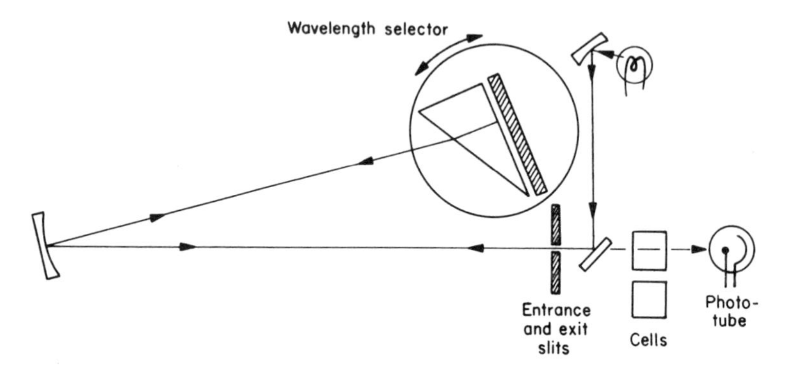

Fig. 6-12. Beckman DU spectrophotometer.

The light source can be either a tungsten-filament lamp or a high-pressure hydrogen (or deuterium) lamp. Focusing is done with concave mirrors, which (unlike lenses) bring all wavelengths to focus at the same point. The principal focusing mirror is tilted slightly so that the dispersed rays leaving the prism do not return exactly along their original path but pass below the small plane mirror beyond the slit and continue through the cell and into the phototube. Two phototubes are mounted in the instrument, one blue-sensitive and one red-sensitive, and one can change from one to the other by moving a handle in or out. Four cells can be placed in the instrument at once, and one of these must always contain the pure solvent or a "blank" solution to obtain the 100 per cent transmittance reading.

The DU spectrophotometer is a null instrument; that is, the galvanometer needle is always brought back to its zero position. Readings are taken by balancing the photocell current with an opposing current from a potentiometer. The galvanometer need not be linear, that is, its deflection need not be proportional to the current; it is sufficient for it to be sensitive. The accuracy of the phototube reading depends not on the galvanometer but on the uniformity of the potentiometer wire, and potentiometers can be very accurate indeed.

To take a reading, select the wavelength desired, adjust the dark current zero, then place the solvent cell in the beam and adjust the 100 per cent transmittance reading. It is not necessary to turn the potentiometer dial (that is, the transmittance-absorbance dial) to 100 per cent to do this; a special position of the selector switch, called "Check," shorts out the potentiometer wire and gives the same effect as if the dial had been set at 100 per cent. This feature is a great convenience, as one must continually check the solvent reading with single-beam instruments. There are two ways to make the galvanometer needle come to zero at the 100 per cent reading. One is to adjust the slit; the slit width is continuously variable. The other is by an electrical sensitivity control at the top left-hand corner of the instrument case. After the 100 per cent reading has been set, bring a sample into the path of the beam, and turn the potentiometer knob until the galvanometer needle comes to zero. Read the transmittance or absorbance from the scale.

If the transmittance reads below 10 per cent, turn the selector switch to 0.1. The transmittance range 0 - 10 per cent is now expanded to cover the entire scale. To obtain the actual transmittance, divide the scale reading by 10; to obtain the actual absorbance, add one unit to the reading on the absorbance scale. This feature allows one to read with good accuracy absorbances from 1.0 up to 2.0 and even higher.

D. The Beckman Model B

This is a popular instrument, medium-priced, with glass optics. Its wavelength range is 340-1000 nanometers; that is, it covers the visible region and little more. It is a good instrument for inorganic photometric analysis, nearly all of which is done above 340 nanometers. It reads transmittances and absorbances directly. A special feature of the Model B is its dispersing element, which is a glass prism with curved sides, a so-called Féry prism. This prism disperses the light and focuses it at the same time. The focal length depends on wavelength; blue light is focused at a shorter distance than red light. To compensate for this effect the wavelength selector dial moves the prism back and forth as well as rotating it about its axis. The general plan of the optical system is like that of the DU (Fig. 6-12) with a plane mirror, instead of a concave mirror, between the slits and the prism.

The slit width is continuously variable, but there is no rheostat to adjust the sensitivity. Instead, there is a sensitivity switch with four positions. Moving this switch by one unit multiplies or divides the transmittance reading by the square root of 10; that is, it reduces or increases the absorbance reading by 0.5 unit.

The cell holder carries four cells, as in the DU. To take a reading, place one cell with the solvent and another with the sample in the holder. Close the shutter (on the right of the instrument) and adjust the dark current control until the needle reads zero on the transmittance scale. (The zero reading should not change as the sensitivity is changed, but if it does, select the sensitivity desired before making the dark current adjustment.)

Select the wavelength desired, set the solvent cell in the light path, and open the shutter. Adjust the slit until the needle reads 100 per cent transmittance. Now move the sample cell into the light path and read the position of the needle on the scale, using the transmittance or the absorbance scale, as desired. If the reading falls below 10 per cent transmittance, increase the sensitivity by one unit or more. The needle will move to the right on the scale, changing the absorbance reading by 0.5 unit for each unit of the sensitivity control.

If the needle cannot be brought to 100 per cent with the sensitivity first selected, one must use a higher sensitivity before adjusting the slit. At high or low wavelengths where the phototube sensitivity is low, one may have to use the highest sensitivity in order to get a 100 per cent balance at all.

Two phototubes are supplied with the instrument, a "blue" tube for use up to 600 - 650 nanometers, and a "red" tube for use above this wavelength. To change phototubes the holder must be removed by loosening four screws. The phototube that is not in use should be kept in a dry atmosphere in the dark.

In a humid climate one often finds, on starting up the instrument, that one cannot bring the dark current reading to zero. The dark current reading drifts slowly downward and may take hours before it stabilizes. The reason is moisture, and the trouble can be avoided by keeping the phototube (in its holder) in a desiccator while the instrument is not in use. (Or the outside of the phototube or its resistor may be dirty; clean by wiping with alcohol.) In the Beckman DU a package of desiccant is mounted in the phototube compartment; this desiccant must be replaced from time to time.

E. The Beckman Model DB-G

This is a double-beam spectrophotometer that uses a vibrating mirror and a single photocell, as shown in Fig. 6-9. It covers the ultraviolet-visible region from 200 to 800 nanometers. A tungsten filament and hydrogen lamp are mounted in the instrument, and one can change from one to the other by turning a mirror. The dispersing element is a diffraction grating. To prevent interference from overlapping orders, filters are automatically moved into the beam as the wavelength dial is moved. The meter reads directly in transmittance and absorbance.

Several features distinguish this double-beam instrument from the single-beam instruments we have described. Two cells are placed in the cell compartment together, one with solvent, the other with the sample. The zero adjustment (dark current) is made by placing an opaque block in the sample beam and adjusting the galvanometer, if necessary, to read zero; once adjusted, the zero is stable and seldom needs to be reset. To make the 100 per cent adjustment both cells are filled with solvent, and a "balance" knob is turned to make the galvanometer read 100 per cent transmittance. The two parts of the divided beam are now in balance and remain in balance when the wavelength is changed. Absorption spectra can thus be scanned by simply turning the wavelength dial and watching the galvanometer needle. The wavelength dial may be turned by a motor at constant speed; there is a choice of two scanning speeds, fast and slow, and the phototube output may be fed into a recorder which traces absorption spectra automatically. The slit width can be set manually if desired, but more often the width is automatically "programmed," that is, changed with the wavelength in a predetermined manner to follow the changes in light source intensity and phototube sensitivity.

For measuring absorbances this instrument is no more precise than the Model B and less precise than the DU. The great convenience of double-beam instruments is in scanning absorption spectra. The DB-G is also easily adapted to flame photometry and atomic absorption spectrometry.

IV. VISIBLE SPECTROPHOTOMETRY: DETERMINATION OF ELEMENTS

The visible part of the spectrum is used chiefly for determining elements, in contrast to the ultraviolet, which is used mainly for identifying and determining organic compounds. Photometric methods have been devised for most, if not all, of the chemical elements. Some of these methods are indirect and many are of little practical use. The technique of atomic absorption spectroscopy (Chapter 7) has supplanted the older photometric methods for many metallic elements, but atomic absorption instruments are not always available, and it is well to be aware of at least the better methods that depend on light absorption in solution.

The best photometric methods are those for the transition elements. In the first long period this means the elements from titanium to copper, inclusive. The ions of these elements contain unfilled d-orbitals to which, from which, and between which electrons can move by absorbing light quanta. One of the easiest and best spectrophotometric determinations is that of the element manganese. It is converted by very strong oxidizing agents, like periodic acid, into the permanganate ion, MnO_4^-, which as every chemist knows is intense purple in solution. It is the intense colors, with molar absorptivity 1000 or more, that are of primary interest in chemical analysis, and it is worthwhile to ask why the color of the permanganate ion should be so intense. The reason seems to be that the light-absorption process involves "charge transfer." An electron in one of the four oxygen atoms absorbs a quantum of light and is excited. Instead of passing to a higher level in oxygen (which would require much energy) the electron passes to one of the higher-energy unoccupied *d*-orbitals of the manganese atom. Crudely one may picture the process like this:

$$Mn^{7+}O^{2-} \xrightarrow{h\nu} Mn^{6+}O^{1-}$$

Processes like this cause a big shift of electric charge and are therefore very efficient at absorbing electromagnetic waves, that is, light. A similar case is the chromate ion, CrO_4^{2-}, which is yellow. This also absorbs by charge transfer. Most of the absorption of the chromate ion occurs in the near ultraviolet.

Charge transfer also accounts for the intense colors of $FeCl_4^-$ (yellow) and $FeBr_4^-$ (red). The ion FeI_4^- is unstable. Charge transfer is so easy that this ion breaks down to Fe^{2+}, I^-, and free iodine, I_2.

The atoms or molecules that are attached to the central metal atom in complexes are called ligands. A simple ligand used in photometric analysis is the thiocyanate ion, CNS^-. This is rather like the halide ions in its chemistry, and it, too, acts as an electron donor in charge-transfer processes. The blood-red of the ferric thiocyanate complexes is well known. It is not much used in photometric analysis because more than one complex is formed and the absorbance of the solutions is not proportional to the concentration of Fe(III) unless conditions are carefully controlled. Thiocyanate is a good reagent for cobalt(II), with which it forms a deep blue complex, extractable by solvents like ether, ethyl acetate, and isoamyl alcohol. Thiocyanate is also a good reagent for molybdenum, tungsten, and other transition elements (see Table 6-1).

Another simple reagent is hydrogen peroxide, H_2O_2. This gives intense yellow solutions with titanium(IV) in acid solutions, and less intense orange-yellow with vanadium(V) and molybdenum(VI). In strongly alkaline solution it gives a yellow color with uranium(VI), but this is not often used for analysis as better reagents are now known for uranium.

TABLE 6-1

Some Colorimetric Reagents for Metals

Reagent	Metals
H_2O_2	Ti, V, Mo, U
Thiocyanate	Fe, Co, Mo, W, U, Nb, Ti
Phenanthroline and derivatives	Fe, Cu, Ru
8-Hydroxyquinoline	Al, Ga, In, Mg, U, V
Dithizone	Cu, Ag, Zn, Cd, Hg, Pb
Sodium diethyldithio-carbamate	Bi, Cu
Dithiol (4-mercapto-4-methyl-1,2-dimercaptobenzene)	Sn, Mo, W

The organic reagents that give colored complexes with metal ions are numerous. Some of the complexes are easily extracted by nonaqueous solvents. Extraction makes the test more complicated, but offers two advantages: first, added selectivity, which means fewer problems from interfering elements; and second, the possibility of raising the concentration of the element sought, by transferring it from a large volume of very dilute aqueous solution to a small volume of the organic solvent. This makes the test more sensitive.

These are a few of the more important organic photometric reagents:

(i) 1,10-phenanthroline and related compounds. The compound 1,10-phenanthroline combines as a ligand with metal ions through the unshared electrons on the two nitrogen atoms, as follows:

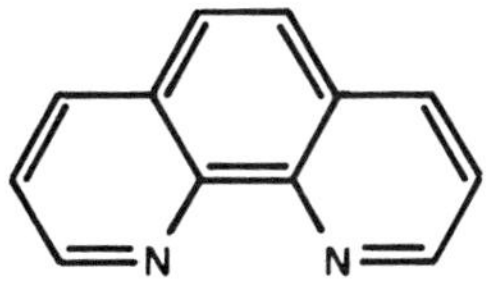

N N

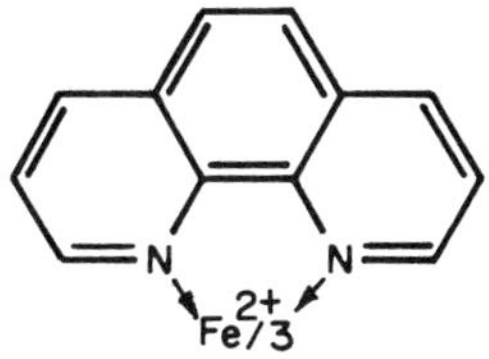

N N
$Fe^{2+}/3$

Three phenanthroline molecules surround one ferrous ion to form a complex which is intensely red, with maximum molar absorptivity ε of 1.1×10^4 at 510 nanometers. A similar complex is formed with ferric ions, but this is green; it is also less stable. Phenanthroline itself is colorless, and so, almost, are ferrous and ferric ions. The intensely colored complexes seem to be of the charge-transfer type.

The important part of the phenanthroline molecule, as far as complex formation is concerned, is the linkage -N=C-C=N-. With the metal ion this forms a five-membered chelate ring. (Chelate means a ring closed by coordinate bonds.) The -N=C- linkages must be part of aromatic rings. Thus, these compounds also form red complexes with ferrous ions:

2,2'-bipyridine, $\varepsilon = 8.65 \times 10^3$

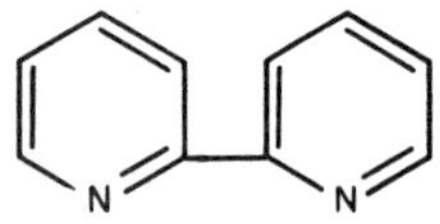

4,7-diphenyl-1,10-phenanthroline, or "bathophenanthroline" $\varepsilon = 2.24 \times 10^4$

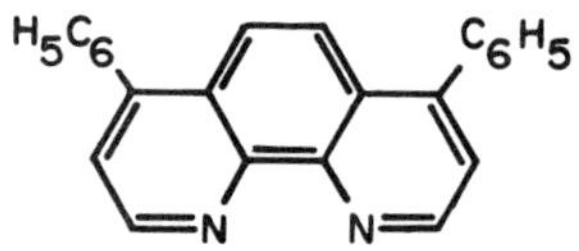

H_5C_6 C_6H_5

2,4,6-tripyridyl-s-triazine, $\varepsilon = 2.4 \times 10^4$

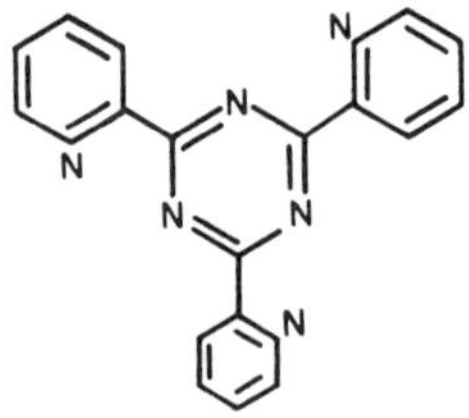

One notes that the larger the molecules are, the greater is the molar absorptivity. The reason is very simple; larger molecules have more chance of intercepting light quanta.

Three phenanthroline molecules around one ferrous ion make a tight fit, and if the positions next to the nitrogen atoms (the positions 2 and 9 in 1,10-phenanthroline) are substituted, say by methyl groups, it becomes impossible to coordinate three molecules around one ferrous ion. The compounds 2,9-dimethyl-1,10-phenanthroline and 2,2'-biquinoline

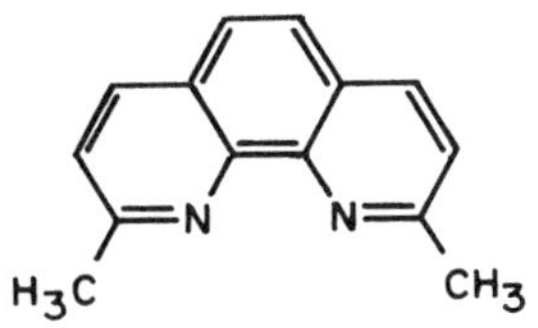

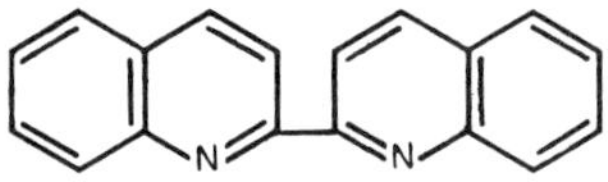

H3C CH3

N N

do not combine with iron but combine very effectively with copper(I) ions, which coordinate two ligand molecules in a tetrahedral configuration. The Cu(I) complexes are intensely colored (yellow and purple, respectively) and are extracted, associated with chloride ions, by isoamyl alcohol. One part of copper in ten million parts of water can be detected and measured with reasonable accuracy by the reagent 2, 2'-biquinoline, which is also called cuproine.

Using these reagents for iron and copper one must remember, first, that a reducing agent must be added to reduce Fe(III) to Fe(II) or Cu(II) to Cu(I), and second, that the pH must be controlled within certain limits. If the solution is too acid, protons become attached to the nitrogen atoms and prevent their combining with metal ions; if it is too basic, the hydroxide of iron or copper is formed instead of the phenanthroline complex. The need for pH control is characteristic of all organic complex-forming reagents. They combine with metal ions by donating electrons. These electrons can be donated to protons instead. In general,

pH too low, $\text{reagent} + H^+ = \text{reagent}\cdot H^+$
pH too high, $(\text{metal ion})^{n+} + nOH^- = \text{metal hydroxide}$

The effective pH range may be narrow or it may be broad. In the case of iron(II) and 1,10-phenanthroline the range is broad, about 3 - 8.

(ii) Diphenylthiocarbazone, or "dithizone." This is a black powder that dissolves in carbon tetrachloride, chloroform, benzene, acetone, and many other organic solvents to give intense green solutions. A concentration of 25 parts per million, or 1×10^{-4} molar, gives a color that is too deep to use in quantitative analysis; 10 parts per million is better. With certain metal ions dithizone combines to form complexes that are an intense red, purple, or yellow. To form these complexes, dithizone loses a proton and forms chelate rings as follows:

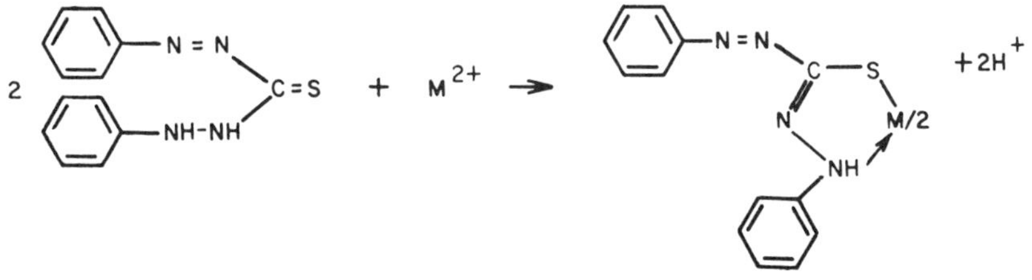

2 N = N C=S NH-NH + M^{2+} → N=N C S N M/2 NH + $2H^+$

The metals that react in this way are: Cu, Ag, Au, Zn, Cd, Hg, Tl, Pb, Bi. To perform the test, a solution of dithizone in an organic solvent is shaken for a minute or two with an aqueous solution of the metal ion, whose pH has been suitably adjusted. The more stable is the dithizone complex, the lower is the pH at which it can be extracted. The most stable complexes are those of mercury, bismuth, and silver.

For quantitative photometric analysis the best way to use dithizone is to shake the aqueous solution with an excess of reagent, then measure the absorbance of the organic solvent layer at two wavelengths, one at which the complex has its maximum absorption (about 550 nanometers, in the green part of the spectrum) and the other at which dithizone has its maximum absorption (about 600 nanometers; this depends somewhat on the solvent). Knowing the molar absorptivities of the complex and of dithizone at each of the two wavelengths, one can solve two simultaneous equations and calculate the concentrations of the complex and of the unreacted reagent; see Chapter 5.

As a reagent, dithizone suffers from the disadvantage that its solutions are unstable. They decompose significantly in the course of a working day. This means that to get accurate results with dithizone, one has to make repeated checks and calibrations. Another characteristic of dithizone, which is an advantage in certain circumstances and a disadvantage in others, is its lack of selectivity; it reacts with more than one metal at a time. Where a quick and sensitive test for "heavy metals" is required, or where one wants to extract traces of "heavy metals" from a solution, dithizone is a very useful reagent. It is used in geochemical prospecting, but less extensively than it once was, for atomic absorption spectroscopy has taken its place. It is also used to determine minor elements, principally lead, in aluminum alloys and petroleum products, and it is used in industrial hygiene to determine lead, copper, and mercury in blood. It can be made more selective by the use of auxiliary complexing agents ("masking agents") and by careful pH adjustment.

(iii) Sodium diethyldithiocarbamate:

$$(H_5C_2)_2N-C(=S)-S^-\,Na^+$$

This is a very useful reagent for determining copper. Copper(II) forms a yellow complex which is extracted at pH 8 by chloroform or carbon tetrachloride. Less than one part of copper in a million of water can be determined in this way, and there are few interferences; the chief are Co, Ni, and Bi. Iron is kept from interfering by complexing it with citrate.

A compound chemically related to this reagent is ammonium pyrrolidine dithiocarbamate:

$$\begin{array}{l} H_2C - CH_2 \\ | \qquad\quad \rangle N - - C(=S)S^- NH_4^+ \\ H_2C - CH_2 \end{array}$$

This compound forms complexes with all the metals that form complexes with dithizone, and a few others besides, and the complexes can be extracted by solvents such as chloroform and methyl isobutyl ketone. The complexes are usually colorless, but the extracted metals can now be determined by atomic absorption.

The Heteropoly Acids. A group of compounds that is very useful in photometric analysis is the heteropoly acids. The prototype of these acids is phosphomolybdic acid, $H_3[PO_4 \cdot 12MoO_3]$. The twelve MoO_3 units are coordinated around the oxygens of the phosphate ion, three molybdenum atoms being attached to each phosphorus. There are six oxygen atoms around each molybdenum atom, and some of these oxygens act as bridges to join two molybdenum atoms. Molybdenum can be replaced by tungsten, and phosphorus, the central atom in the complex, can be replaced by arsenic, germanium, or silicon. Heteropoly acids containing boron and vanadium are also known.

The color of the acids is an intense yellow. Absorption starts at 450 nanometers, approximately, and continues to increase as the wavelength is lowered, the maximum being somewhere in the ultraviolet region. The wavelength used for spectrophotometric determination is generally 400 nanometers. The absorption is higher at lower wavelengths, but the light intensity of the tungsten lamp falls rapidly as the wavelength is lowered, and 400 nanometers represents a good compromise. One must remember that one is measuring on the side of an absorption band (Chapter 5) and that the Beer's law calibration curve will show negative deviations unless the slit width is kept very small.

The heteropoly acid color is produced by adding ammonium molybdate in considerable excess to an acid solution and is used to determine phosphorus, silicon, and arsenic. Where phosphorus and silicon occur together, as in soils, fertilizers, treated water supplies, and some detergents, care is needed to prevent one element from being mistaken for the other, and it is best to perform a preliminary separation, for example by evaporating to fumes with perchloric acid, which dehydrates and precipitates silica while leaving phosphoric acid in solution. To determine phosphate in fertilizers and phosphate rocks, a favorite method uses a mixture of ammonium molybdate and ammonium vanadate, forming a yellow-orange phosphovanadomolybdate. The compound contains phosphorus and vanadium in the atomic ratio 1:1, with about 18 atoms of molybdenum to each atom of phosphorus. Its molar absorptivity, based on phosphorus, is about 1800 at 400 nanometers.

Phosphomolybdic, silicomolybdic, and arsenomolybdic acids are reduced by a variety of reducing agents to form a blue substance of uncertain composition that is called "molybdenum blue." The color is very intense; the molar absorptivity based on phosphorus is about 25,000. (In other words, the blue solution prepared from phosphate and containing 1×10^{-4} gram-atom of P per liter would have an absorbance of 2.5 in a 1-centimeter cell.) The maximum absorption is in the red, at about 850 nanometers, but the peak is broad and flat, so that the choice of wavelength is not very important and a wide slit can be used.

"Molybdenum blue" is used for determining traces because of its great sensitivity, but there are disadvantages to its use. It is not a stoichiometric compound, and its absorption spectrum depends on the way in which it is prepared. Its solutions are colloidal and change with time, eventually precipitating, so that spectrophotometric readings must be taken within an hour or two after mixing. Another reason for reading the color after a fixed and relatively short period is that the excess molybdate is reduced too, though much more slowly than the heteropoly acid. In strongly acid solution (1 molar or more in hydrogen ions) phosphomolybdic acid is reduced much faster than silicomolybdic, which helps to reduce interferences from silicate. Many reducing agents have been used, the commonest being stannous chloride, ferrous sulfate, hydrazine, and 1,2,4-aminonaphthol sulfonic acid.

V. ULTRAVIOLET SPECTROSCOPY OF ORGANIC COMPOUNDS

Table 6-2 presents a quick view of how organic molecules absorb radiation in the ultraviolet.

Saturated hydrocarbons absorb in the "hidden" part of the spectrum. It takes a lot of energy to excite these compounds because all the outer electrons are involved in bonding, and one cannot drive an electron to a higher energy level without breaking the chemical bonds. Large energy quanta are required, and this means radiation of short wavelength.

Compounds with oxygen atoms, like alcohols and ethers, have electrons on these atoms that are not involved in bonding and can therefore be excited to higher energy levels without disrupting chemical bonds. These compounds absorb at higher wavelengths (smaller energy quanta) than saturated hydrocarbons, but the wavelengths are still below the range of ordinary analytical instruments.

The same is true of compounds with carbon-carbon double bonds. In the carbon-carbon double bond one pair of electrons is like the pair that forms a single bond (these are called <u>sigma</u>-electrons) and the other pair is in a different type of orbital, shaped something like a pair of sausages above and below the plane of the molecule, called a <u>pi</u>-bond (π). These electrons can be promoted to a higher energy level without breaking the sigma-bond that joins the two carbon atoms. However, the orbitals now have a different shape (see Fig. 6-13) and are called "anti-bonding orbitals," symbol π^*. The electron transition is shown as $\pi \rightarrow \pi^*$. The charge distribution of the molecule changes considerably, and this transition is therefore very effective at absorbing radiation. The molar absorptivity is very high, as is seen from Table 6-2. However, the absorption of ethylene is still below the range of commercial instruments.

Compounds like acetone, which have a double bond between carbon and oxygen, show intense $\pi \rightarrow \pi^*$ absorption and another absorption band, much weaker but occurring at wavelengths well within the range of analytical instruments, which is associated with movement of an unpaired electron from the oxygen atom to an antibonding π^* orbital. This band is too weak to be of much use in the analytical determination of these compounds, but quite strong enough to prevent the use of acetone as a solvent for ultraviolet spectroscopy.

Compounds with <u>conjugated double bonds</u> have intense $\pi \rightarrow \pi^*$ absorptions that occur at wavelengths well within the range of commercial spectrophotometers. It is for these compounds that ultraviolet spectroscopy is a useful analytical tool. The conjugated bonds may be C=C-C=C or

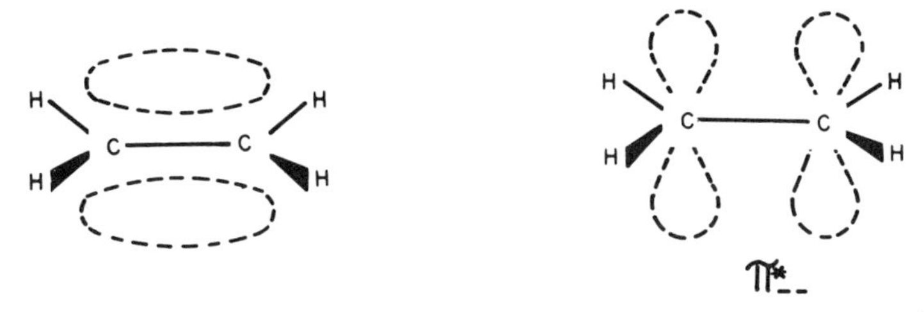

Fig. 6-13. Carbon-carbon double bond: bonding and antibonding pi-orbitals.

TABLE 6-2

Absorption of Ultraviolet Light

Compound	Wavelength of max. absorption	Molar absorptivity	Transition
C_3H_8	135	—	$\sigma \rightarrow \sigma^*$
CH_3OH	183	150	$n \rightarrow \pi^*$
C_2H_4	171	15,530	$\pi \rightarrow \pi^*$
CH_3CHO	160	20,000	$n \rightarrow \sigma^*$
	180	10,000	$\pi \rightarrow \pi^*$
	290	17	$n \rightarrow \pi^*$
CH_3COOH	208	32	
$CH_3COOC_2H_5$	211	57	
C_6H_6	184	46,700	
	202	6,900	
	255	225	

C=C-C=O. The ultraviolet absorption gives a qualitative test for conjugation, distinguishes between conjugated and nonconjugated compounds (note the absorptions of mesityl oxide and its isomer), and also permits quantitative determinations at very low concentrations.

The longer the conjugated chain, the more intense is the absorption; this can be attributed to the greater cross-sectional area of the light-absorbing electron system. The wavelength also increases, indicating that the energy levels in the ground and excited states get closer together. This is predicted from quantum mechanics. The simplest calculation that can be made from the Schrodinger equation, that of the energies of a particle in a one-dimensional box, shows that the energies lie closer together the longer the box. [The relation is: $E = (n^2 h^2)/8ma^2$, where E is the energy, n an integer, h is Planck's constant, m is the mass of the particle, and a the length of the box.] An electron in a conjugated system is not quite the same thing as a particle in a box, but there is a connection, and Table 6-3 shows that the longer the conjugated chain, the longer is the wavelength of maximum absorption. Chemists who make dyestuffs have known this fact for nearly a century, and naturally colored

TABLE 6-3

Ultraviolet Absorption by Conjugated Systems

Compound	Max. wavelength	Molar absorptivity
C_2H_4	171	15,530
$H_2C{=}CH{-}CH{=}CH_2$	217	20,900
$H_3C(CH{=}CH)_xCHO$:		
x = 1	217	15,600
2	270	27,000
3	312	40,000
4	343	40,000
5	370	57,000
6	393	65,000
trans-beta-carotene, $C_{40}H_{56}$	452	139,000
	478	122,000
C_6H_5COOH	230	11,600
$C_6H_5CH{=}CHCOOH$: cis-	268	10,700
trans-	272	15,900
$C_6H_5CH{=}CHC_6H_5$, trans-	244	25,000
	280	10,000

substances have long conjugated chains. Carotene, the orange-red coloring matter in carrots which also occurs in green leaves, is a hydrocarbon with eleven double bonds in a conjugated chain. It absorbs at 452 and 478 nanometers with molar absorptivities of 139,000 and 122,000.

Aromatic compounds represent a special class of conjugated compounds, and they all absorb in the ultraviolet above 200 nanometers. Benzene has two very strong bands below this wavelength, and a series of weaker bands, with molar absorptivity around 200, between 240 and 265 nanometers. The fine structure of these bands is due to vibrational energy changes. These bands appear in the spectra of many compounds that carry the phenyl group, like phenylacetic acid, the amino acid phenylalanine, and phenethylamine; see Fig. 6-14. The benzene spectrum is modified more drastically if there is an electron-donating group immediately next to the benzene ring, as in the compounds phenol and aniline. It is very interesting to note the

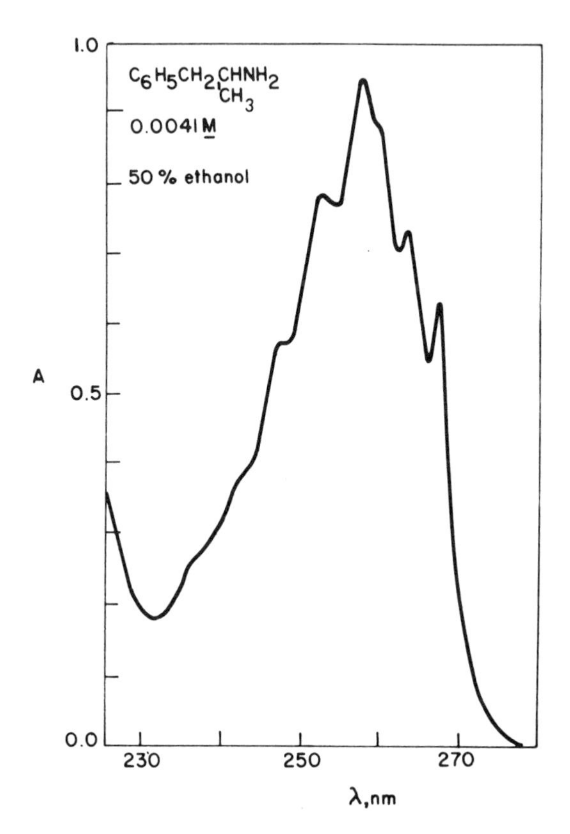

Fig. 6-14. Absorption spectrum of amphetamine.

effect of protonation on these compounds; phenol, C_6H_5OH, has a band at 270 nanometers with ε = 1450; the phenolate ion, $C_6H_5O^-$, which has two more nonbonding electrons on its oxygen, absorbs at 287 nanometers with ε = 2600. Aniline, $C_6H_5NH_2$, absorbs at 280 nanometers with ε = 1430, whereas the anilinium ion, $C_6H_5NH_3{}^+$, which has no nonbonding electrons on the nitrogen atom, absorbs at 254 nanometers with absorptivity ε of only 170. The effect of protonation on the absorption of phenolic compounds and aromatic amines is, of course, used in acid-base indicators, where the molecules are large enough, and sufficiently conjugated, to absorb in the visible region.

Table 6-3 shows the effect of adding double bonds to the aromatic rings. It also shows the very interesting effect of stereoisomerism. The π-orbitals of conjugated double bonds overlap better if the carbon atoms all lie in the same plane. The trans isomers of stilbene and cinnamic acid are more nearly coplanar than the cis isomers and absorb more intensely.

Polynuclear aromatic compounds like naphthalene, anthracene, and phenanthrene have broad and intense absorption bands extending to wavelengths that are higher, the more extensive is the ring system. Heterocyclic compounds of aromatic character, like pyridine and thiophene, absorb in a manner similar to benzene. Many compounds of biological importance have heterocyclic rings and absorb in the ultraviolet. Examples are the purine and pyrimidine bases that spell out the "genetic code" in DNA (deoxyribonucleic acid), vitamin B1, and alkaloids, such as caffeine. The ultraviolet spectra serve for identification as well as quantitative determination, and in favorable cases, like caffeine and phenacetin, mixtures of two components can be analyzed by measuring absorbances at two wavelengths.

Ultraviolet absorption bands are generally sharper and more detailed than bands in the visible region and are therefore better for identification and analysis of mixtures. They are not as sharp as infrared bands, however, and infrared spectroscopy is a much more useful tool for identification and structural determination.

VI. SOLVENTS FOR ULTRAVIOLET SPECTROSCOPY

As we have seen, there are many compounds that have no ultraviolet absorption above 200 - 220 nanometers. Thus there is little difficulty in finding solvents for ultraviolet spectroscopy. Good solvents, transparent down to 220 nanometers or below, are: water; methyl, ethyl, and isopropyl alcohol; hexane, cyclohexane, octane; diethyl ether, dioxane; acetonitrile.

The solvent often affects the absorption of the dissolved compound. Thus the fine structure of the benzene bands that is observed in benzene vapor is easily visible in cyclohexane but is blurred and shifted slightly to longer wavelengths in polar solvents like alcohols.

Very important considerations in ultraviolet spectroscopy are purity and cleanliness. Small traces of highly absorbing impurities may make the study of ultraviolet spectra impossible unless these impurities can be removed. Solvents should be of specially purified "spectral grade." The cells must be scrupulously clean; fingerprints or films of grease on the outside of the cells may be invisible to the eye but may absorb strongly in the ultraviolet.

EXPERIMENTS

Experiment 6-1

Absorption Spectra of Potassium Permanganate and Potassium Dichromate; Analysis of a Binary Mixture

Materials. Potassium permanganate, 5.0×10^{-4} M and 2.5×10^{-4} M; potassium dichromate, a solution 2.5×10^{-4} M in $K_2Cr_2O_7$ and 0.5 M in sulfuric acid. This solution may be made by dissolving 73.5 mg of $K_2Cr_2O_7$ in water, adding 15 ml concentrated sulfuric acid, mixing well, and making up to 1 l. An "unknown solution" should also be prepared which is about 2×10^{-4} M in both potassium permanganate and potassium dichromate and is also 0.5 M in sulfuric acid.

A spectrophotometer is needed that is sensitive down to 340 nm. This may be a Spectronic-20, Beckman Model B or Beckman Model DU, or any other commercial instrument. A set of 1-cm glass cells is also needed.

Objects of Experiment. To learn to use the spectrophotometer; to plot absorption spectra; to illustrate Beer's law and apply it to the analysis of a two-component mixture; and if the instrument permits, to study the effect of slit width.

First, measure the absorbance and transmittance of the three standard solutions at wavelength intervals of 25 nm. With the dichromate, start at 500 nm and work downward. With the permanganate solutions, start at 600 nm and work down.

The permanganate spectrum shows more detail, which can be observed if the slit of the spectrophotometer is sufficiently narrow. It cannot be observed with the Spectronic-20 as the slit (band pass = 20 nm) is too wide. It can be observed very well with the Beckman DU if the sensitivity knob is kept turned to its maximum, and the slit width to its minimum. The curves of Fig. 6-15 were obtained with a Beckman DU, first with the sensitivity kept at a maximum and the needle brought to zero by carefully adjusting the slit, and second, with the sensitivity kept at the minimum and the needle again brought to zero by carefully adjusting the slit. It is evident that the fine structure of the spectrum between 520 and 580 nm is seen with the narrow slit but lost with the wide slit. (The advantage of the wide slit is, of course, the greater stability that comes with more light energy falling on the phototube. We could say in general that a nar-

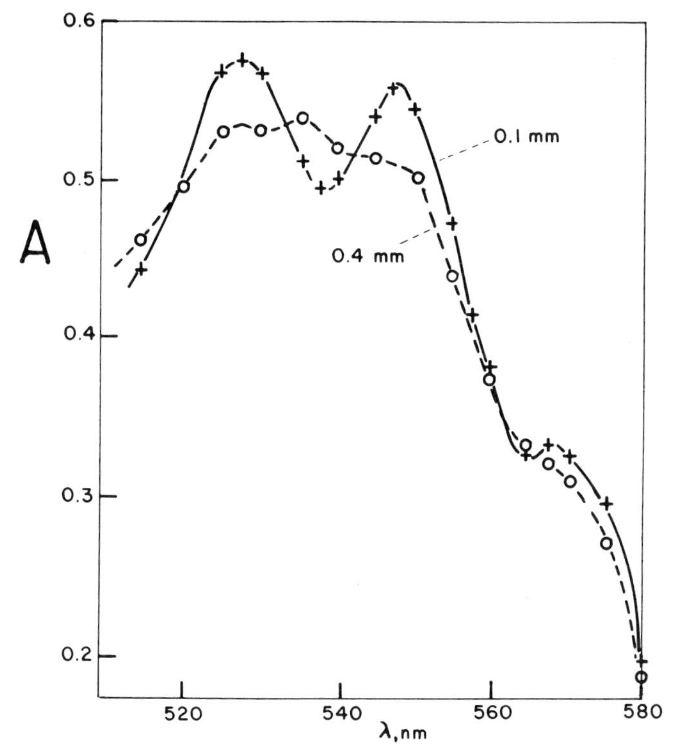

Fig. 6-15. Spectrum of potassium permanganate, 2.5×10^{-4} M: effect of slit width. Beckman DU: slits 0.1 and 0.4 mm.

row slit is best for mapping spectra, but a wide slit is better for quantitative analysis.)

Plot the permanganate spectrum for both solutions in as much detail as you can. With the Beckman Model B, make one set of measurements at the highest sensitivity (therefore, lowest slit width) and another set at low sensitivity (wide slit).

In preparation for analyzing the "unknown," choose one wavelength at which the dichromate absorption is a maximum (about 350 nm) and determine as carefully as you can the absorbance of the permanganate solutions, as well as that of the dichromate, at this wavelength. (If the dichromate absorbance is too high, that is, very much above 1.0, at its maximum you may have to compromise and choose a wavelength at which the absorption is not quite as great.) Choose another wavelength at which the permanganate absorption is maximum (about 550 nm) and measure the absorbance of the permanganate and the dichromate solutions at this wavelength.

Now take the "unknown" and measure its absorbance at each of the wavelengths chosen. Calculate the molar concentrations of manganese(VII) and chromium(VI) in the solution. Note that the calculation is quite easy, because the dichromate does not absorb at all at 550 nm. One can use the absorbance at 550 nm to calculate the permanganate concentration, then calculate the absorbance by permanganate at 350 nm, subtract this absorbance from the absorbance that you measured, and use the difference (which is the absorbance due to dichromate) to calculate the dichromate concentration. It is not necessary to go through the "determinant" procedure described in Chapter 5.

In using this method to determine chromium the acidity is important, because hydrogen ions repress the hydrolysis of dichromate:

$$Cr_2O_7^{2-} + H_2O = 2CrO_4^{2-} + 2H^+$$

But hydrogen ions also combine with $Cr_2O_7^{2-}$ to form $HCrO_4^-$ and species like $Cr_3O_8^{2-}$. It is interesting to study the effect of acidity on the absorption spectrum of dichromate between 330 and 400 nm.

Experiment 6-2

Iron by 1,10-Phenanthroline

This is a simple and useful determination that is capable of very high precision.

Materials. A solution of 1,10-phenanthroline, made by dissolving 0.2 g of the crystals in 100 ml of hot water; hydroxylamine hydrochloride, 5 g in 100 ml; sodium acetate, 1 M, or 14 g $CH_3COONa \cdot 3H_2O$ in 100 ml; and a standard solution containing 100 ppm iron, that is, 100 mg of iron per liter of solution. Make this by weighing 0.702 g ferrous ammonium sulfate, $FeSO_4 \cdot (NH_4)_2SO_4 \cdot 6H_2O$, formula weight 392.14, dissolving in water, adding about 0.1 mol sulfuric acid, and making up to 1.000 l. The acid is added to prevent ferric oxide from precipitating when the solution comes in contact with air.

You will also need pH paper and a buret or pipet that can measure 1.00 ml with good accuracy.

The instrument may be any spectrophotometer, or a filter photometer with a green filter.

For an "unknown" we suggest a fertilizer, a limestone, sandstone rock (finely ground), or a soil.

First construct a calibration curve. Take four 100-ml beakers and place in them 1.00 ml, 2.00 ml, 3.00 ml, and 4.00 ml of the 100-ppm iron standard, measured as accurately as possible. Add roughly 20 ml water, then 2 ml of hydroxylamine [which reduces any Fe(III) to Fe(II)], 5 ml of 1,10-phenanthroline and 5 ml of 1 M sodium acetate. Test the pH with indicator paper. Any pH value between 3 and 8 is satisfactory. Add ammonia, if necessary, to bring the pH above 3. Stir, transfer the solutions to 100-ml volumetric flasks, and make up to the mark.

For highest accuracy the solutions should be left for 2 or 3 hr to develop the color, but for ordinary purposes a wait of 15 to 30 min is enough. Measure the absorbances at 505 nm and plot a graph of absorbance against micrograms of iron. The graph should be a straight line if a spectrophotometer is used, but may be curved if a filter photometer is used.

In analyzing "unknowns" spectrophotometrically it is desirable to have an idea of the concentration of the element sought, so that a sample can be taken that will give an absorbance somewhere between 0.1 and 1.0. If one has absolutely no idea of the concentration one must make one or two preliminary trials to find the right range. Fertilizers commonly contain about 0.5% of iron. To determine this iron content we may proceed as follows:

Weigh accurately about 0.5 g of fertilizer (do not dry, as drying may vaporize ammonium salts) and place in a small beaker. Add a few milliliters of water and 2-3 ml concentrated hydrochloric acid. Stir, dilute to 25 ml, and filter, collecting the filtrate in a 100-ml flask. Wash the residue on the filter paper with water and test it with a few drops of concentrated hydrochloric acid. A yellow color will show that the paper still contains some iron. Transfer the iron as completely as possible to the volumetric flask and make the volume to 100 ml; mix well.

If we guessed correctly that the iron content was 0.5%, a 10-ml portion of this solution contains 0.25 mg, or 250 μg of iron, which will fall nicely in the middle of our calibration curve. So, pipet 10.0 ml of the solution into a beaker and add hydroxylamine, 1,10-phenanthroline, and sodium acetate in the same way as was done with the standards. It may be necessary to add more ammonia this time to adjust the pH. Transfer to a 100-ml volumetric flask, make up to the mark, and read the absorbance as before.

If the absorbance is very high or very low, take a different sized portion of the fertilizer solution and repeat the process.

Limestone samples may be dissolved in the same way as the fertilizer; sandstones may have to be heated for some time with 6 M hydrochloric acid, or even fused with potassium bisulfate in a platinum crucible if the iron oxide that they contain is in an insoluble form. Soil samples may be treated as described in Experiment 6-3.

Calculate the per cent of iron in the sample.

Experiment 6-3

Phosphate by the Vanadomolybdate Method

This is a simple, straightforward method of medium sensitivity and high precision. The main difficulty is that silica interferes and must be eliminated completely. Where more sensitivity is needed, the heteropoly blue method (Experiment 6-4) is preferred.

Materials. (a) Ammonium metavanadate solution; dissolve 2.5 g ammonium metavanadate in about 400 ml hot water; wait until the salt has all dissolved, then add 10 ml concentrated nitric acid, and make to 500 ml by adding water. (b) Ammonium molybdate, 5 g per 100 ml. (c) Nitric acid, 6 M. (d) A standard solution containing 100 ppm phosphorus; dissolve 0.4394 g of potassium dihydrogen phosphate, KH_2PO_4, formula weight 136.1, in water and make to 1.000 l.

Perchloric acid (60% or 72%) will be needed to prepare the samples, and so, possibly, will other reagents; see below.

Any spectrophotometer or filter photometer may be used.

The "unknown" may be a fertilizer, phosphate rock, or a soil. It can be a detergent, but in this case the solution of the sample should be heated with dilute nitric acid for 1 hr to convert meta- and pyrophosphates to orthophosphate.

To construct the calibration curve, take 5.0, 10.0, and 15.0 portions of the 100-ppm standard; dilute each with water, add in this order 10 ml 6 M nitric acid, 10 ml ammonium metavanadate, and 10 ml ammonium molybdate. Transfer to 100-ml volumetric flasks and make up to the mark;

mix well. Prepare another solution with no added phosphate, but with the same reagents, to serve as a "reagent blank." After 30 min measure the absorbance at 400 nm and also at 460 nm. Plot a graph of absorbance against concentration for the two wavelengths. The "reagent blank" is smaller at 460 nm and the graph will be more nearly straight at this wavelength, especially if a filter photometer is used. (Use a blue filter.)

Treat the unknown as follows:

(a) Fertilizer. Most phosphate fertilizers contain about 10% phosphate expressed as P_2O_5. We want a solution to put in the spectrophotometer that will contain about 1 mg phosphorus in 100 ml. One mg P = 2.3 mg P_2O_5 = about 25 mg fertilizer. Therefore, proceed as follows:

Weigh accurately about 250 mg fertilizer (do not dry, as ammonium salts might be vaporized), place in a small beaker, add 5 ml 6 M nitric or perchloric acid, and evaporate carefully to dryness. This treatment converts any soluble silicates to insoluble SiO_2. Add another 2-3 ml 6 M acid (to dissolve insoluble phosphates) and a few milliliters of water; transfer through a filter paper into a 100-ml volumetric flask, washing the paper well, and make up to the mark. Now pipet 10.0 ml of this solution into a second 100-ml volumetric flask, add reagents as was done with the standards, make up to the mark. Read the absorbance at 400 nm, or at 460 nm if you prefer, depending on your calibration curve. If the absorbance is low, take a larger portion of the filtered solution, say 25 ml instead of 10 ml; if the absorbance is too high, take a smaller portion. Remember that for best photometric accuracy the absorbance should be between 0.2 and 1.0, more or less; see Chapter 5.

Calculate the phosphorus content of the fertilizer as P_2O_5. (This is usual practice in the fertilizer industry.)

Phosphate rock may be analyzed in the same way as the fertilizer. It should be ground finely and will be harder to dissolve than fertilizer. If it contains much iron, the iron should be removed by extraction with ethyl acetate in the manner described in part (b). The phosphate content will be higher [42% P_2O_5 for the compound $Ca_5(PO_4)_3OH$], so take a smaller portion for analysis.

(b) Soil. A good agricultural soil contains about 50 ppm of phosphorus (as P) that is "extractable," that is, that can be dissolved at pH 3. The amount is extremely variable, and the total phosphorus may be considerably greater. Soils vary greatly in their characteristics, and it is impossible to give a procedure that will be suitable for all soils. Sampling is a problem; so is the destruction of organic matter. The following procedure worked well for desert soils low in organic matter:

Screen the soil to remove particles larger than 2 mm diameter; weigh a 2-g sample, place in a small beaker with 2 ml water and 5 ml 72% perchloric acid. Evaporate to dryness on a hot plate, with a good hood to remove the perchloric acid fumes. Do not overheat, as this would make iron and aluminum phosphates hard to redissolve. Cool, add 20 ml 6 M hydrochloric acid, filter through medium porosity filter paper, wash the paper with small portions of 6 M hydrochloric acid. Transfer the filtrate to a separating funnel, add about 20 ml ethyl acetate, and shake. The iron (as ferric chloride) is extracted almost completely by the ethyl acetate. Run the aqueous layer (the lower layer) into a small beaker; discard the ethyl acetate layer, which contains iron but no phosphorus.

Note: it is not necessary to remove the iron completely, and it may not be necessary to remove it at all. One part of iron gives a color in the final solution equivalent to 0.01 part of phosphorus. If iron need not be removed, the procedure can be shortened.

Evaporate the aqueous layer just to dryness, add 0.5-1 ml 72% perchloric acid, then water; transfer to a 25-ml or 50-ml volumetric flask, make up to the mark; take a convenient aliquot part of this solution for the spectrophotometric determination. If the phosphorus content is low, use all the solution without taking an aliquot; if it is very low, use the heteropoly blue method (Experiment 6-4). The same solution may be used for the determination of copper or other trace metals; see Experiment 6-5.

If the soil is rich in organic matter it is unwise because of the danger of explosion, to entrust the treatment to perchloric acid alone. Mix the soil with 1-2 ml water, then 10 ml concentrated nitric acid, and add 5 ml perchloric acid. Evaporate as described.

This treatment gives the total phosphorus, not the amount that is available to plants.

Experiment 6-4

Phosphorus by Heteropoly Blue

This method has ten times the sensitivity of the phosphovanadomolybdate method and is less affected by silicate. It is better for filter photometers because the peak in the absorption spectrum is broad and flat. The drawback is that the blue compound is unstable. It must be prepared under closely controlled conditions and the color must be read within a short interval of time.

Materials. (a) Ammonium molybdate solution: dissolve 15 g ammonium molybdate, $(NH_4)_6Mo_7O_{24} \cdot 4H_2O$, in 300 ml warm water, cool, add slowly, with stirring, 350 ml concentrated hydrochloric acid, dilute to 1 l. (b) Stannous chloride: dissolve 10 g $SnCl_2 \cdot 2H_2O$ in 25 ml concentrated hydrochloric acid. This solution oxidizes rapidly in air and must be kept in a small, closed bottle; it should be prepared fresh every week or two. For use, dilute 2 ml of this solution with 100 ml of distilled water. (c) A standard solution containing 50 ppm or 100 ppm of phosphorus, prepared from potassium dihydrogen phosphate in the manner described in Experiment 6-3.

Spectrophotometer readings are best taken at 830 nm to minimize interference by reduced ammonium molybdate. This wavelength requires a red-sensitive photocell and, with a grating instrument like the Spectronic-20, a red filter. The wavelength 660 nm may also be used.

To construct the calibration curve, take 0.50-, 1.00-, 1.50-, and 2.00-ml portions of the 100-ppm standard, or twice these volumes of a 50-ppm standard. Place in 100-ml volumetric flasks, dilute with water; add 10 ml ammonium molybdate and 1 ml of the diluted stannous chloride. Make up to the mark; read the absorbance after 5 min, but not more than 20 min after mixing. If the curve fails to obey Beer's law, the most likely reason is that the stannous chloride has become oxidized or that an insufficient amount was added.

The same "unknowns" may be used as in Experiment 6-3 and treated in the same way. However, the final solution may contain up to 100 ppm of silica (in the form of soluble silicate) without interference; silicomolybdic acid is reduced much more slowly than phosphomolybdic acid in the strongly acid solution that is used here. Iron(III) interferes badly, because it reacts with the stannous chloride. It may be removed as described in Experiment 6-3, or it may be reduced to Fe(II) by passing the sample solution through a short column of granulated, amalgamated zinc. Fifteen ppm of Fe(III) may be tolerated in the final solution.

Biological samples like wheat germ or flour may be analyzed for their phosphorus content; the organic matter is destroyed by evaporating with a mixture of nitric and perchloric acids. First heat with a few milliliters of concentrated nitric acid, then, when part of the organic matter has gone, add a 1:1 mixture of concentrated nitric and perchloric acids, and evaporate nearly to dryness. Make a preliminary experiment to determine the best sample size.

It is good practice to run one or two standards at the same time that the unknowns are run, because of the instability of the blue product and as a check on the stannous chloride reagent.

Experiment 6-5

Copper by Sodium Diethyldithiocarbamate

In this method the colored substance is extracted from water into a nonaqueous solvent, which may be carbon tetrachloride, chloroform, or isoamyl acetate. The method is extremely sensitive and can measure 10 µg of copper without difficulty. The characteristics of the reagent were discussed on page 166.

Materials. (a) Solution of sodium diethyldithiocarbamate, 0.1 g in 100 ml of water; this should be prepared fresh every 3 days. (b) Ammonium citrate, 5%. (c) Standard copper sulfate solution, containing 5.0 ppm Cu. Prepare this by dissolving 200 mg pure $CuSO_4 \cdot 5H_2O$ in water and making to 1 l., or 50 mg to 250 ml. The factor $Cu:CuSO_4 \cdot 5H_2O$ is 1:3.929, roughly 1:4. This solution will contain about 50 ppm Cu. Prepare a 5-ppm solution by diluting this solution quantitatively in the ratio 1:10. While diluting, add enough sulfuric acid to make the solution about 0.1 M in hydrogen ions. This is a routine precaution in preparing dilute solutions of heavy metal salts; the purpose is to prevent hydrolysis and the adsorption of metal ions on glass. From the weight of copper sulfate taken, calculate the copper concentration to three significant figures and note it on the bottle.

Other materials needed are ammonia (5 M), pH indicator paper, carbon tetrachloride or chloroform, and perhaps solid ammonium citrate and solid disodium salt of EDTA (ethylenediamine tetraacetic acid); see below. Separating funnels, about 125-ml capacity and preferably pear-shaped, are needed; these should have Teflon plugs, or if glass plugs are used, they should be free from grease.

To make the calibration curve take four accurately measured portions of the 5-ppm copper standard, with maximum volume 10 ml; volumes of 2.5, 5.0, 7.5, and 10.0 ml are suggested. To each solution add about 10 ml water and 10 ml 5% ammonium citrate, then add ammonia drop by drop until the pH is between 6 and 8. Transfer the solutions, one at a time, to the separating funnel and add 1.0 ml 0.1% sodium diethyldithiocarbamate. (It is best to add the reagent in the funnel, rather than in the beaker, because the copper diethyldithiocarbamate is sparingly soluble and tends to stick to glass.) Now add a measured volume of 10.0 ml of the solvent (chloroform or carbon tetrachloride); shake for 1 min by your watch. (Extraction is fast, but not instantaneous.) Let the layers separate, and run some of the yellow-brown solvent layer into a test tube or cell for spectrophotometric measurement. There is a technique for doing this. First, an ungreased glass stopcock is hard to handle. Keep a light touch so that the stopcock turns but does not fall out. Second, it is essential to prevent droplets of water from entering the spectrophotometer cell. Take a piece of filter paper, roughly 2 cm by 5 cm, roll it into a loose cylindrical plug 2 cm long, and place this plug in the outlet tube of the separating funnel. The solvent leaving the funnel flows past this plug, which absorbs little drops of water. Reject the first few drops of solvent, as they may be contaminated by trace metal impurities from the paper.

Measure the absorbance of the solution at 435 nm. If a filter photometer is used, use a blue filter.

Run a "reagent blank," that is, a test in which no copper standard is added but all the reagents are mixed and treated in the same way as the standards. This will show whether the reagents (in particular the ammonium citrate) contain copper.

For an "unknown" use a fertilizer or soil. If a fertilizer is used, take a weighed quantity of 1 to 5 g and treat it with some 5 ml of 6 M nitric or hydrochloric acid, enough to dissolve whatever is going to dissolve. Filter into a 50-ml volumetric flask, make up to the mark, and take a 10-ml aliquot portion for analysis, mixing this with citrate and diethyldithiocarbamate in the same way that the standards were treated. Fertilizers vary considerably in their trace metal contents; adjust the sample size to give an absorbance between 0.2 and 1.0, if possible.

Soils normally contain between 5 and 50 ppm of copper (1 ppm = 1 μg of metal per gram of soil). Treat the soil as described in Experiment 6-3 to destroy organic matter and extract the copper. The same extract that was used for the phosphorus determination may be used to determine copper if the copper content is high enough. If the copper content is low, start again with a 5-g sample of soil. The ethyl acetate extraction may be omitted if the iron content is low. Small amounts of iron are complexed by the ammonium citrate and do not interfere. If much iron is present, which is the case with red soils, the iron should be extracted from 6 M hydrochloric acid as described, or complexed as follows:

After evaporating with perchloric acid, take up the residue in 1 ml concentrated hydrochloric acid and 10 ml water, and filter into a small beaker. Add 2 g of solid ammonium citrate, then add ammonia to pH 8. Now add 1 g of EDTA (disodium salt). Readjust the pH, if necessary, by adding more ammonia; transfer to a 50-ml volumetric flask, make up to the mark, take an aliquot portion, transfer to the separating funnel, and add sodium diethyldithiocarbamate. If the copper content is low, do not take an aliquot, but place the whole solution in the separating funnel after adding the EDTA, then add diethyldithiocarbamate.

Report the copper content of the soil in parts per million.

The city water may contain enough copper to be determined by this method. Copper gets into water supplies from copper piping and from copper sulfate used to control algae in reservoirs. The copper concentration in potable water should, however, be below 0.1 ppm. For analysis, take 100 - 200 ml of the city water, add reagents as described.

Experiment 6-6

The Ionization Constant of an Indicator

This experiment combines acid-base equilibria and pH measurement (Chapter 1) with spectrophotometry. It also illustrates the "isosbestic point."

Materials. Acetic acid and sodium hydroxide, each 0.1 M; dilute hydrochloric acid; a solution of bromocresol green, made by dissolving 100 ml of the solid indicator in about 3 ml 0.1 M sodium hydroxide, adding water, and making up to 100 ml. The concentration of the indicator should be about 0.1% but need not be accurately known.

A pH meter and standard buffer will be needed, also a spectrophotometer and glass cells and a 2-ml pipet.

Prepare five solutions, as follows: (i) 10 ml 0.10 M NaOH; (ii) 10 ml 0.1 M NaOH plus 15 ml 0.10 M acetic acid; (iii) 10 ml 0.10 M NaOH plus 20 ml 0.10 M acetic acid; (iv) 10 ml 0.10 M NaOH plus 25 ml 0.10 M acetic acid; (v) about 1 mmol of hydrochloric acid, for example, 10 ml 0.10 M or 1 ml 1.0 M HCl. This amount need not be measured exactly. To each of the five solutions add exactly the same volume of the 0.1% indicator solution; 2.00 ml is a suitable volume, but whatever volume is chosen, it must be the same for all. Transfer each solution to a 100-ml volumetric flask and make up to the mark with water.

Measure the absorbance of each one of the five solutions at intervals of 20 nm from 400 to 620 nm, and plot the values on the same graph paper. A series of curves will be obtained which ideally should all intersect at a point called the isosbestic point. For a family of curves like these to show an isosbestic point, two conditions are necessary: (a) the total indicator concentration must be the same in every case; (b) the indicator must exist in two, and only two, forms whose proportion changes as the conditions are changed (in this case, as the pH is changed). A good isosbestic point is also an indication of good experimental technique.

It is now necessary to measure the pH of each of the solutions (ii), (iii), and (iv). Do this with the pH meter, and be careful to standardize the meter with a known buffer.

Now calculate the ionization constant of bromocresol green from these pH measurements and the absorbance curves. Read the absorbances at two wavelengths, one the wavelength of maximum absorbance of the acid form of the indicator (λ_1), the other the wavelength of maximum absorbance of the base form (λ_2). Let the absorbances at λ_2 be P for the acid form, R for the base form and Q for the buffer. In the buffer the concentrations of base and acid forms are in the ratio (P - Q) to (Q - R). The ionization constant is therefore

$$K_a = \frac{[H^+][In^-]}{[HIn]} = \frac{[H^+][P - Q]}{[Q - R]}$$

$[H^+]$ is known from the pH reading, hence K_a can be calculated.

Example. Three solutions of an indicator having the same total concentration of indicator have absorbances at a certain wavelength that are 0.10, 0.65, and 0.90 in solutions of pH 2, 7.40, and 12, respectively. What is the ionization constant of the indicator?

In the solution of pH 7.70 the proportion of In^- to HIn is (0.90 - 0.65) to (0.65 - 0.10), that is, 0.25:0.55. The hydrogen ion concentration is 3.0×10^{-8}. The ionization constant is therefore $0.25 \times 3.0 \times 10^{-8}/0.55$ = 1.36×10^{-8}.

From your three "buffer" curves you can make six calculations of the ionization constant. Instead of bromocresol green you can use methyl red, which has almost the same ionization constant. Other indicators may be chosen, but it may be necessary to use another buffer system, instead of acetic acid - sodium acetate, to give pH values that are close to the pK_a of the indicator. An interesting extension of this experiment is to vary the ionic strength. The solutions (i) - (iv) above all have ionic strength 0.01, but this can be increased by adding sodium or potassium chloride. Methyl red and bromocresol green will show different ionic strength effects, and the difference can be easily explained from their chemical structures. The color change of methyl red corresponds to a change: $HX^+ = H^+ + X$, whereas the color change of bromocresol green accompanies charge separation: $HY = H^+ + Y^-$. Ionic strength affects the second equilibrium far more than it does the first.

Experiment 6-7

Ultraviolet Absorption Spectrum of Acetylsalicylic Acid (Aspirin) and Its Anion

Materials. Acetylsalicylic acid, recrystallized; hydrochloric acid and sodium hydroxide, each 1 M or another convenient, approximately known concentration; chloroform, cyclohexane. Volumetric flasks, 100-ml; pipets, 5-ml and 10-ml; a set of silica (quartz) cells of 1-cm path length; spectrophotometer for use in the ultraviolet. This instrument may be of the single-beam or double-beam type.

Objects of Experiment. To gain experience in ultraviolet spectrophotometry; to observe typical features of absorption spectra of organic compounds, including the effect of the solvent; to apply one's observations to the determination of acetylsalicylic acid in commercial products (aspirin tablets).

Note on the handling of silica cells. Cells of silica (quartz) are necessary for making measurements below 340 nm, which is the limit of transmission for glass. Silica transmits to 200 nm and below.

Silica cells are expensive and easily broken. Their polished optical surfaces are easily scratched. Invisible films of dirt and grease, including fingerprints, can absorb ultraviolet light.

Handle the cells by their roughened sides, never by the smooth surfaces, and keep them scrupulously clean. After filling a cell and before placing it in the spectrophotometer, rinse the smooth outer surfaces with a few drops of distilled water and wipe them lightly with lens paper, white cleaning tissue, or a clean handkerchief, laundered but not starched. If the cells have covers, be careful. Some covers are not square; they fit when turned one way but not the other. Never try to force them into place, or you will split the cell.

As soon as you have finished your work, rinse the cells thoroughly and dry them inside by rinsing with acetone, then draining. When they are dry, put them back in their case.

Silica cells usually carry markings to identify them, but if there is any doubt about whether they are silica or glass, place an empty cell in the spectrophotometer and compare its absorbance below 300 nm with that of air. If the instrument is of the double-beam type, place the empty cell on the "sample" side and no cell at all on the "reference" side. You will soon see if the cell transmits light or not. If it does, it is made of silica.

(a) The spectra of aqueous solutions. In this part of the experiment you will note the differences between the spectrum of the undissociated acid and that of its anion. Acetylsalicylic acid (hereafter abbreviated to ASA) has an ionization constant about 1×10^{-4}. In the very dilute solutions that will be used, ASA is more than half dissociated into its ions. To keep at least 99% of it in the undissociated acid form we shall add enough hydrochloric acid to provide a hydrogen-ion concentration of 0.05 M.

Weigh accurately about 25 mg of acetylsalicylic acid (ASA) and transfer it to a 100-ml volumetric flask. Add water, shake until the acid has dissolved, then add more water to make up to the mark, and mix well. Label this solution "A".

Solution "A" is too concentrated for us to measure its absorbance directly. Take two more 100-ml volumetric flasks; label them "B" and "C". Pipet 10.0 ml of "A" into each of these flasks. To "B" add 5 ml 1 M sodium hydroxide; to "C" add 5 ml 1 M hydrochloric acid. Add water to both flasks to make up to 100 ml and mix.

Now plot the spectra of solutions B and C from 300 nm down to 215 nm. The procedure will depend on the type of instrument, whether it is single-beam or double-beam and whether it has automatic recording. If it is a single-beam, non-recording instrument, like the Beckman DU or Zeiss PMQ-II, place three cells together in the cell carrier with pure water, solution B and solution C. Use the water cell to set the 100% transmission adjustment; this must be reset each time the wavelength is changed. Be sure to check the zero adjustment. Some of the absorbances will be high, greater than 1, and a small error in the zero setting will make a big error in the measured absorbance. Keep the slit width to a minimum by using a high sensitivity setting. Take readings every 2 nm between 220 and 240 nm.

The spectra of B and C will be different, but each shows two maxima, one around 230 nm, the other at 280 nm or higher. The lower-wavelength peak is stronger. It is called the primary peak. The higher-wavelength or secondary peak is weaker. If the maximum absorbance of this peak is below 0.2 it is a good idea, especially with a double-beam recording spectrophotometer, to prepare a more concentrated solution, taking a fresh portion of solution "A" and diluting it in a smaller ratio, keeping the acid (or base) concentration the same as before. For example one could take 10.0 ml of "A" and 1.25 ml 1 M hydrochloric acid and make the volume to 25.0 ml with water. Run the spectrum of this solution between 260 and

300 nm. The molar absorptivities should be the same as those of the more dilute solution, but the values for the more concentrated solution should be more precise.

The formula weight of ASA is 180. Convert your absorbance readings to molar absorptivities, and plot the molar absorptivities of the acid and base forms against wavelength on the same sheet of graph paper.

(b) Spectrum of a nonaqueous solution. Acetylsalicylic acid is very soluble in chloroform. However, chloroform absorbs ultraviolet light strongly below 230 nm, and the absorption of the solvent almost hides the absorption peak of ASA itself. We shall therefore use a mixture of chloroform and cyclohexane.

Prepare a solution of ASA in pure chloroform of accurately known concentration, about 40 mg in 100 ml. Pipet 5.00 ml of this solution into a dry 100-ml volumetric flask and make up to the mark by adding cyclohexane. In another 100-ml volumetric flask place 5.00 ml of pure chloroform, without ASA, and add cyclohexane to the mark. Use this solvent mixture in the reference cell of the spectrophotometer. In the sample cell place the solution of ASA. Plot its absorption spectrum from 300 nm downward to 230 nm or less.

If you use a single-beam spectrophotometer you will notice that the solvent begins to absorb appreciably below 240 nm. You will have to open the slit wider and wider to set the 100% transmittance reading. Keep the electrical sensitivity adjustment near its maximum, use the slit control to set the 100% adjustment, and record the slit width for each wavelength. This record will show how the absorbance of the solvent increases with decreasing wavelength. With a double-beam instrument one does not notice the absorption of light by the solvent until the light energy becomes so weak that the recorder pen moves erratically.

Plot the molar absorptivity of ASA against the wavelength on the same graph paper that you used for the aqueous solutions. You will see the difference in the spectra. The primary band in cyclohexane is narrower and more intense than that observed in water, reflecting the weaker interaction between the solvent and solute.

An interesting study is that of the effect of the chloroform:cyclohexane ratio on the absorption spectrum. For this study it is helpful to have cells with 2-mm or 1-mm path lengths, to reduce the effect of the absorption of light by the chloroform itself.

(c) Determination of ASA in an aspirin tablet. Aspirin tablets usually contain starch as a binder, and may contain sugar as well. The proportion of acetylsalicylic acid may be as low as 50%. From the information obtained in parts (a) and (b) of this experiment, devise your own procedure for determining ASA in aspirin tablets, bearing in mind the following points:

(i) The ASA in aspirin tablets is usually the undissociated acid, but one should not rely on this. Decide whether you want to measure the absorption of the acid or base form, and adjust your procedure accordingly.

(ii) Disperse the tablet in cold water, then filter off the starch. In hot water the starch granules swell and form a jelly that cannot be filtered. In cold water, on the other hand, ASA dissolves very slowly.

(iii) In alkaline solutions ASA hydrolyzes slowly to give the salicylate ion. Solutions may, however, be left overnight.

(iv) It may be better to use the secondary maximum, rather than the primary maximum, for your measurements, as you will not have to dilute the sample as much.

Experiment 6-8

Simultaneous Determination of Caffeine and Acetylsalicylic Acid by Ultraviolet Absorption Spectroscopy

Objects of Experiment. To apply the principle, discussed in Chapter 5, of the simultaneous determination of two substances in the same solution that have different absorption spectra; to note the problem that arises when one substance is much more concentrated than the other; to analyze a commercial drug preparation.

Tablets sold for the relief of headaches commonly contain acetylsalicylic acid (ASA) and caffeine. They may contain other drugs as well, like phenacetin (p-ethoxyacetanilide) and p-hydroxyacetanilide, and these compounds absorb in the ultraviolet. If more than two ultraviolet-absorbing substances are present in the same solution it is very difficult to analyze the mixture by absorption spectroscopy alone. Separations are needed; see Chapter 10. For this experiment, choose a drug preparation that contains only aspirin (ASA) and caffeine as active ingredients.

Such preparations commonly contain twenty times as much ASA as caffeine. Absorbances must therefore be measured with great care. Small errors in absorbance measurements will cause large errors in the calculated percentages.

Caffeine, 1,3,7-trimethylxanthine:

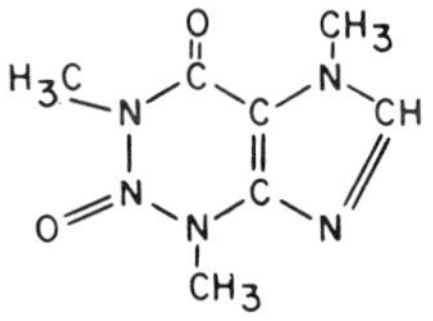

formula weight 195, is a weak base. Its ionization constant, K_b, is about 10^{-14}. In 1 M acid, therefore, caffeine is present; half as the uncharged base and half as the protonated cation. The spectrum of caffeine in 1 M hydrochloric acid has maximum absorption near 268 nm, compared with 273 nm in water, and the cation absorbs less strongly than the uncharged base.

In this experiment you will first plot the absorption spectrum of caffeine dissolved in water, then, comparing this spectrum with those obtained in the last experiment for ASA, you will devise and test a method for determining the proportions of caffeine and ASA in an analgesic drug tablet. Only minimal directions will be given.

(a) The absorption spectrum of caffeine. The molar absorptivity of caffeine at 273 nm is about 10,000. Therefore, use a solution of concentration close to 1.0×10^{-4} M, if you are using 1-cm cells. This means about 20 mg caffeine per liter. It will be best to prepare this solution in two stages, as you did with ASA; first, prepare a solution of 200 mg/l or more, then pipet a known volume of this solution into a volumetric flask and dilute with water to a known volume. As in all ultraviolet work, make sure that the distilled water is clean and of good quality, and make quite sure that your glassware is clean also. Run the absorption spectrum between 220 nm and 300 nm.

If you wish, run the spectrum of an acid solution; but remember that even in a strongly acid solution, part of the caffeine will exist as the free base, and if you plan to use this spectrum in part (b), be sure that you know the acid concentration and can reproduce it to within 5%. (The differences in the spectra are not so great as to need more accuracy than this.)

(b) Analysis of a drug tablet. Take a weighed amount of the drug, preferably at least 100 mg to minimize sampling errors, dissolve it in cold water (allowing time for the ASA to dissolve) and filter to remove the starch; wash the filter with distilled water. Bring the filtrate and washings up to a known volume in a volumetric flask, then make another quantitative dilution to bring the concentration into the range (about 20 mg/l) that will give conveniently measurable ultraviolet absorbances. In the course of this dilution, add acid or base, as you wish, to determine the forms, anionic, cationic or uncharged, whose spectra you wish to measure.

For example, one possibility is to add enough hydrochloric acid to make the final hydrogen-ion concentration 0.05 M. At this acid concentration the ASA, the major constituent, will be present to more than 99% as the uncharged acid, while caffeine will be present to the extent of 90-95% as the uncharged base. The differences in the spectra are greatest for the uncharged species.

Choose two wavelengths, at one of which the caffeine absorbs much more strongly than ASA, while at the other, ASA absorbs much more strongly than caffeine. Measure the absorbance of your solution at each of these wavelengths, and as a check, make measurements at two other wavelengths, separated as much as possible from the first two. You now have the data to make four independent calculations of the concentrations of caffeine and ASA. Make these calculations, using the method described in Chapter 5. Estimate the precision of each calculation, and finally make your best estimate of the weights of caffeine and ASA in 100 mg of the drug tablet, noting the probable error.

A more accurate way to determine the caffeine in analgesic tablets is to extract the caffeine with chloroform from an aqueous solution of pH 8-9. At this pH, ASA exists as its anion which is not extracted into the nonpolar solvent. The concentration of caffeine in the chloroform extract is determined by ultraviolet absorption, using, of course, solutions of caffeine in chloroform as standards. It is wise to measure the absorbance at more than one wavelength to see whether any ASA has been extracted.

QUESTIONS

1. Sunlight has its maximum intensity in the yellow part of the spectrum. Calculate the temperature of the surface of the sun.

2. In determining cobalt by extracting the blue thiocyanate complex it is important (a) to eliminate iron, (b) to use a considerable excess of thiocyanate. Explain the reasons for (a) and (b).

3. Traces of "heavy metals" in soils and powdered rocks may be measured, quickly and approximately, by extracting with ammonium citrate and adding dithizone solution until the red color first formed starts to turn a permanent green that persists on shaking. In such a test, 0.50 g rock = 3.0 ml of a solution containing 40 mg dithizone per liter. Calculate the "heavy metal" concentration of the rock in parts per million (milligrams metal per kilogram rock), assuming all the heavy metal to be lead.

4. Would acetone be a good solvent for ultraviolet spectroscopy? Explain your answer.

5. Absolute alcohol is made from 95% alcohol (the constant-boiling mixture of alcohol and water) by adding benzene and boiling out the ternary azeotrope, benzene-water-alcohol. Which is best to use as a solvent for ultraviolet spectroscopy, absolute alcohol or 95% alcohol?

6. A bottle of cyclohexane shows some absorption in the ultraviolet. Instead of throwing it away, you purify it by passing it through a column of dried, activated aluminum oxide. The solvent is now transparent in the uv. What has happened? (See Chapter 9)

7. The diffraction grating of a spectrophotometer has 2,400 lines/cm. A beam of light, wavelength 600 nm, strikes the grating at right angles. At what angle to the incident beam is the first-order diffracted beam?

8. How many millimeters wide is the exit slit of this spectrophotometer if it passes a band of wavelengths 20 nm wide and is 25 cm away from the grating? (Assume that the entrance slit has zero width and that the rays striking the grating are parallel; they are focused after they leave the grating.)

9. What is the minimum volume of a solution of 1,10-phenanthroline, 1.0 g/l, needed to combine with 0.50 mg of Fe(II)?

10. A fertilizer contains 8% phosphorus expressed as P_2O_5. What is the percentage of the element phosphorus?

11. To "open up" rocks for analysis one sometimes fuses them with sodium carbonate, sometimes with $KHSO_4$. For what type of rock would you use one substance, and for what type would you use the other?

12. Perchloric acid is a useful reagent for destroying organic matter, but serious explosions may occur when perchloric acid is heated with organic matter. Well-known and well-understood procedures exist for the safe use of perchloric acid in destroying organic matter. Describe these procedures and explain why they are effective.

13. In many spectrophotometers the width of the slit can be adjusted. What are the advantages of (a) a narrow slit, (b) a wide slit?

14. Why are high absorbance readings (above 1.5 on most instruments) not to be trusted? What instrumental error may occur?

Chapter 7

INFRARED SPECTROSCOPY

I. INTRODUCTION

The "infrared" region of the spectrum is the region "beyond the red," that is, with wavelengths greater than 0.8 microns. We noted in Chapter 5 that the infrared region extends to wavelengths of several hundred microns. When an analytical chemist speaks of infrared spectroscopy, however, he usually means the range from 2.5 to about 20 or 25 microns. This is the range of wavelengths that gives him the most information about the vibrations of molecules, and hence about the structure of molecules. Infrared spectroscopy is used primarily for organic compounds and primarily for identification and structure determination. It is not used very much for the quantitative analysis of mixtures, though it may be so used in special cases. Certainly it does not have the precision of visible and ultraviolet spectroscopy for quantitative measurements. On the other hand, it is far more useful for qualitative identification, and an infrared spectrophotometer is an indispensible tool in a modern organic chemical laboratory. Instruments are available at moderate cost that are very simple to use. The infrared spectrum not only characterizes the compound but indicates its degree of purity and often provides clues to the identity of impurities that may be present.

II. INSTRUMENTATION

A. The Near-Infrared

Before describing the common form of infrared spectrophotometer we should mention that the wavelength range from 0.8 microns, the end of the visible spectrum, to 2.5 or 3 microns is accessible to instruments of the type described in the last chapter. This wavelength range is transmitted by quartz and glass and is emitted by a tungsten-filament lamp. Photocells do not respond, however, and it is necessary to use a solid-state lead sulfide photodetector for wavelengths above 1 micron. Thus most ultraviolet-visible spectrophotometers do not read beyond 1 micron, and the instruments that do read into the near-infrared are more expensive. However, they use the same cells as would be used in the visible region, and absorbance measurements can be made that are reasonably precise.

Absorption in the range 1 to 3 microns is caused mainly by "overtone" vibrations of molecules, analogous to the higher harmonics of a vibrating string. The bands are narrow and well-defined, though the molar absorptivities are much lower than in the visible region. The principal analytical use is for hydroxy and amino compounds. Water has a narrow absorption band at 1.89 microns, with molar absorptivity 30, and this band can be used for the quantitative determination of water.

B. The Medium Infrared

This is the range commonly used for chemical analysis. It starts at 2.5 microns, the wavelength at which glass becomes opaque. The upper wavelength limit depends on the instrument. All infrared spectrometers in common use read up to 16 microns, the limit of transmission of a sodium chloride prism, and some go to much higher wavelengths. For the characterization of organic compounds the range 2.5 - 16 microns is quite adequate. In frequency units (see Chapter 5) this range is 4000 - 625 wave numbers or 4000 - 625 cm^{-1}. Infrared spectrometers are often calibrated in wave numbers rather than wavelengths, and one should be able to convert one kind of units into the other without difficulty. (Note that 2.5 microns goes into 1 centimeter 4000 times.)

Infrared spectrometers differ from visible-ultraviolet spectrometers in several respects. First, materials must be used for cells and windows that are transparent to infrared rays. Glass cannot be used, and quartz is opaque above 4.5 microns. Rock salt, or sodium chloride, is the material most commonly used. The upper wavelength limit at which a 2-millimeter thickness of the substance transmits 10 per cent of the radiation is 25 microns for sodium chloride, 40 microns for potassium bromide, and 70 microns for cesium iodide. Prisms, of course, are much thicker than 2 millimeters, and a rock-salt prism can be used only up to 16 microns.

Prisms of such materials as sodium chloride and cesium bromide are very expensive and are easily harmed by water vapor. Most modern instruments use gratings. Diffraction gratings are not harmed by atmospheric humidity and provide uniform dispersion over a broad range of frequencies. However, the infrared spectrum is too broad to be handled effectively by a single grating; the range 2.5 - 16 microns (4000 - 625 cm^{-1}) covers nearly three powers of 2, compared to less than one factor of 2 for the visible region. Therefore, infrared spectrometers use at least two gratings, and provision must be made in the mechanism for exchanging one grating for another as the spectrum is scanned. In some instruments the grating is changed automatically, while in others the spectrum must be run in two stages. The instruments also incorporate filters to exclude unwanted higher-order diffraction spectra; see Chapter 6.

The source of radiation in an infrared spectrometer is, in the cheaper instruments, a heated Nichrome wire and, in the more expensive ones, a heated rod of silicon carbide or a "Nernst glower," a heated rod of rare-earth oxides. The temperature of emission is low, and the energy output is very low compared with the output of a tungsten-filament lamp. The detection method must therefore be very sensitive. Detection is done with a thermocouple. The thermocouple junction that absorbs the radiation must be small and have a very low heat capacity, so that it can follow the changes in intensity as the spectrum is scanned.

All modern infrared spectrometers are double-beam instruments and have automatic recording.

III. CELLS, SAMPLE HOLDERS

There are no convenient solvents for infrared spectroscopy as there are for the visible-ultraviolet range. No liquid transmits throughout the whole infrared, though a few liquids, such as carbon tetrachloride, can be used over limited ranges. Liquid samples are run without solvents, and the cells must therefore be very thin. The thicknesses most often used are 0.1 mm and 0.025 mm. The cell walls are usually of rock salt or sodium chloride. Polished plates of sodium chloride about 2 millimeters thick are mounted in a steel frame, as shown in Fig. 7-1, with a spacer to keep them the correct distance apart. Some cells are demountable, that is, they can be taken apart and reassembled, changing the spacer if a different cell thickness is required, but they are not very satisfactory; the permanently

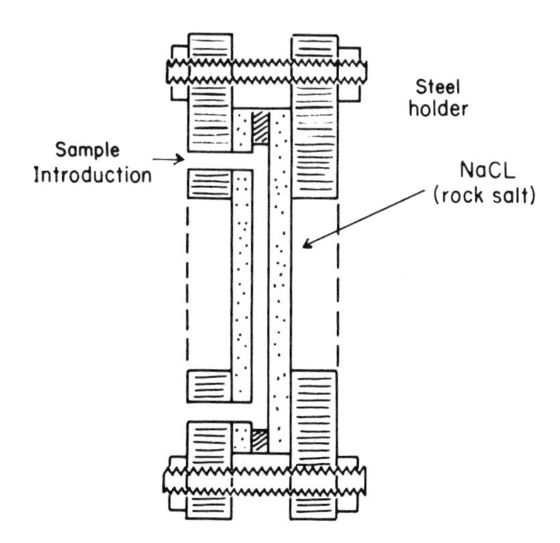

Fig. 7-1. Infrared cells for liquids.

assembled cells are better. Because sodium chloride is soluble in water the cell windows are easily etched. One must take great care to use dry samples and to rinse the cell with a volatile liquid like benzene and acetone after use. Cells should be kept in a desiccator when not in use. In spite of all precautions it is difficult to avoid some erosion of the internal cell surfaces. One can seldom be really sure of the thickness of an infrared cell unless it is new, and this is one reason that quantitative analysis in the infrared is less precise than in the visible or ultraviolet region. If the sides of the cell are smooth and parallel, the thickness can be measured by an interference method; see Section 7-IV.

Plates of sodium chloride that have become roughened can be polished by rubbing with a fine abrasive, and kits for doing this can be bought commercially.

Infrared spectra can be run on solid samples. The best technique for doing this is to grind the solid very finely in mortar together with about 100 times its weight of powdered potassium bromide. The mortar should be of a hard material like agate or Mullite. The finely ground powder is now compressed into a transparent disk about 1 millimeter thick. Strong pressure is needed, about 2000 kilograms for a disk 1 centimeter in diameter, and various kinds of dies and presses are sold for the purpose. The most elaborate ones make provision for evacuating air from the powder during compression. A very satisfactory arrangement, and an inexpensive one, consists of a steel cylinder with a screw thread going through it, and two bolts; one is screwed a short distance into the threaded hole, then the sample powder is poured into the hole, and the second bolt is screwed down onto the powder; see Fig. 7-2. One must take care not to force the screw too tightly or the threads may be stripped. After compressing the sample one removes the bolts and places the steel cylinder with the sample disk inside it in the path of the beam of the infrared spectrometer; the cylinder is made of such a size that it can be easily mounted in the spectrometer. Or, if desired, the compressed sample disk can be removed from the press by cutting around its circumference with a pointed spatula.

Another technique for solid samples is the "Nujol mull." "Nujol" is a paraffin oil sold for medicinal purposes that consists of straight-chain high-molecular-weight hydrocarbons. The same product is sold under other trade names. The oil is transparent over most of the infrared range but has bands at 3.5, 6.9, 7.2, and 14 microns, the first three corresponding to carbon-hydrogen bonds and the fourth to twisting of the long carbon chain. Absorption bands of the sample that happen to coincide with these bands will be hidden, but others will be clearly seen. The solid sample is placed in a small Mullite mortar with a drop or two of paraffin oil and ground thoroughly to a smooth paste, which is then spread on a sodium chloride plate or disk. This is mounted in the path of the infrared beam and the spectrum is run.

Polyethylene film can be substituted for the sodium chloride plate, and it is much cheaper. Chemically, polyethylene and paraffin oil are similar. The molecules of polyethylene are longer, but the spectra are practically identical. Squares of polyethylene film are cut and mounted in the 2- by

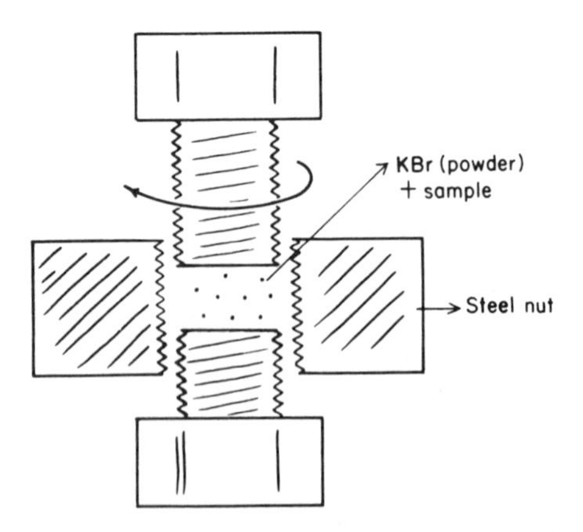

Fig. 7-2. Device for pressing potassium bromide disks. The powder is introduced as shown, then the upper screw is tightened until the powder is compressed into a thin disk. The steel nut is mounted in the spectrophotometer beam without removing the disk.

2-inch cardboard mounts used for color photographs (Fig. 7-3); see Experiment 7-1. Cells made of polyethylene are sold for spectra in the "far-infrared," from 700 to 200 cm^{-1}.

IV. INTERFERENCE WAVES: MEASURING FILM AND CELL THICKNESS

Figure 7-4 shows an absorption spectrum recorded with polyethylene film. The absorption bands are clearly seen, and there is another feature: a regular wave pattern of light intensity plotted against frequency which appears in the region between absorption bands. The rise and fall of intensity is not due to absorption but to interference between the light waves that go straight through the film and those that are reflected between the front and back surfaces of the film, as shown in Fig. 7-5. Ray (a) goes straight on; ray (b) is reflected back through the film and traverses the thickness of the film two more times. Its path is longer than that of the main ray by a distance 2d, where d is the thickness of the film. For ray (a) and ray (b) to reinforce one another and give maxima in the intensity-frequency graph (Fig. 7-4), the distance 2d must be a whole-number

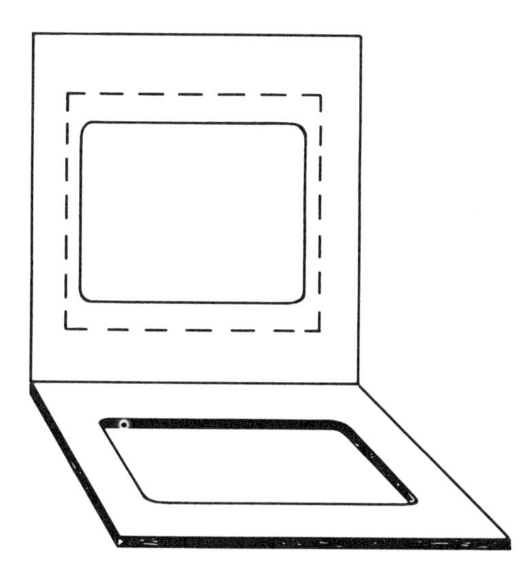

Fig. 7-3. Cardboard mount for 35-mm photographic transparencies. The sample of film to be examined is placed in the cardboard holder; this is folded and sealed with a hot iron.

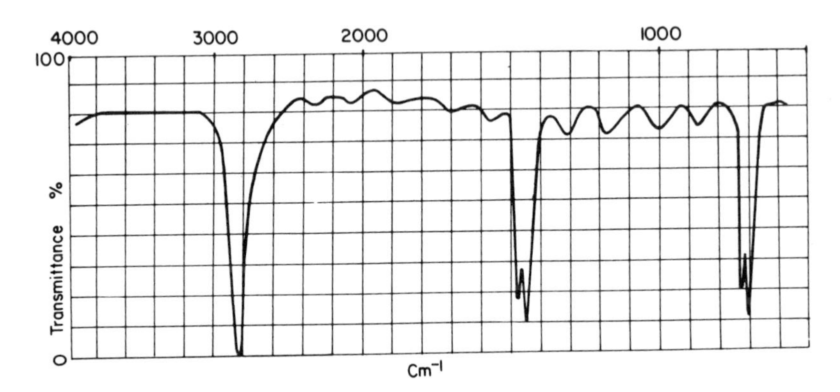

Fig. 7-4. Infrared spectrum of polyethylene film. The C-H stretching vibration is seen at 2850 cm^{-1}; C-H bending at 1450 cm^{-1}; the carbon chain twisting vibration at 700 cm^{-1}. The interference waves are clearly seen below 1400 cm^{-1}.

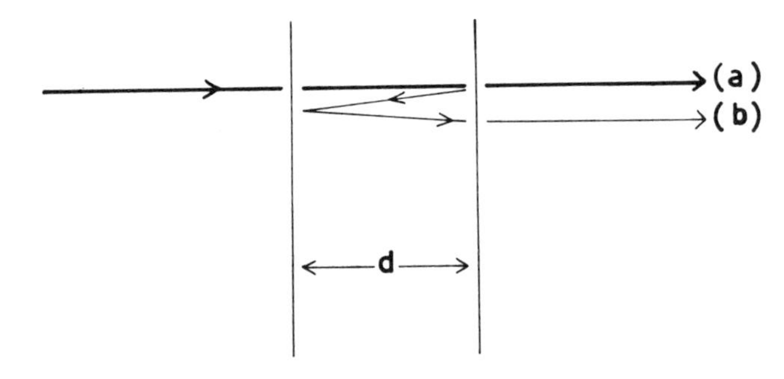

Fig. 7-5. Origin of interference waves in a film or thin cell. Ray (b) travels a distance 2d farther than ray (a); for reinforcement, 2d must be an integral number of wavelengths.

multiple of the wavelength of the light within the film: $2d = n\lambda$. Using wave numbers, $\lambda = 1/\tilde{\nu}$, and $2d = n/\tilde{\nu}$. Here, n is a whole number, the number of waves in the distance 2d. If this relation holds for one number n, it must also hold for the next intensity maximum, which will occur for n + 1 waves. Therefore

$$2d = \frac{n}{\tilde{\nu}} = \frac{n+1}{\tilde{\nu} + \Delta\tilde{\nu}}, \quad n\,\Delta\tilde{\nu} = \tilde{\nu} = \frac{n}{2d}$$

or

$$\Delta\tilde{\nu} = \frac{1}{2d} \tag{7-1}$$

The quantity $\Delta\tilde{\nu}$ is the difference in wave-number frequency between one intensity maximum and the next. <u>However</u>, these are wave numbers <u>in the film</u>, not the numbers read off the scale of the spectrometer. The spectrometer gives the wave numbers, the numbers of waves per centimeter, <u>in vacuo</u>, or in air, for there is practically no difference. There is quite a lot of difference, however, between the wavelength in a vacuum and the wavelength in the polyethylene film; the ratio is the refractive index of polyethylene. Let us call the refractive index ε; then equation (7-1) may be rewritten:

$$\Delta\tilde{\nu}\ \text{(read on spectrometer)} = \frac{1}{2d\varepsilon} \tag{7-2}$$

If we know the refractive index, therefore, we can use the interference waves to measure the thickness of the film. The catch is that we must use the refractive index for the particular frequency or wavelength corresponding to the spectrum, and this is not the same as the value for visible light. However, the interference-wave method provides a very convenient way to compare the thicknesses of different polyethylene films or other plastic films.

The same method is used to measure the width of the cells used for liquids in the infrared. As we noted, these cells are of the order of 0.1 mm wide inside. All we have to do to measure the thickness of a cell is to run its infrared spectrum when it is empty, and note $\Delta\tilde{\nu}$, the frequency difference between the interference maxima. The refractive index of the air in the cell is so close to unity that one can use equation (7-1) directly. The inside surfaces of the cell must be smooth and parallel to give good interference waves.

V. INFRARED SPECTRA AND MOLECULAR VIBRATIONS

The absorption of infrared radiation causes a molecule to vibrate. However, not every vibration of a molecule produces absorption in the infrared. Only those vibrations that are accompanied by a change in the electric dipole moment cause absorption of infrared. Vibrations about a center of symmetry, which do not change the dipole moment, cannot be excited by infrared radiation. Thus it comes about that symmetrical diatomic molecules, like O_2 and N_2, do not absorb in the infrared. This is fortunate, for otherwise we would have to evacuate the air from infrared spectrometers. The carbon dioxide and water vapor in air do absorb, carbon dioxide at 4.2 and 15 microns, water vapor around 2.7 and between 6 and 7 microns, but these absorptions do not affect records taken on a double-beam instrument, as they are fairly weak and cancel out between the sample beam and the reference beam.

The appearance or nonappearance of lines in the infrared tells us something about the symmetry of the molecules of our sample. Of the two 1,2-dichloroethylenes, for example, the trans form has a center of symmetry and the cis form does not. A vibration about the center of symmetry of the trans form (see Fig. 7-6) cannot absorb infrared light, and the spectrum is correspondingly simpler. Likewise para-disubstituted benzenes have simpler spectra with fewer lines than those of ortho or meta derivatives; see Fig. 7-7.

The intensity of absorption depends on the square of the change in dipole moment. Thus there are great differences in the strengths of infrared bands. The carbonyl group, :C=O, has a large dipole moment that changes greatly when the vibration changes. The corresponding infrared band at 1725 cm^{-1}, or 5.8 microns, is very intense and easy to recognize in an infrared spectrum. The infrared absorption of water is intense for the same reason.

The vibration of a diatomic molecule is relatively simple. A vibrating molecule can have energies that are given, approximately, by

$$E = \frac{h}{2\pi}\left(v + \frac{1}{2}\right)\sqrt{\frac{k}{\mu}} \qquad (7\text{-}3)$$

where h is Planck's constant, v an integer (1, 2, 3 ..), k the force constant, and μ the "reduced mass." The reduced mass is defined by

$$\frac{1}{\mu} = \frac{1}{m_1} + \frac{1}{m_2} \qquad (7\text{-}4)$$

where m_1 and m_2 are the masses of the two atoms. Equation (7-3) is accurate only if the molecule vibrates in simple harmonic motion, with k being a true constant independent of the distance between the two atoms. This is never quite true. Now, the quantum theory says that v, the "vibrational quantum number," can change by only one unit at a time when the molecule absorbs or emits radiation. The frequency ν of the radiation is then simply

$$\nu = \frac{1}{2\pi}\sqrt{\frac{k}{\mu}} \qquad (7\text{-}5)$$

"Overtones," or changes of more than one unit in v, are permitted only if the vibration is not strictly harmonic. Since all molecules have some anharmonicity (k does change somewhat with the distance between atoms) overtones are allowed and are observed, though their intensities are low.

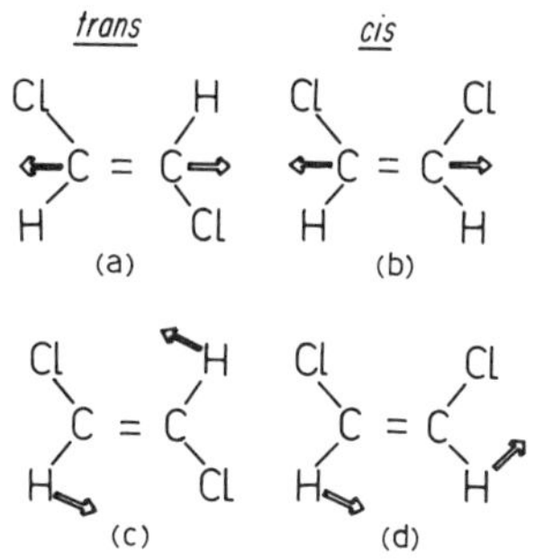

trans *cis*

(a) (b)

(c) (d)

Fig. 7-6. Vibrations of cis and trans isomers. Vibrations (a) and (c) produce no change in dipole moment, and therefore do not absorb infrared radiation.

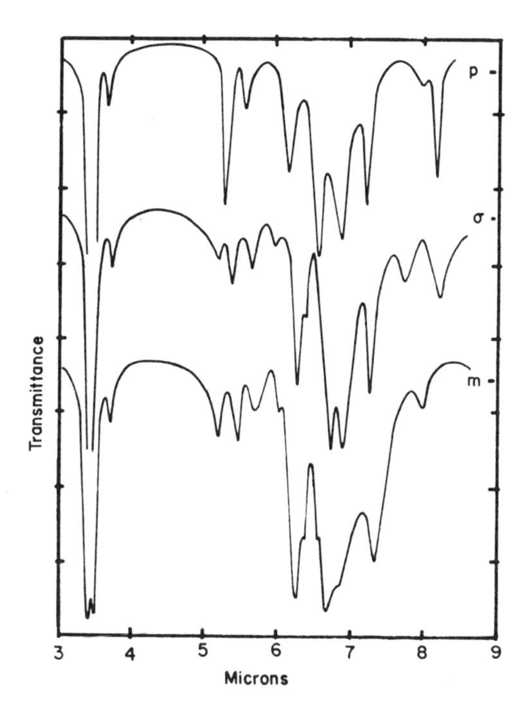

Fig. 7-7. Spectra of ortho-, meta-, and para-xylene. Note the sections between 5 and 6 microns (2000 and 1666 cm^{-1}); these are characteristic "fingerprints" of ortho, meta, and para isomers. Note also the relative simplicity of the spectrum of para-xylene.

These are the vibrations whose frequencies come in the "near-infrared," below 3 microns.

The vibration of polyatomic molecules, the molecules we usually meet in chemical analysis, is much more complicated. A molecule with n atoms has 3n - 6 "normal modes" of vibration. The molecule of acetone, therefore, can vibrate in (3 x 10 - 6) = 24 fundamentally different ways. Fortunately we can consider that, to a first approximation, different parts of a molecule can vibrate independently. Thus we can resolve the vibration of the acetone molecule into a vibration of the :C=O group, a vibration by stretching of the C-H bonds of the methyl groups, another vibration by bending of the H-C-H bond angle, another (like that of a tuning fork) by bending of the C-C-C bond angle, and so on. The concept of "group frequencies" is basic to the understanding of infrared spectra and their use in identification, but one must always bear in mind that the group frequencies are not independent of one another and are influenced by all the vibrations of the molecule.

Let us accept the idea of a "group frequency" for the moment, and apply equation (7-5) to vibrations of two masses m_1 and m_2 connected by an imaginary spring of strength (force constant) k. We see that the tighter the spring, the higher the frequency (just as in a vibrating violin string). Thus atoms connected by double bonds should vibrate faster than the same atoms connected by single bonds. It is easier to "bend" a bond, or change the angle between two bonds, than it is to stretch it, therefore "stretching" vibrations should have a higher frequency than "bending" vibrations. These predictions are fulfilled by the following characteristic group frequencies:

C-C	800 - 1200 cm^{-1}	C-N	1050 cm^{-1}
C=C	1680 - 1620 cm^{-1}	C=N	1650 cm^{-1}
C≡C	2100 cm^{-1}	C≡N	2250 cm^{-1}

C-H stretch, 2900 cm^{-1}; C-H bending, 1370, 1460 cm^{-1}

The effect of mass is seen very plainly in vibrations in which one partner is the hydrogen atom. For the C-H pair, the reduced mass μ is 12/13, or 0.92. For C-Cl, μ is 8.9. Thus vibrations of bonds with hydrogen are relatively fast, and the heavier the atoms that are connected, the slower the vibration. This is seen in the following set of frequencies:

C-H	2900 cm^{-1}
C-F	1000 - 1400 cm^{-1}
C-Cl	600 - 800 cm^{-1}
C-Br	500 - 600 cm^{-1}

Extensive tables of group frequencies have been compiled and can be found in reference books. An abbreviated table is given here; Table 7-1. Using this table of group frequencies, let us analyze the spectrum of acetone (Fig. 7-8) as follows:

Band at:	Corresponding vibration:
3.4 microns = 2950 cm^{-1}	C-H stretch
5.8 " = 1720 "	:C=O stretch
7.0 " = 1430 "	$-CH_3$ bending
7.35 " = 1360 "	
8.2 " = 1220 "	C-C-C bending (tuning-fork)
9.2 " = 1090 "	C-C stretch
11.1, 12.8 microns	vibrations of molecule as a whole

Division of the vibration frequencies of a molecule into the contributions of individual groups and bonds works fairly well as long as the frequencies of two adjacent bonds are not close together. The closer the two frequencies are, the more the vibrations interact, and the more difference there will be between the observed frequencies and those tabulated as group frequencies. And, naturally, the more complicated the molecule the more complex is its spectrum. No two compounds have the same infrared spectrum, and the infrared spectrum of a compound that is known to be pure

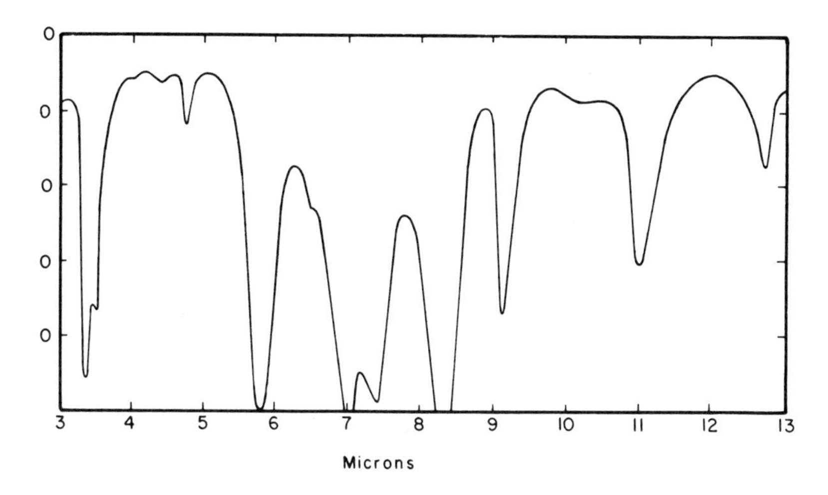

Fig. 7-8. Infrared spectrum of acetone.

(by chromatography, for example; see Chapter 12) is like a fingerprint and provides positive identification. Catalogs are available that give in-

TABLE 7-1

Group Frequencies

Wavelength, μ	Frequency, cm^{-1}	Vibration
2.75	3640	O-H stretch; alcohols, phenols N-H stretch
2.95-3.1	3400-3225	Hydrogen-bonded O-H and N-H stretch, ≡C-H stretch
3.2-3.3	3125-3000	=C-H stretch, aromatic C-H
3.4-3.5	2940-2850	Saturated C-H stretch
3.45-3.7	2900-2700	Aldehyde C-H stretch
3.85-3.9	2600-2560	S-H stretch
4.4-4.75	2270-2105	C≡C stretch
5.0-6.0	2000-1670	Aromatic substituted C-H overtones
5.4-5.75	1850-1740	C=O, anhydrides
5.55-5.8	1800-1725	C=O, esters
5.75-5.95	1740-1680	C=O, ketones, aldehydes (aliphatic), carboxylic acids
5.85-6.0	1710-1665	C=O, conjugated and aromatic
5.95-6.15	1680-1625	C=C stretch, unconjugated
6.15-6.25	1625-1600	C=C stretch, conjugated
6.8-6.9	1470-1450	CH_3 and CH_2 bending
7.20	1390	CH_3 bend (symmetrical); $C(CH_3)_2$
7.25-7.3	1380-1370	CH_3 bend (symmetrical); $C(CH_3)_2$
7.7	1300	C-H bend in trans-disubstituted olefin
7.2-7.7	1390-1300	COO^- carboxylate ion stretch
7.1-7.6	1410-1315	C-OH, tertiary alcohols and phenols
7.4-7.9	1350-1265	C-OH, primary and secondary alcohols
7.5-8.0	1330-1250	C-N, aromatic amines
8.7	1150	O-H bend, tertiary alcohols
9.1	1100	O-H bend, secondary alcohols
9.5	1050	O-H bend, primary alcohols
10.1-10.4	990-960	C-H bend, olefines (out-of-plane vibration)
10.5	950	O-H bend, acids
11.0	910	C-H bend (out-of-plane), vinyl olefines

TABLE 7-1 (continued)

Wavelength	Frequency, cm^{-1}	Vibration
12.0-12.5	830-800	C-H bend (out-of-plane), trisubstituted olefines
12.5	800	para-Disubstituted benzenes
13.0	770	meta-Disubstituted benzenes
13.5	740	ortho-Disubstituted benzenes
13.6-13.8	735-725	Four or more linked $-CH_2-$ groups
12.5-16	800-625	C-Cl stretch
14.5	690	Monosubstituted benzenes

frared spectra of tens of thousands of compounds, and they are indispensable tools of the research laboratory.

Infrared spectroscopy is one of three or four methods that are used very widely to identify organic compounds and determine their structure. The other important methods are nuclear magnetic resonance, mass spectroscopy, and for certain classes of compounds, ultraviolet spectroscopy (Chapter 6).

EXPERIMENTS

Operation of Infrared Spectrometers. The simpler, lower-priced infrared instruments, like the Perkin-Elmer Model 700, are very easy to operate. There is a switch to turn on the light source and electronics; it is best to turn this on and leave it on, especially in a humid climate, for as many days as the instrument is being used. There are buttons to start and stop the recorder, a lever to raise and lower the recorder pen, and a knob to adjust the pen for 100% transmittance. On some models there are switches that give a choice of scanning speeds (slow to show the greatest detail; fast to get the result as soon as possible), and often there is a switch to choose the wavelength or frequency range (for example, in the Perkin-Elmer Model 337, one range, with one grating, covers 4000 to 1200 cm^{-1}, while the other range, with another grating, covers 1333 to 400 cm^{-1}; other instruments have other ranges).

If the instrument is a double-beam instrument, and most instruments are, the sample cell is placed in one beam (labeled as "sample") and nothing is placed in the other beam; that is, the sample is run against air as a reference. Obviously other cells or materials can be placed in the reference beam if desired. Often, to help with samples whose absorbance is high, one places in the reference beam a kind of black metal "comb" or occluder that shuts out part of the radiation in the reference beam; the fraction of radiation that is obstructed can be changed by changing the angle of the "comb."

One thus has two ways to set the pen to read 100% transmittance. One is with the electronic adjustment knob, the other is with the "comb." By using the "comb" one blocks out part of the radiant energy and makes the instrument less responsive to fine details of the spectrum. One should not use it, therefore, unless the sample absorbs so strongly that the spectrum would be pushed down toward the zero transmittance line; see Fig. 7-9.

The 100% transmittance control should be set so that the pen, in the course of its travel, rises as near as possible to 100% without actually going off the scale. One places the sample in the "sample" beam and adjusts the 100% control to make the pen read about 90% transmittance at the lowest end of the wavelength scale before starting the scan. (Most instruments start at the low wavelength, or high frequency, end of the scale and move toward the high wavelength end.) Now it frequently happens that the transmittance rises just after the beginning of the scan, between 2.5 and 3.0 μ. One must therefore be prepared to readjust the 100% setting, if necessary, to keep the pen on scale; see Fig. 7-9.

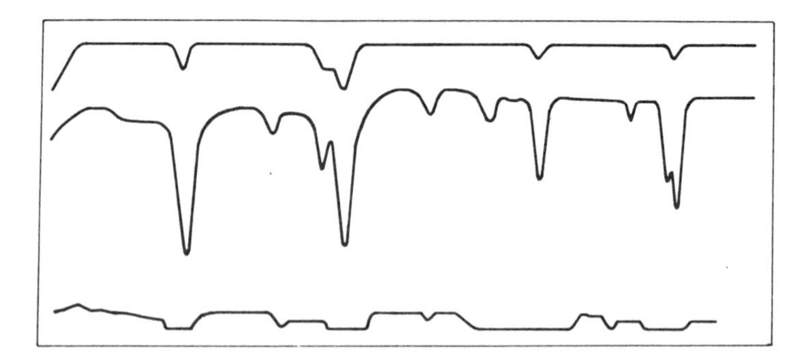

Fig. 7-9. Correct and incorrect use of occluder or "comb": (a) too much occlusion, 100% transmittance set too high; (b) correct; (c) not enough "comb," 100% transmittance set too low.

Spectra are generally recorded on special chart paper supplied by the instrument maker. The abscissae may be linear in wavelength, or they may be linear in frequency (wave number). One must watch for a possible change of scale around the middle of the chart. The Perkin-Elmer Model 700, for example, records the whole spectrum on one chart, from 4000 to 625 cm^{-1}. The main divisions are 400 cm^{-1} apart from 4000 to 2000 cm^{-1}, then 200 cm^{-1} apart from 2000 to 625 cm^{-1}. Filters are changed automatically during the run to select the proper order spectrum from the grating (see Chapter 6).

The ordinates are usually per cent transmittance but may be absorbance. Absorbances are used less in infrared spectroscopy than in the visible and ultraviolet because of the uncertainty in quantitative interpretation.

Chart paper is expensive, and one may substitute ordinary unruled paper if one calibrates it. Even with chart paper one should not take the wavelength-frequency scale for granted without calibration. One should check the scale from time to time by running the spectrum of a polystyrene film; see Experiment 7-1.

In some spectrometers the pen carrier moves across the paper; in others the pen carrier is stationary and the chart paper is moved underneath it. The chart carrier, or the pen, is moved back to its original position after the spectrum has been scanned. One caution is necessary: never try to move the pen or carrier by hand unless the scanning mechanism is turned off. And, when you do move it, move it gently.

Experiment 7-1

Examination of Plastic Films

Equipment. Infrared spectrophotometer, samples of plastic film, cardboard mounts (the mounts 2 x 2 in., or 5 x 5 cm, used for color transparencies are the most convenient, but one can make one's own with cardboard and scissors).

Objectives. Film samples are the easiest to run in the infrared. They need no special preparation, and one can concentrate on manipulating the spectrometer itself. One object of this experiment is to calibrate the wavelength scale of the instrument. Another is to measure film thickness by interference.

Obtain samples of the following films: polystyrene, polyethylene, polyvinyl chloride, Cellophane. Polystyrene film is usually supplied by the instrument maker, mounted in cardboard, for calibrating the wavelength scale. Polyethylene and Cellophane are widely used for wrapping purposes. Polyvinyl chloride is sold under the name "Saran Wrap."

Obtain other samples if possible. The films must be thin or they will not be sufficiently transparent to infrared rays.

Mount samples of the films in the cardboard holders. This can be done by sealing the holders, or mounts, with a hot iron, as one would do if mounting a photographic film, or it can be done with a paper stapler. Then run the spectra of the films, starting with the polystyrene. This spectrum is complex and has many bands, corresponding to the many possible vibrations of the benzene rings and the chain that connects them:

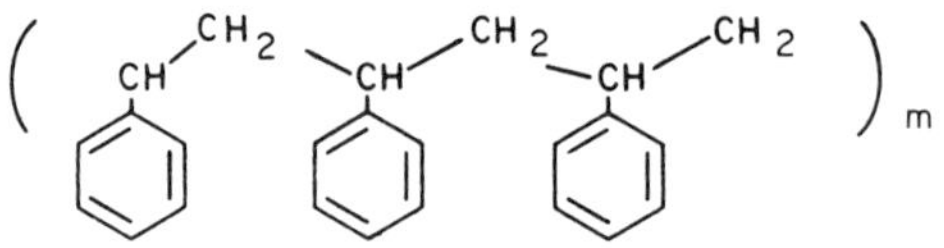

It is shown in Fig. 7-10. The sharp peaks at 3.51, 6.24, and 11.03 μ (2850, 1600, and 906 cm^{-1}) are those used to calibrate the wavelength-frequency scale of spectrometers. Note the positions of these peaks, and compare the chart readings with the true wavelengths or frequencies. Note also the pattern of four bands between 2000 and 1700 cm^{-1}, which is characteristic of monosubstituted benzene rings.

Now run the polyethylene. This spectrum is simple because the molecule is simple:

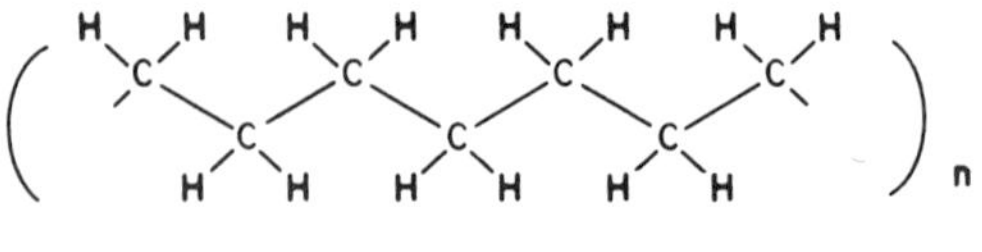

It should look like Fig. 7-4. If other bands are seen, they indicate that the film has other substances incorporated into it. Sometimes plasticizers are added; they are usually esters and show the characteristic, telltale band of the -C=O group at about 5.75 μ, or 1740 cm^{-1}.

The region between 7 and 13 μ should be free from absorption bands and should show the regular sine-wave profile that indicates interference of waves reflected between the front and back of the film; see Section 7-IV.

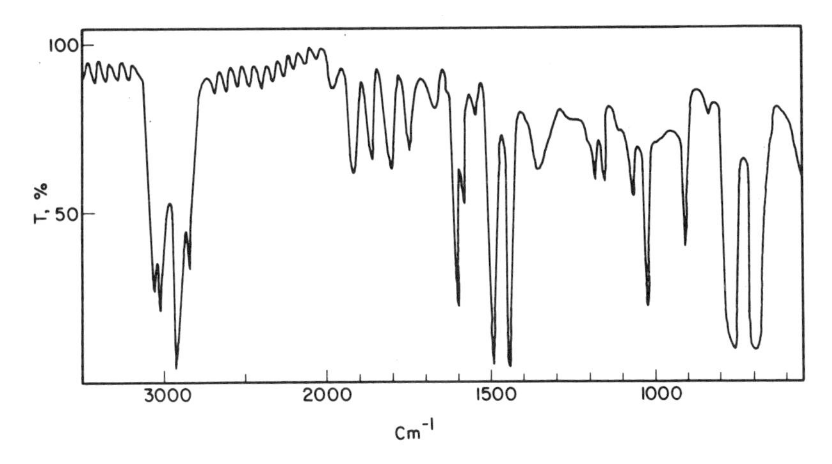

Fig. 7-10. Infrared spectrum of polystyrene film.

Measure the frequency difference between one maximum and the next. For maximum accuracy measure the length of, say, five waves and divide by the number of waves to get the frequency span of one wave, and use this value, $\Delta\tilde{\nu}$, to calculate the thickness of the film, taking the refractive index as 1.4. If possible, get samples of polyethylene film of two thicknesses, the very thin film used for plastic bags and the somewhat stronger film used for constructing model airplanes, and compare the thicknesses of the two films. The thicker film (about 40 μ thickness) will give better-defined interference waves.

Run the other film samples, and identify what features you can. Characteristic of the polyvinyl chloride are the C-Cl stretching and bending vibrations at relatively long wavelengths. Identify unknown films if you can.

Experiment 7-2

Examination of Paraffin (Nujol) Mulls

Equipment. Small mortar, about 5 cm diameter, of hard material like Mullite or agate; rock salt disks or polyethylene film. If polyethylene film is used, choose, if possible, a film that is somewhat thicker than the very thin films used to wrap goods in the grocery store.

The main purpose of this experiment is to get acquainted with the "mull" technique for running infrared spectra of solid samples. A wide variety of samples can be chosen. For example, benzoic acid and sodium benzoate can be studied; the differences between benzoic acid and phthalic acid (or potassium hydrogen phthalate) will show characteristic differences between mono- and disubstituted benzenes, although the pattern of low-intensity overtone bands between 2000 and 1700 cm^{-1} is partly obscured by the absorption of the Nujol itself. A good pair of substances to compare is salicylic acid and acetylsalicylic acid. Acetylsalicylic acid may be prepared from salicylic acid by warming 1 g of the acid with excess acetic anhydride, then purifying the product by recrystallizing from hot water.

To prepare the "mull," grind the solid sample very thoroughly with a drop or two of the paraffin oil to obtain a thin paste that can be spread smoothly with a spatula. Spread it on a mounted polyethylene film (as noted above, it is best to use a film that is not too thin) or on a disk or plate of sodium chloride, and run the spectrum. Run the spectrum of the paraffin oil alone for comparison. Also try running the sample "mull" with a film of paraffin oil alone, on polyethylene or on rock salt as the case may be, in the reference beam of the infrared spectrometer.

Experiment 7-3

Use of Potassium Bromide Pellets

Equipment. A press to make pellets (see page 198); small mortar, as used in Experiment 7-2; powdered potassium bromide, spectroscopic grade if this is available; sample holder for spectrometer.

As was mentioned in Sec. 7-III, this technique is the best for solid samples, because potassium bromide is transparent through the entire range used in most infrared work. Potassium bromide that is not of spectral grade may contain impurities that absorb in the infrared, and a "blank" spectrum should be run on a pellet of potassium bromide alone.

The solid sample should be ground to a fine powder with 200 times its weight of potassium bromide. Convenient quantities are 2 mg of sample to 400 mg potassium bromide. It is very important to keep the potassium bromide dry, because water absorbs strongly in the infrared. The sample must be dry too. More than a trace of water not only spoils the spectrum, it makes it impossible for the powder to cohere into a pellet. A moist sample will remain a powder even after strong compression.

Press the mixed powder into a pellet, or a disk, about 1 cm in diameter and 1-2 mm thick, using the technique appropriate to the type of press you have; see page 197. Mount the disk in the sample beam (using the holder recommended for the instrument) and run its spectrum.

As in the last experiment, a great variety of samples can be chosen. To observe the effect of isomerism, maleic and fumaric acids can be run; their spectra are substantially different. Assign as many as possible of the observed bands to particular vibrations; in other words, interpret your spectrum.

Experiment 7-4

Spectrum of Liquid Samples: Quantitative Analysis

Equipment. Cell for liquids with sodium chloride windows, thickness 0.05 mm (other widths may be used); glass syringe; pure liquids including benzene, toluene, and isopropanol.

Objectives. To investigate structural effects on infrared spectra and to use infrared absorption to analyze simple mixtures.

The type of cell used for liquid samples is illustrated in Fig. 7-1. It has an inlet and an outlet tube which are closed by Teflon plugs or stoppers. To fill the cell with a liquid sample, remove both plugs, draw some of the liquid up into the syringe, and drive it gently into the cell. Let some of the liquid flow through the cell to rinse it before inserting the plugs, and make sure there are no air bubbles in the cell. A good way to proceed is to use *two* syringes, as shown in Fig. 7-11. Press the liquid out of one syringe and into the other, and have the cell slightly inclined, with the inlet hole lower than the outlet. (The syringes are the common variety sold in pharmacy stores, without the needles.)

Place the filled cell in the sample beam, and run the spectrum against air on the reference side. Record spectra of the following liquids: benzene, toluene, ethyl benzene, *o*-xylene, *m*-xylene, *p*-xylene, isopropanol, and as many other substances as you wish. Interpret each spectrum as well as you can, correlating each band with a specific vibration where possible. Note especially the spectra of benzene, toluene, ethyl benzene, and the three isomeric xylenes between 5 and 6 μ (2000 and 1600 cm^{-1}). The bands in this region are characteristic of mono- and disubstituted benzenes, and ortho-, meta-, and para-disubstituted derivatives have characteristic

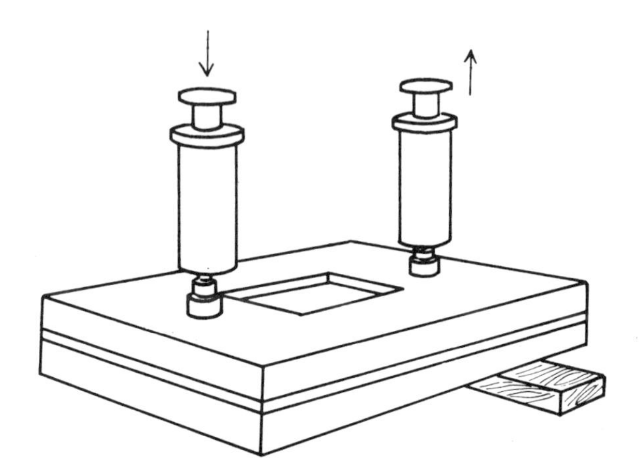

Fig. 7-11. Using two syringes to fill an infrared cell.

forms that can be used for identification. Note also that it is easily possible to distinguish ethyl benzene from m- and p-xylene; these three compounds are not only isomeric but have very similar boiling points and may be confused with one another in gas chromatography (Chapter 12).

A good exercise is to record the spectrum of a number of different compounds containing the carbonyl group, including a ketone (methyl ethyl ketone and acetone could be compared), an aldehyde, an ester (ethyl acetate and ethyl benzoate could be compared), and an amide (such as dimethylformamide). In every case there is a strong band at or near 5.8 μ or 1730 cm^{-1}.

Quantitative Analysis. To show how infrared can be used in quantitative analysis, we shall use mixtures of benzene and isopropanol. Such mixtures, containing some 10-20% of isopropanol, are used in thin-layer chromatography (Chapter 11) and as solvents for nonaqueous acid-base titrations (Chapter 1), as well as for many commercial applications, such as paint removers. (Of course, commercial products may contain other substances. It would be interesting to take a commercial paint remover and examine it by infrared spectroscopy and gas chromatography.)

Prepare standard mixtures with 10%, 20%, and 30% of isopropanol by volume, that is: 1.0 ml isopropanol plus 9.0 ml benzene; 2.0 ml isopropanol plus 8.0 ml benzene, and so on. Run their spectra in the infrared, along with the spectrum of pure benzene. Be sure that all the spectra are run at the same scanning speed and under conditions as nearly identical as possible. Also run an "unknown," provided by the instructor, with a proportion of isopropanol that is in the same range as the standards.

Use the absorption band at 12.2 μ (815 cm^{-1}) to measure the proportion of isopropanol. This band appears in the isopropanol spectrum but not that of benzene. Measure the minimum transmittance at the bottom of this band (call this _t_), and compare it with the transmittance reading obtained by drawing a straight line across the top of the band and reading the same wavelength as the band minimum; let this be t_o; see Fig. 7-12. The _absorbance_ caused by the isopropanol is $\log(t_o/t)$. Plot a graph of this absorbance against the volume percentage of isopropanol; ideally it should be a straight line. Use this graph to find the fraction of isopropanol in the unknown.

For this technique to be effective the values of _t_ should be greater than 0.1 (10%) and less than 80% of t_o. If the values of _t_ are too small, use a narrower cell, or restrict your analysis to smaller proportions of isopropanol. If they are too great, use a wider cell or more concentrated solutions.

Test the applicability of this method to mixtures of isopropanol with liquids other than benzene, for example, toluene, _o_-xylene, or ethyl benzene, which do not absorb significantly at 815 cm^{-1}. The absorbance of the isopropanol should be almost independent of the other liquid or "solvent"; it should depend only on the volume fraction of isopropanol.

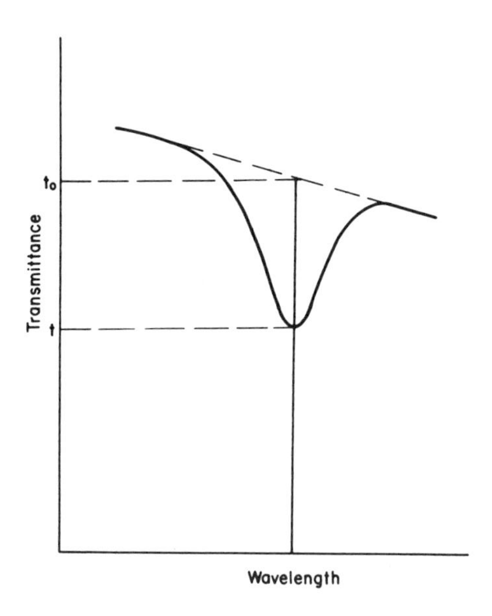

Fig. 7-12. Quantitative analysis by infrared.

QUESTIONS

1. Compounds containing the methylene group, $-CH_2-$, show infrared absorption bands at 3.4 and 6.8 μ. How would these bands be shifted if deuterium atoms were substituted for the atoms of ordinary hydrogen?

2. What bands would you expect phosgene, $COCl_2$, to show in the infrared? Phosgene is formed, among other products, when carbon tetrachloride vapor is heated strongly with air, and it is very toxic. How could you use infrared to measure the concentration of phosgene in air containing the vapor of carbon tetrachloride and its various oxidation products?

3. A compound has the molecular formula C_3H_6O. List the various structural formulas that correspond to this molecular formula, and indicate how you could distinguish between them by infrared spectroscopy.

4. The compounds m-xylene, p-xylene, and ethyl benzene have boiling points very close together and are hard to distinguish by gas chromatography (see Chapter 12, Experiment 12-2). Show how you would distinguish these compounds by their infrared spectrum.

5. A sample of polyethylene film gives interference maxima 40 cm^{-1} apart between 514 and 674 cm^{-1}. A rectangle 3.0 x 4.0 cm weighs 98 mg; the density of polyethylene is 0.92 g/cm^3. What is the refractive index?

6. Why is potassium bromide used for pellets for infrared spectra (Experiment 7-3) rather than KCl or K_2SO_4?

Chapter 8

ATOMIC ABSORPTION SPECTROSCOPY

I. INTRODUCTION

Atomic absorption spectroscopy is a new technique in chemical analysis. It started in 1955 with the work of the Australian scientist, A. Walsh, and his paper entitled "The Application of Atomic Absorption Spectroscopy to Chemical Analysis." Yet, in another way we could say that atomic absorption has been known for a long time. The dark lines in the spectrum of the sun were observed as early as 1802 and were described by Joseph Fraunhofer in 1814. They are called the "Fraunhofer lines," and are caused by the absorption of certain wavelengths of light by atoms in the sun's atmosphere, which is relatively cool compared to the intensely hot center of the sun from which most of the sun's light comes. Of course this explanation was given long after Fraunhofer lived.

Atomic absorption spectroscopy is a method for detecting and measuring chemical elements, particularly metallic elements. Compounds have to be broken down into their atoms, and this is done by spraying them into a hot flame. The arrangement is shown in Fig. 8-1. Light of a certain wavelength, produced by a special kind of lamp, is passed through the long axis of a flat flame and into a spectrophotometer. Meanwhile the sample solution is aspirated into the flame. Before it enters the flame the solution is dispersed into a mist of very fine droplets, which evaporate in the flame to give first the dry salt, then the vapor of the salt. At least a part of this vapor must be dissociated into atoms of the element to be measured.

The atoms absorb light over very narrow wavelength bands, so narrow that one usually calls them "lines." This behavior is characteristic of atoms and molecules in gases. In liquids the molecules are close together and influence each other's energy levels in a random way; the result is an absorption spectrum consisting of broad bands hundreds of angstrom units

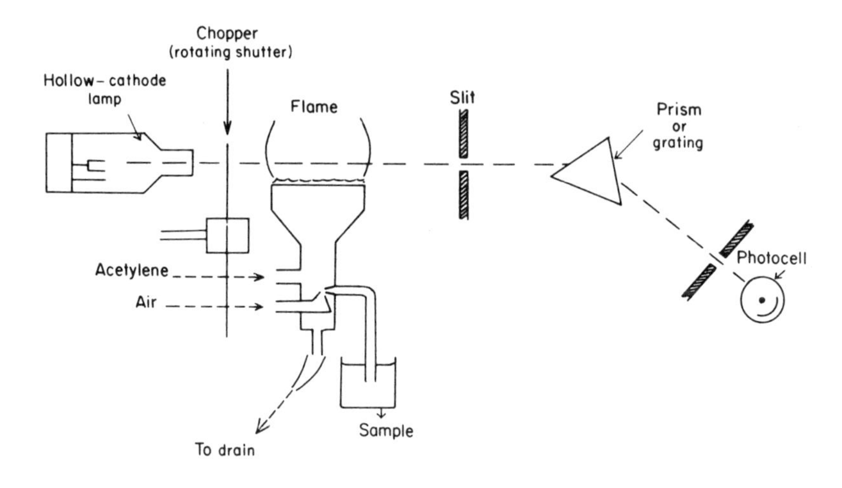

Fig. 8-1. Arrangement for atomic absorption spectroscopy.

wide. In gases, on the other hand, the atoms and molecules are far apart. Each absorbs independently of its neighbors, and the energy levels are sharply defined. This means that the frequencies of absorption are also sharp.

Because the absorption bands are so narrow, the light source must also give very narrow bands. If it gave broad bands or continuous radiation most of the light would pass through the flame without being absorbed. Furthermore, the light source must emit exactly the same wavelengths that are absorbed by the flame. What made atomic absorption spectroscopy possible was the hollow-cathode lamp. In this lamp the cathode is made of, or lined with, the same element that is to be determined in the flame. It gives out light in a spectrum of sharp lines that are characteristic of the element and have exactly the right wavelengths to be absorbed.

II. EQUIPMENT

In this section we shall describe the various individual elements of the apparatus shown in Fig. 8-1.

A. The Hollow-Cathode Lamp

The construction of the lamp is shown in Fig. 8-2. The hollow cathode and the anode wire are mounted side by side in a glass tube that has a quartz window fused into the end; quartz is necessary because nearly all the useful spectral lines are in the ultraviolet. The tube is filled with argon or neon under low pressure. A potential of some 400 volts is placed between the electrodes. When the lamp is switched on the inert gas ionizes, that is to say, the electric field tears electrons out of some of the atoms and converts them into positive ions. These strike the metal surface of the cathode and cause some metal to vaporize. The metal atoms in the vapor become excited, that is, electrons are raised out of their normal orbitals into other orbitals having higher energies. When an electron returns to a lower orbital the excess energy is released as a quantum of light.

It is important to realize that not all the lines emitted by the hollow-cathode lamp are absorbed by atoms in the flame. The atoms in the flame are nearly all in their ground state, the state of minimum energy. Only those emission lines will be absorbed that correspond to transitions from the ground state. Light quanta corresponding to transitions from one excited state to another will not be absorbed by the flame. Emission lines caused by transitions from the ground state of an atom are called resonance lines.

After a hollow-cathode lamp is switched on it takes some time, perhaps as much as an hour, for it to "warm up" and give light of constant intensity. The warm-up period is longer the greater the current. Normal operating currents are about 10-15 milliamperes. The maximum operating current is marked on the lamp itself and should be noted before the lamp is used. The higher the current the more intense the light, but the shorter is the life of the lamp.

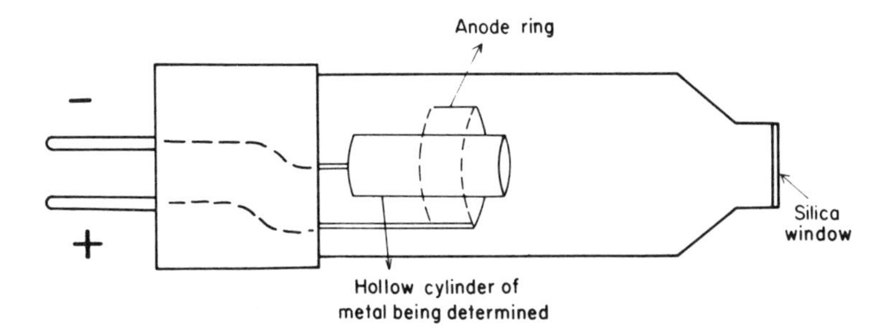

Fig. 8-2. Hollow-cathode lamp.

Normally a separate lamp is used for each element to be determined, and to save time, provision is made in most instruments to have three or more lamps warmed up and running at the same time. One can then exchange one lamp for another with minimum delay. Another way to save time is to use multielement lamps, in which the cathodes are made of alloys of two or more elements or of segments of different pure elements. The drawback to multielement lamps is that the metals do not vaporize equally. In the course of time the most volatile metal settles out and coats the other metals, giving what is effectively a single-element lamp. Therefore, in the long run one does not save money by buying a multielement lamp in place of several single-element lamps.

B. The Chopper

Between the hollow-cathode lamp and the flame is interposed a rotating sector wheel that breaks the steady light from the lamp into an intermittent or pulsating light. This gives a pulsating current in the photocell. Superimposed on this current is a steady current caused by light that is emitted from the flame. Only the pulsating (or alternating) current is amplified and recorded, and in this way, the absorption of light is measured without interference from the light emitted by the flame itself.

C. The Burner

The flame must be hot to produce the necessary atoms. Most often it is fed by acetylene and compressed air. Natural gas and bottled gas (liquefied propane and butane) can be used instead of acetylene, but the flame is not as hot, so the sensitivity is less and the instrument readings depend greatly on the experimental conditions. A hotter flame than the air-acetylene flame is obtained by using nitrous oxide and acetylene. With the nitrous oxide - acetylene flame several elements can be measured, notably aluminum, that cannot be measured in the air-acetylene flame.

The acetylene is called the *fuel*; air or nitrous oxide is called the *support gas*. The kind of burner normally used in atomic absorption spectroscopy is not designed to take oxygen as support gas, and the user is warned that oxygen will cause an explosion. The burner is normally of the type called a *laminar flow burner*. There is another type called a *total consumption burner* which is not often used because it is noisy and hard to control; total-consumption burners do use oxygen, with hydrogen or acetylene, and give very hot flames.

In the laminar flow burner the sample solution is aspirated through a narrow capillary tube and converted into a spray by the air or other support gas; the air and fuel are then mixed and fed together into the flame. Not all the sample solution gets into the flame. Only the finer droplets are carried into the flame; the larger, heavier droplets fall to the bottom of the mixing chamber and are drained out. Thus the conditions of aspiration must be carefully controlled if consistent results are to be obtained.

Once they get into the flame the fine droplets of solution evaporate; in some types of burner the air-fuel mixture can be preheated to evaporate the solvent before the mixture reaches the flame, and this arrangement gives increased sensitivity. Next, the solid particles from the solution evaporate, and then they dissociate into atoms. The processes of evaporation and dissociation are not instantaneous, but proceed at a measurable rate as the gases flow upward through the flame. The population of atoms in the flame is not uniform but varies with the height. Another complication is that some elements form very stable oxides, and the proportion of oxide depends on the position in the flame and also on the proportions of fuel and support gas that were used. The oxide molecules do not absorb light, but only the free atoms.

Thus it is important to be able to adjust the height at which the light beam passes through the flame. This is done by moving the burner up and down. In some models the burner can also be rotated to shorten the light path within the flame and so facilitate readings with more concentrated solutions.

The design of the burner and mixing chamber is of great importance, and much attention has been given to it in commercial atomic absorption spectrometers.

D. The Monochromator

This is exactly the same as an ultraviolet-visible spectrophotometer (Chapter 6) except for the light source and sample cells. It has entrance and exit slits and a grating or prism as dispersing element. The wavelength setting is adjusted to select the wavelength that one wishes to observe. As a rule, the slit width is independently adjustable. Careful wavelength setting and slit adjustment are necessary if the element one is observing has two lines close together but of different intensities, or if another element has a resonance line very close to the line one wishes to observe.

E. Readout and Recorder

In making a measurement of atomic absorption one first sets the meter needle to read zero absorbance or 100 per cent transmittance with no sample in the flame, or with pure water being aspirated into the flame; then one places the cup of sample solution in its holder, lets the solution aspirate into the flame, and reads the absorbance. Ideally the absorbance reading rises to a maximum and remains constant until the sample is used up. In practice the reading always fluctuates a little, and it may be important to know whether the fluctuation is random or whether it indicates a steady drift. For this reason the photocell output is often fed into a recorder. The record of absorbance against time should look like Fig. 8-3a. If it looks like Fig. 8-3b, with the absorbance decreasing as the solution aspirates, it is probable that the capillary needs cleaning.

A recorder is a convenience, but by no means necessary.

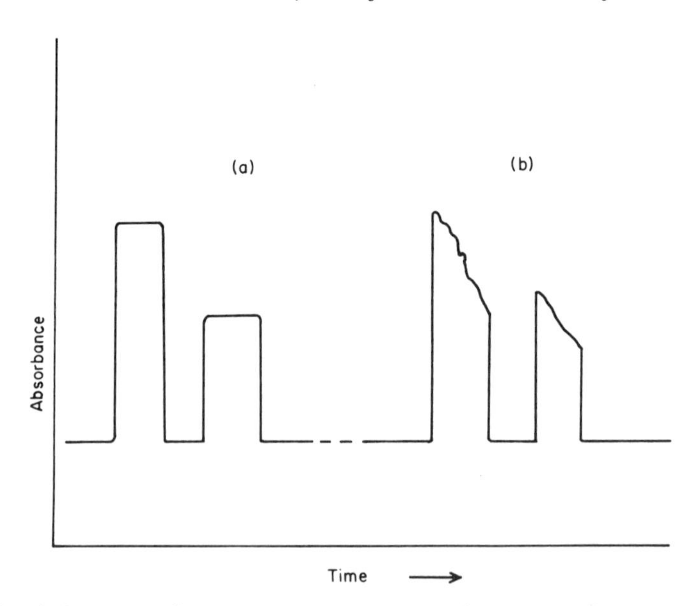

Fig. 8-3. Recorder trace at constant wavelength: (a) burner working well; (b) burner dirty, aspirator needs cleaning.

F. Single and Double Beam Instruments

Most atomic absorption instruments use only a single beam. Once the lamp is warmed up and steady, and when the flame is burning smoothly, the photocell output with no sample, that is, the base line of Fig. 8-3, remains steady. However, one gets increased stability by using a double beam. The chopped beam from the hollow-cathode lamp is divided into two, one part goes through the flame while the other part bypasses it. This is like an ordinary spectrophotometer where one beam passes through the sample and the other passes through the solvent. Double-beam operation in atomic absorption saves time when the lamp is warming up. With a single beam one has to wait until the lamp output becomes steady before taking readings; with a double beam one can start taking readings within a few minutes after turning on the lamp. This is an advantage in a routine analytical laboratory that handles many samples, but it is not so important in university laboratories.

G. Capability for Flame Emission

Most atomic absorption instruments have adjustments that permit one to measure the emission of light by the flame, that is, to use the instrument as a flame photometer. The hollow-cathode lamp and the chopper are not used, the light is received by a sensitive photomultiplier tube, and the output is read on the transmittance scale, with the needle set to read zero transmittance with no sample in the flame.

Flame emission is specially useful for measuring the alkali metals, sodium, potassium, and lithium. This technique is compared with atomic absorption in the next section.

III. ATOMIC ABSORPTION COMPARED WITH FLAME EMISSION

Flame emission spectroscopy, or flame photometry, is an older analytical technique than atomic absorption. It was introduced in a qualitative way by Bunsen and Kirchhoff in 1860 and was used to identify the elements rubidium and cesium. It was developed as a quantitative tool after 1930.

The sample solution is sprayed into a flame and the intensity of emitted light of certain wavelengths is measured. The emission of yellow light by sodium, of wavelength 5890 and 5896 angstroms, and of red light by potassium of wavelength 7665 angstroms, is particularly strong. Instruments for measuring sodium and potassium are on the market. They are simple and inexpensive. No elaborate gratings or prisms are needed to distinguish the sodium and potassium emissions; colored glass filters are sufficient.

Before atomic absorption was developed, much work went into the investigation of flame photometric methods. By using very hot flames like cyanogen and oxygen it was possible to detect and measure thirty or more elements by their flame emission. Now, for all elements but a very few, atomic absorption has proved to be a much more sensitive method. The reason is simple. For an atom to emit light, it must be excited above its ground state; energy must be supplied to move one of its electrons into a high-energy orbital. Even at the high temperature of a flame, only a few atoms, relatively speaking, have enough energy to emit light. Absorption of light, however, is done by atoms in the ground state. Almost all the atoms in the flame are in the ground state and able to absorb light. Atomic absorption is thus a much more sensitive property than flame emission.

The proportion of atoms in the excited state can be calculated from the Boltzmann distribution equation. The energy of the excited state is given by the wavelength or frequency of the resonance line. For the zinc line at 2139 angstroms, the fraction of atoms having enough energy to emit this line in a flame at 3000°K is only 5.6×10^{-10}. Similar fractions can be calculated for other lines and other temperatures.

Atomic absorption has another advantage over flame emission in quantitative analysis. The relation between absorbance and concentration is nearly linear, that is, Beer's law is obeyed over a wide concentration range. This is not true of flame emission. The intensity of light emitted by sodium atoms in a flame is anything but linear in concentration; the graph of intensity against concentration is strongly curved, the emission rising less and less with concentration as the concentration increases. One reason is that the light emitted by atoms in the middle of the flame is absorbed again by ground-state atoms before it can get out of the flame. Flame emission, therefore, needs more careful calibration than atomic absorption, and it is more subject to interferences.

IV. INTERELEMENT EFFECTS IN ATOMIC ABSORPTION

Atomic absorption spectroscopy is remarkably free from interferences, and in general, one element absorbs light without regard to the presence of other elements. Thus, elements can often be determined without having to separate them from other elements, which gives an immense saving in time. However, interelement effects do exist, and separations are necessary for the most precise work. The sensitivity of atomic absorption varies widely from one element to another (see Table 8-1), and in the determination of trace concentrations it is often necessary to concentrate the trace element before it can be measured.

The effects of one element on the absorption by another are complex and not well understood, but certain types of interference can be recognized. Thus phosphate ions interfere with the determination of magnesium and calcium, making the absorption weaker than if phosphate were absent. The reason is almost certainly the formation of stable phosphates of calcium and magnesium that do not easily break up into atoms in the flame. The interference is less at higher flame temperatures, and it may be reduced by adding a salt of lanthanum or thorium to the sample solution; these elements combine preferentially with phosphate ions.

Ideally, atomic absorption spectroscopy should be performed on dilute solutions, 0.01 molar or less, and if the constituent sought is accompanied by other constituents at much higher concentrations, a preliminary separation is desirable. For one thing, high concentrations of salts tend to clog up the burner. If, for reasons of speed and convenience, one wants to pass concentrated salt solutions into the burner, one must be careful about the calibration; one should not rely on calibration curves obtained from dilute solutions but should compare the unknowns with standards made to be as nearly like the unknowns as possible. One can also use the method of standard additions; that is, one runs the unknown, then adds a known proportion of the element sought and runs it again, noting the increase in the signal. To check linearity of response one should always add two or more known proportions of the element, and compare the readings.

When using concentrated salt solutions one must always run plenty of distilled water, or, perhaps, dilute hydrochloric acid through the burner to keep it clean. Acid should be followed by pure water. This is common sense, but operators often overlook these simple precautions.

V. NONAQUEOUS SOLVENT IN ATOMIC ABSORPTION

The samples do not necessarily have to be solutions in water; other solvents can be used if desired. There are two practical reasons for this.

First, one may want to measure the concentration of a metal in a liquid like gasoline or alcohol. Second, one may separate metal ions from one another, and concentrate them from dilute aqueous solutions, by extracting them with organic reagents dissolved in an organic solvent. A reagent that is often used for this purpose is ammonium pyrrolidine dithiocarbamate (see Chapter 6, Sec. IV). Copper, silver, zinc, cadmium, iron, cobalt, nickel, and lead are extracted by this reagent and may be concentrated from large volumes of dilute solutions. The reagent has been used, for example, to determine trace metals in sea water. The organic solvent most used in atomic absorption work is methyl isobutyl ketone.

The choice of organic solvent is important. Most solvents burn with a smoky flame when aspirated into the air-acetylene flame. They burn better if the air flow is increased and the acetylene decreased. In some burners provision is made for a supplementary air flow in addition to the air used to aspirate the sample, and this feature is a distinct advantage when working with organic solvents. Solvents vary considerably and unpredictably in their burning characteristics. Methyl isobutyl ketone burns well, and so does normal octane.

Generally a metal in an organic solvent shows a much higher absorption in the flame than it would if it were dissolved in water. One reason for this effect is the higher flame temperature. Water aspirated into a flame cools it, while a combustible organic solvent makes the flame hotter. Another reason is that many organic solvents have a lower surface tension than water, and the aspirated solution breaks more easily into fine droplets. The solvent may also be less viscous than water and flow more rapidly into the burner.

To correlate absorption with concentration in nonaqueous solvents it is convenient to be able to prepare standard solutions of metal salts in these solvents. Inorganic salts like chlorides or nitrates are insoluble in organic solvents. Compounds of metals with organic acids are soluble to some extent, and such compounds are now available commercially in high purity. Examples are the cyclohexanebutyrates of lead, zinc, cadmium, and copper.

VI. LINE WIDTH AND TEMPERATURE

A spectral "line," whether of absorption or emission, is never a mathematical line with zero width, but always extends over a band of wavelengths, though the band may be small. A factor that always limits the sharpness of a spectral line is the Heisenberg uncertainty principle: the lifetime of the excited state, multiplied by the uncertainty in the energy of the quantum of light emitted or absorbed, must equal Planck's constant, 6.6×10^{-27} erg-second. However, the uncertainty in the wavelength is no more than 10^{-4} angstrom and can be neglected. A more important cause of line broadening is the Döppler effect, and this is related to the temperature of the atoms that are absorbing or emitting light. The hotter they are, the more rapid is the movement of the atoms back and forth along the line of the light beam. An atom that is moving in the same direction as the light will absorb at a slightly lower frequency than an atom at rest. This effect causes the intensity of absorption to vary with wavelength in the manner shown in Fig. 8-4. At 3000°K the width of the line, or band, is about 0.02 angstrom. The higher the temperature the broader the band, and the lower is the intensity of absorption at the center or maximum of the band. If the light source gives narrower lines than the absorption lines - and this is the case with the hollow-cathode lamp - the observed intensity of absorption will fall somewhat with rising flame temperature <u>if</u> the element that is absorbing is completely in the form of its atoms. If the compounds in the flame are only partly dissociated, which is usually the case, then the absorption is likely to increase with increasing flame temperature.

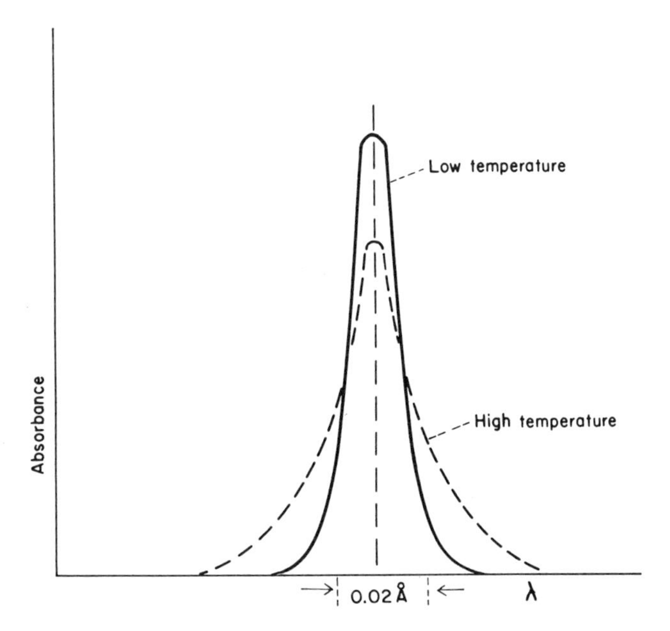

Fig. 8-4. Absorbance vs wavelength in a typical atomic absorption "line."

The Döppler broadening of spectral lines is responsible for the phenomenon of "reversal" seen so often in emission spectroscopy, where the cooler atoms at the edge of an arc or flame absorb light emitted by the hotter atoms in the middle and may cause the line to look hollow, with less intensity in the middle than at the sides. This effect complicates the determination of sodium by flame photometry, as was noted above.

VII. SUMMARY OF FACTORS AFFECTING ATOMIC ABSORPTION INTENSITY

It will now be evident that atomic absorption in flames is affected by many factors, so many that it is a wonder that the technique is as good as it is for quantitative analysis. The absorption intensity is affected by the following factors:

(a) Rate of aspiration of the sample into the flame. This depends on the air pressure, the size of the capillary, and the viscosity of the solution.

(b) Degree of dispersion or atomization of the solution; only the finer drops get into the flame, while the larger ones fall out and escape to the drain. The proportion of fine droplets depends on the air pressure, the temperature of the nozzle where atomization takes place, and the surface tension of the solution.

(c) Flame temperature. This affects the degree of decomposition of compounds into atoms and affects the width of the absorption line.

(d) Position of the light beam within the flame. The atom population changes with height in the flame in a complicated way. If the decomposition into atoms is slow, the atom population rises as one goes upward into the flame until near the top of the flame, where it drops again as the flame gets cooler. If the decomposition is rapid, atom population closely follows the flame temperature, which is a maximum just above the blue cone at the bottom of the flame.

If atmospheric oxygen forms stable oxides, the proportion of atoms bound to oxygen will depend on flame position in a complicated way.

(e) Interelement effects, the most pronounced of which are caused by chemical combinations within the flame. The combination of calcium and phosphate has been mentioned; calcium and aluminum also combine to form calcium aluminate.

(f) Solvent effects; these are complex and only partly related to flame temperature. In aqueous solutions containing much organic matter, for example, in sugar solutions, the organic solute will probably affect the atomic absorption.

It will be seen that considerable care and thought is necessary to get good quantitative results in atomic absorption. Fortunately, standard known solutions can be run with great ease, and the careful worker will check his calibrations several times in the course of a day's work.

VIII. SENSITIVITY OF ATOMIC ABSORPTION FOR VARIOUS ELEMENTS

Most elements in the periodic table can be measured by atomic absorption. Primarily it is used for determining metals. Nonmetals can sometimes be determined indirectly; for example, phosphorus may be measured by its effect in lowering the absorption of magnesium or calcium. Among metals the sensitivity varies very widely. Magnesium, calcium, zinc, and cadmium can be detected and measured in extremely small amounts. Metals that form stable involatile oxides, like lanthanum, zirconium, and thorium, are hard to detect, and for these elements atomic absorption spectroscopy is virtually worthless. Generalizations are risky, however, and published tables showing limits of detection must be accepted with caution. These limits depend greatly on the experimental technique. Flame temperature is an important factor, and certain elements, such as aluminum and silicon, can be measured in the nitrous oxide - acetylene flame but not in the air-acetylene flame.

Table 8-1 lists the analytically useful absorption lines for a number of elements and gives an idea of the relative sensitivity.

IX. COMPARISON WITH OTHER METHODS

Atomic absorption is a fast, simple analytical method that is especially useful when a small number of elements have to be determined in a large

TABLE 8-1

Wavelengths and Sensitivities for Atomic Absorption[a]

Element	Wave-length	Sensi-tivity	Flame	Element	Wave-length	Sensi-tivity	Flame
Ag	3281	A		Mg	2852	AA	
Al	3093	Low	N_2O	Mn	2795	A	
As	1937	Low		Mo	3133	Low	N_2O
Au	2428	C		Na	5890	AA	
Ba	5535	C	N_2O	Ni	2320	B	
Be	2349	A	N_2O	Pb	2170	C	
Bi	2231	Low		Rh	3435	B	
Ca	4227	B		Sb	2176	Low	
Cd	2288	A		Se	1960	Low	
Co	2407	B		Si	2516	Low	N_2O
Cr	3579	B		Sn	2246	Low	N_2O
Cu	3248	A		Sr	4607	A	
Fe	2483	B		Ti	3643	Low	N_2O
In	3039	Low		Tl	2768	Low	
K	7665	A		Zn	2139	AA	
Li	6708	A		Zr	3601	Low	N_2O

[a]Wavelengths are given in angstrom units. Sensitivities are estimated as follows:

AA Below 0.01 ppm
A 0.01-0.05 ppm
B 0.05-0.10 ppm
C 0.1-0.2 ppm
Low More than 0.2 ppm

These are the concentrations in micrograms per milliliter that give 1% absorption (99% transmittance) in a typical commercial instrument. Flame is air-acetylene unless noted.

number of samples with moderate accuracy. It is not a method of high precision, and where high-precision analyses are desired at low concentrations, the ordinary spectrophotometric methods are often better. For example, if iron is to be determined at concentrations around 1-10 parts per million (milligrams/liter), photometric determination with 1,10-phenanthroline (Chapter 6) will give better accuracy than atomic absorption and may be preferred to atomic absorption in practice. If large concentrations of iron must be measured with maximum precision, the preferred method may be titration with standard dichromate or ceric sulfate, or perhaps coulometric titration (Chapter 2). The analytical chemist learns how to choose the best method for the task at hand.

Atomic absorption is not good for multielement analysis, because a separate lamp must be used for every element. To detect and measure a large number of elements simultaneously in the same sample, emission spectroscopy is much more convenient.

X. FLAMELESS ATOMIC ABSORPTION: MERCURY AND OTHER ELEMENTS

The element mercury is easily vaporized without using a flame, and the vapor consists of free atoms. Moreover, lamps are readily available, the so-called mercury-vapor lamps, that give the resonance lines of the mercury spectrum in high intensity. Thus, mercury can be determined by atomic absorption in simple equipment at very low concentrations. Traces of mercury are separated from the sample in an appropriate way; for example, they can be collected from water samples by absorption on silver wire or gauze (silver is higher in the activity scale than mercury and displaces it from its ions and dissolved compounds); the mercury is then vaporized, and the vapor is passed into a long cell made of quartz, which transmits ultraviolet light. The absorption is measured at 2537 angstroms. The method is so sensitive that 10^{-8} gram of mercury can be detected.

Measurement of trace concentrations of mercury is important for two reasons. First, mercury is toxic, and excessive concentrations of mercury are sometimes found in foodstuffs and in air. Second, the presence of traces of mercury in air, in water, and in rocks is used in geochemical prospecting to look for ore deposits. Very often the presence of ores of such metals as lead, zinc, and copper is indicated by accompanying small amounts of mercury, and mercury is easy to detect, even in concentrations so low as to be harmless to the health.

Another kind of "flameless atomic absorption" consists of vaporizing the metals and turning their compounds into atoms in a hollow graphite rod heated electrically to a very high temperature. Fantastic sensitivities are obtained in this way, because the vapor stays in position and is not carried away as it would be in a flame. Equipment for this kind of analysis is available commercially.

EXPERIMENTS

General Instructions. There are many types of atomic absorption equipment on the market. Detailed directions will not be given, because specific controls vary from instrument to instrument. The following general directions will apply to all instruments and all determinations.

Lamp Selection. Select the lamp, or lamps, needed for the element or elements to be determined. Place them in position in the apparatus, first noting the maximum current at which they are to be operated. If the maximum lamp current is marked as 20 mA, one can usually get good results with 15 mA, and the lower current means that the lamp will last much longer. Turn on the filament first, then increase the dc voltage as instructed by the instrument maker. See that the lamp is aligned properly, and that the beam of visible light it emits (this is generally the red light of neon) falls on the entrance slit of the monochromator (or spectrophotometer).

Wavelength Selection, Slit Adjustment. From Table 8-1 select the wavelength to be used in your determinations. Open the slit fairly wide, if the slit is adjustable, and set the wavelength setting of the monochromator to correspond with the desired wavelength. The needle of the instrument should deflect, showing that light is falling on the photocell. (If there is a selector switch for reading absorbance or transmittance, use the transmittance setting for the preliminary adjustment.) If necessary, increase or decrease the "gain" so that the needle reads between 20% and 80% transmittance, roughly.

The wavelength scale of the instrument is seldom highly accurate, and the width of atomic absorption lines is very small. Thus it is very unlikely that you will hit the correct wavelength setting the first time. Move the wavelength control back and forth very slowly and carefully, watching the galvanometer needle, until the "per cent transmittance" reads a maximum. If the needle goes off the scale, reduce the gain. Now cut down the slit width and keep adjusting wavelength and gain until you have the slit width desired, and have the wavelength adjusted as closely as you can. Finally adjust the gain or balance control until the needle reads 100% transmittance. You can switch over to the "absorbance" mode now, if there is a switch to do this. The needle will drift slowly, if the instrument is a single-beam instrument, and will need readjusting from time to time. Make sure you really have a resonance line; other lines are not absorbed.

The proper slit width to use depends on the experiment. The narrower the slit, the less stray light, the less the chance of observing two emission lines at once, and the closer is the obedience to Beer's law. On the other hand, the narrower the slit the less light enters the photocell, the higher must be the "gain" of the amplifier to make the needle read 100% transmittance, and the worse is the "noise" or random variation of the transmittance reading. Experiment 8-2 will show the effects of varying the slit width.

Sample Preparation and Layout. The sample solutions may conveniently be prepared and arranged while the lamp is warming up. It is desirable to have all the samples ready before the flame is lighted.

Prepare a standard solution of the metal that is to be determined, in the form of any convenient salt; make the solution slightly acid, say 0.1 N or less, preferably with hydrochloric acid. This will prevent hydrolysis and adsorption of metal ions on the surface of the bottle used to store it. The concentration of the first standard depends on the sensitivity of the atomic absorption. Usually it will be in the range 10 - 100 ppm (milligrams metal per liter).

Using pipets and small volumetric flasks (50 ml is a convenient size), prepare a series of more dilute standards. If you know in advance what concentrations are needed to give absorbances between 0.1 and 1.0 in your flame, you are fortunate; prepare solutions of these concentrations, taking three or four concentrations to cover the range. Often, however, one does not know exactly what concentrations he is going to need, and one should have on hand a few volumetric flasks, small pipets or graduated cylinders, and distilled water, ready to make further dilutions if necessary. The aim is to have standards with absorbances of about 0.1, 0.2, 0.5, and 0.8 to make a good calibration chart.

Once work has begun, it is desirable to be able to aspirate one solution after another into the flame, with minimum delay or waiting time between samples. We recommend placing the samples in small beakers, say 5 ml capacity (in some instruments there is a clip, or holder, to hold 5-ml cups while they are being aspirated) and laying these on a large piece of paper beside the instrument, with the concentration of each solution written on the paper in front of the beaker; see Fig. 8-5. One of the beakers should contain distilled water, and the aspirator should be rinsed with distilled water at frequent intervals.

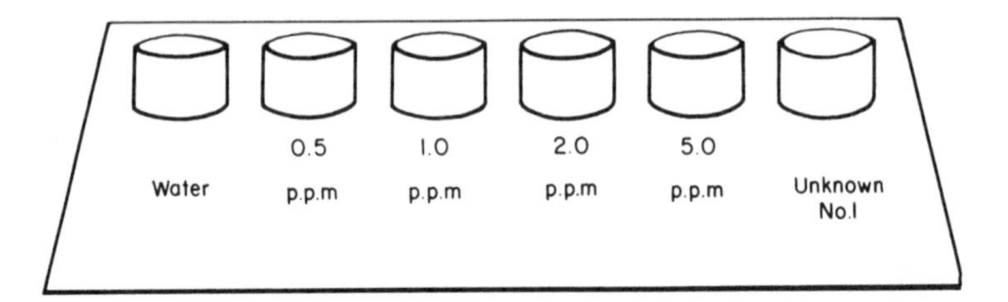

Fig. 8-5. Layout of samples for atomic absorption. The samples are placed in small beakers on a piece of paper with the concentrations written on the paper, as shown.

Flame Light-up and Adjustment. When the lamp is warmed up, the wavelength setting adjusted, and the samples ready, light the flame. Carefully follow the directions supplied with the instrument. Always turn on the air (support gas) first, and open the valve wide enough to give a pressure sufficiently high that the fuel valve will open. This will be between 15 and 20 $lb/in.^2$. All atomic absorption burners have, as a safety feature, a spring-loaded valve that shuts off the fuel if the air pressure is insufficient. So, after turning on the air (or nitrous oxide if this is used), turn on the acetylene to the pressure recommended (usually 3-4 lb) and immediately light the burner.

A handle or crank is provided to raise and lower the burner. Make sure that the burner is at the correct level so that the beam from the hollow-cathode lamp crosses it at the correct hight, usually about 1 cm above the top of the burner. In some determinations the height adjustment is very important.

After the flame is lit, it usually takes a few minutes for it to settle down and become stable. The first readings may be somewhat erratic, but one can use these few minutes to find out if one's standard solutions are in the proper concentration range.

Preheater. Some instruments have provision for heating the mixing chamber, so that the droplets of solution are vaporized to a fine powder before the material enters the flame. Preheating may raise the sensitivity by as much as a factor of 10. If preheating is to be used, allow up to 1/2 hr for the mixing chamber to heat up. Make sure the cooling water (to keep the outside of the chamber, the aspirator nozzle, and the burner head cool) is running according to the manufacturer's instructions.

Taking the Readings. When all is ready, place the standards and samples, one after the other, in position with the inlet tube of the aspirator dipping into them, so that the solutions are sucked up into the flame. Watch to see that the galvanometer needle has come to a constant value before recording the reading. This may take half a minute or more. When you have taken the reading, remove the beaker of solution from the aspirator, let air pass into the aspirator for several seconds, then put the next solution into position. Rinse with water from time to time, and if any of the samples are very concentrated, rinse with water immediately afterward.

Be on the watch for any blocking of the aspirator tube. This is more likely to happen in some instruments than others. It will be made manifest, first, by a falling of the absorbance reading as aspiration proceeds, and second, by an observed slowness of the aspiration. If the tube gets blocked it can be cleaned by using the fine wire _provided for the purpose_; not just any wire, as a wire that is too wide will scratch the tubes and spoil their performance.

Turning Off the Instrument. Note these points: after a series of tests, wash the aspirator thoroughly by drawing several portions of distilled water through it; and, when you turn off the flame, shut the fuel valve first, then let the air pass for a minute or two to cool the burner. Finally, turn off the air, and make sure all valves are tightly closed.

Experiment 8-1

Determination of Zinc: Study of Solution Variables

Solutions and Materials. Hollow-cathode lamp for zinc; standard zinc sulfate solutions, sodium chloride, glycerol, fertilizer, soil (for analysis).

Prepare a standard solution containing 10 ppm of zinc, as follows: Weigh out 0.110 g of zinc sulfate hydrate, $ZnSO_4 \cdot 7H_2O$, dissolve it in water, and add about 1 ml 6 _M_ hydrochloric acid. Transfer the solution to a 250-ml volumetric flask and make up to the mark with distilled water. Mix thoroughly. This solution contains 100 ppm of zinc. Make a solution that is 10 ppm by diluting this solution in the ratio 1:10, for example by taking 25 ml with a pipet and diluting to 250 ml in a volumetric flask. Keep the solution slightly acid by adding 0.5-1 ml 6 _M_ HCl during this dilution, that is, before making up to the mark.

From the 10 ppm standard, prepare other, more dilute solutions, of concentrations 1.0, 2.5, and 5.0 ppm, by diluting the 10-ppm solution with distilled water. Prepare two more solutions of concentration 5 ppm; let one contain 10% of glycerol, approximately, and the other contain 10% of sodium chloride. For example, take two 100-ml volumetric flasks; in one, place 10 g of sodium chloride (weighed roughly); in the other, place 10 ml glycerol (measured roughly); to each flask add 50 ml 10 ppm zinc solution (or 5.0 ml 100 ppm zinc solution); then add distilled water to each, mixing well, and making up to the mark. These solutions are to test the effect of salt and the effect of the viscosity of the solution on the absorbance in the flame.

For "unknowns," use soil or fertilizer. For soil, take 5.0 g of oven-dried soil, shake it well for 5 min with 20 ml 5% ammonium citrate solution to extract the heavy metals, filter, wash with a minimum of water, and make the filtrate and washings up to 50 ml or 100 ml. For fertilizer, take 2.0 g, warm in a beaker with 2 ml concentrated hydrochloric acid and 10 ml of water, warm until the material has disintegrated, then dilute, filter into a 250-ml volumetric flask, and make to the mark with distilled water. NOTE: if the proportions suggested do not give a convenient absorbance reading, use other proportions; for example, if there is so much zinc in the fertilizer that the absorbance reading is above 1.0, or the needle goes off the scale, dilute the solution quantitatively with more distilled water.

When all the solutions are ready, run them in turn, as described on page 234, starting with the standards. Run at least four standards, choosing them so that their absorbances cover the range 0.1 to 0.8. You might start with solutions of 1.0, 2.5, 5.0, and 10.0 ppm; if the 10-ppm absorbance goes over 1, make a more dilute standard, say 0.5 ppm, and measure its absorbance in the flame. The aim is to plot a calibration curve, absorbance against concentration, with absorbances ranging up to 0.8 - 1.0.

You have two extra solutions with 5 ppm zinc; one contains glycerol, which increases the viscosity and makes the solution feed into the flame more slowly. The result is that the concentration of zinc atoms in the flame is smaller than normal, and one might think that the solution contained less zinc than it actually does. This test, and the test with 10% sodium chloride present, will show that one must use care and common sense in interpreting atomic absorption readings.

Run the "unknown" solutions. In addition, run samples of distilled water and of your city tap water. The distilled water should not contain any zinc, but it is wise to test the distilled water when working with such low concentrations. The city tap water may contain enough zinc to be detected, for traces of zinc are very widespread in nature.

Using your calibration curve, calculate the proportion of zinc in the soil or the fertilizer. Express your result in parts per million (milligrams per kilogram or micrograms per gram).

Experiment 8-2

Determination of Iron: Effect of Slit Width

Solutions and Materials. Hollow-cathode lamp for iron, standard iron solutions, unknowns (fertilizer, soil, wheat flour).

Prepare a "master" standard solution containing 200 ppm of iron. This may be made from the salt ferrous ammonium sulfate, which is obtainable in high purity and contains almost exactly one-seventh its weight of iron. Weigh 0.350 g of ferrous ammonium sulfate, dissolve it in water, add about 1 ml of 6 M hydrochloric or sulfuric acid to prevent hydrolysis (this is important; ferrous salt solutions rapidly oxidize in air and precipitate yellow hydrous ferric oxide if acid is not added), transfer to a 250-ml volumetric flask, and make up to the mark.

From this solution prepare more dilute standards, for example 100, 50, 25, 10, and 5 ppm. Note that the sensitivity of atomic absorption for iron is at least a factor of 10 less than the sensitivity for zinc.

The "unknown" solutions of soil and fertilizer can be prepared as they were in Experiment 8-1. The iron content of these materials varies very widely, and the proper size sample and the proper dilution must be found by trial. Wheat flour may contain about 50 ppm of iron, more or less. To measure the iron in wheat flour the flour must be "wet-ashed" (see Experiment 6-4). Take a 2-g sample, place in a 100-ml beaker, add 10 - 20 ml concentrated nitric acid, and heat gently until the acid boils. Then stir in, carefully, about 2 ml concentrated sulfuric acid. Heat on a hot plate or over a small flame until the nitric acid has all evaporated and dense white fumes of sulfuric acid start to appear. By now the black carbonaceous matter that was first formed should have disappeared, and the liquid should be almost colorless. There may be a white solid present, but there should be no dark material. If it is still dark, keep heating, and add a few drops of nitric acid, stirring well. If nitric acid does not clear the solution, add a few drops of perchloric acid. Evaporate the clear solution to about 1 ml, cool, add water, and transfer to a 10-ml volumetric flask. Wash any white precipitate into the flask; this may contain ferric sulfate. Make up to the mark with distilled water, mix well. Use this solution for the atomic absorption reading.

Calibration: Effect of Slit Width. Reference to Table 8-1 shows that there are two absorption lines of iron whose wavelengths are very close together, 2483.3 Å and 2488.1 Å. They are absorbed to different extents in the flame, and if the instrument reads the intensities of both lines at once, we shall find negative deviations from Beer's law (Chapter 5). By using a very narrow slit we can observe just one line at a time and obtain a graph of absorbance against concentration of iron that is nearly linear. The penalty we pay is that very little energy enters the photocell, and as a result the galvanometer needle does not remain steady but fluctuates erratically back and forth. We call this phenomenon "noise." If we use a wide slit, we read both lines at once, and the graph of absorbance against concentration is not linear but flattens out at high concentrations. Examples of calibration graphs at narrow and wide slit widths are shown in Fig. 8-6.

The first part of this experiment, therefore, is to construct calibration graphs for narrow and wide slits. If your instrument has a fixed slit, you will have to omit this part of the experiment.

In the second part of the experiment, use the calibration graphs to measure the concentration of iron in your unknown. Report the result in parts per million.

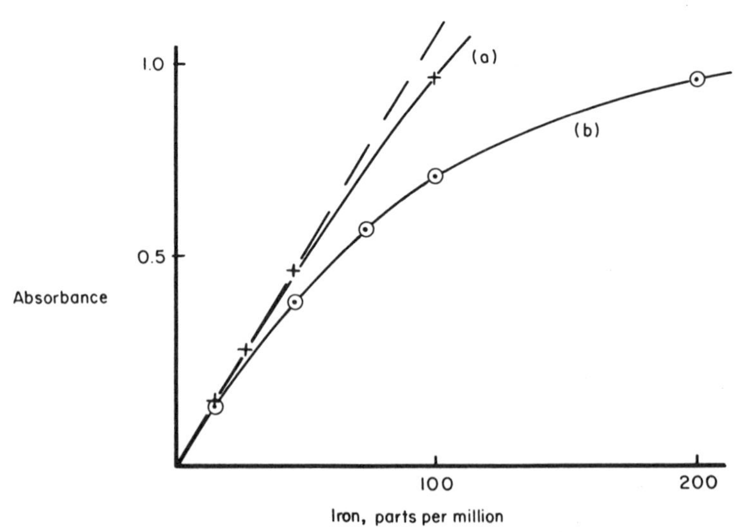

Fig. 8-6. Calibration curves for iron at 2483 Å: (a) narrow slit; (b) wide slit that also admits the line at 2488 Å.

QUESTIONS

1. Why is atomic absorption much more sensitive than flame emission spectroscopy for most elements?

2. Why does the sensitivity of atomic absorption vary greatly from one element to another?

3. If we examine the emission spectrum of a hollow-cathode lamp we find that it consists of a large number of lines, yet only a few of these lines are absorbed by a flame containing the element that is in the lamp. Why is this?

4. The effect of flame temperature on absorbance is generally small. It may be positive or negative, and in some cases the absorbance increases considerably with increasing flame temperature. Explain why this is so.

5. If a substance raises the viscosity of a solution, what effect will it have on the readings obtained in atomic absorption spectroscopy?

6. A solution containing a certain element gives absorbance 0.45. For calibration by the standard addition method, 5.0 ml of the test solution are mixed with 1.0 ml of a standard that contains 100 mg/l. of the element sought. The absorbance is now 0.80. What is the concentration of the test solution?

7. How many volumes of air are needed to burn one volume of acetylene? How many volumes of nitrous oxide, N_2O, are needed?

8. What data would you need to make an estimate of the temperatures of the air-acetylene and nitrous oxide - acetylene flames?

9. The most sensitive absorption line of nickel has wavelength 2320.0 Å. However, there is a non-resonance line at 2319.8 Å and the two cannot be resolved. Many workers therefore prefer to use the line at 3414.8 Å, which is only a tenth as sensitive. Discuss the reasoning behind this choice.

Chapter 9

CHROMATOGRAPHY

I. INTRODUCTION

Chromatography is the name given to a family of methods for separating complex mixtures and determining the amounts of their constituents. It is one of the most important areas of chemical analysis and is much used in organic chemistry and biochemistry.

The word "chromatography" means "color writing." It was introduced by Mikhail Tswett, the Russian botanist who developed chromatography in the year 1905. He separated the colored substances in green leaves, the two chlorophylls (green), carotin (red) and xanthophyll (yellow), by extracting the dried leaves with petroleum ether, then pouring the dark-colored extract into a vertical glass tube packed with powdered calcium carbonate. The colored substances were absorbed by the calcium carbonate and remained as a dark green band at the top of the tube. Then he poured more petroleum ether down the tube. The colored substances were gradually washed down the tube, but at different rates; orange-red carotin moved the fastest and formed a characteristically colored band in the white calcium carbonate when it moved ahead of the others. Above this band, moving more slowly, appeared a yellow band of xanthophyll, then two green bands of chlorophyll. The pattern of colored zones he called a "chromatogram." He found that the zones moved faster if he mixed a little alcohol with the petroleum ether. As he continued to pass the solvent, the colored substances came out of the bottom of the column, one by one, the carotin emerging first; they could be collected in separate vessels (see Fig. 9-1).

Of course the technique is not restricted to colored substances. Colorless substances can be separated too, and we can make use of the separation if we have a way to detect the substances when they come out of the column. Chemical color-producing reagents can be applied to the solution, or various physical properties can be used, such as changes in refractive index or electrical conductivity. Today, one seldom looks at colored zones formed within the tube; more commonly one examines the solution as it comes out of the tube. Then it is not necessary for the solid absorb-

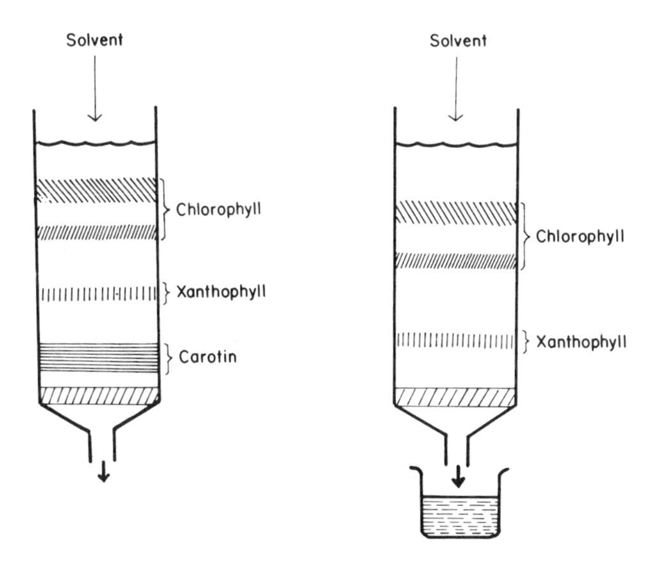

Fig. 9-1. Tswett's experiments on leaf pigments.

ent in the tube to be white, nor is it necessary to make the tube of glass. In spite of the changes that have been made in the technique, we still call the method "chromatography."

The essentials of chromatography are a <u>fixed</u> or <u>stationary phase</u>, usually a finely divided granular solid; a <u>moving phase</u>, which can be a gas or a liquid; and the <u>substrates</u> or <u>solutes</u>, two or more substances that are to be separated. The substrates are first dissolved in the solvent or introduced as vapors. When they meet the stationary phase they are partitioned between the two phases, the stationary and the moving phase. Their molecules move rapidly back and forth between these two phases. The more time they spend in the stationary phase, the more slowly they move. If they are not held by the stationary phase at all, they move along with the moving phase at the same speed as the molecules of the moving phase.

As a rule the stationary phase is confined within a tube or <u>column</u>, as it was in Tswett's pioneer work. The column is not necessary, however. The stationary phase may be the fibers of filter paper (paper chromatography) or it may be an absorbent powder coated on the surface of a glass plate (thin-layer chromatography).

We noted that the moving phase can be a liquid or a gas. The commonest kind of chromatography, that is, the kind by which most analyses are actually performed, is gas chromatography, in which the samples are carried as vapors in a stream of an inert gas like helium or nitrogen, the carrier gas. The fixed or stationary phase in gas chromatography is generally a granular solid which is coated with a thin film of a nonvolatile liquid, and it is this liquid film that dissolves the substrate vapors to different degrees and makes the chromatographic separation possible. This kind of chromatography is often called gas-liquid chromatography.

Though the experimental arrangements used in different kinds of chromatography vary widely, all forms of chromatography have this feature in common, they depend on differences in the partitioning, or distribution, of the different substrates between the two phases, the moving and the stationary phase. Thus there is a general theory of chromatography that is valid for all kinds of chromatography. We shall develop the theory in an elementary way, considering first the case of elution chromatography. (Another kind of chromatography called displacement chromatography is mentioned in Sec. VIII.) Here, a very small amount of substrate is introduced at the entrance to a column (or at the starting line of a thin-layer plate) and is made to move down the column (or along the plate) by passing the moving phase, the solvent or carrier gas. To fix our ideas, let us think of a column and a moving liquid; however, the same reasoning will apply to a thin layer or to a moving gas.

Originally the substrate is concentrated in a very thin, sharp zone at the top of the column. As solvent is passed, the substrate moves, and at the same time the zone becomes diffuse. Some of the substrate molecules move faster than the average, while others move more slowly. Drawing a graph of substrate concentration against position in the column, we get Fig. 9-2. A graph of substrate concentration in the liquid coming out of the column plotted against the volume of liquid passed looks like Fig. 9-3. Two different substrates are shown, one moving twice as fast as the other. We note that the faster moving substrate comes out of the column first (a

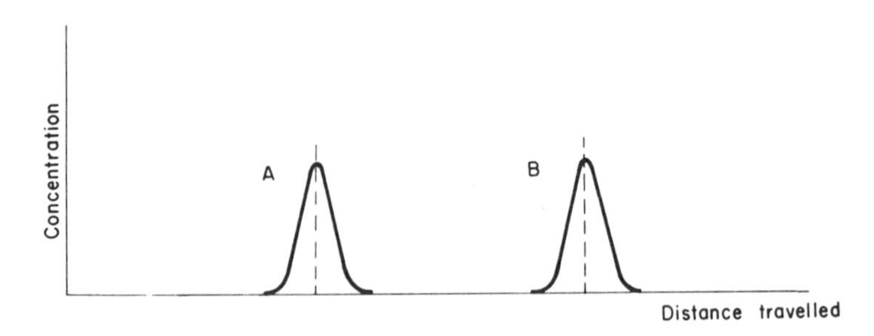

Fig. 9-2. Concentrations of the substrates A and B plotted against the distance along the column at a fixed time.

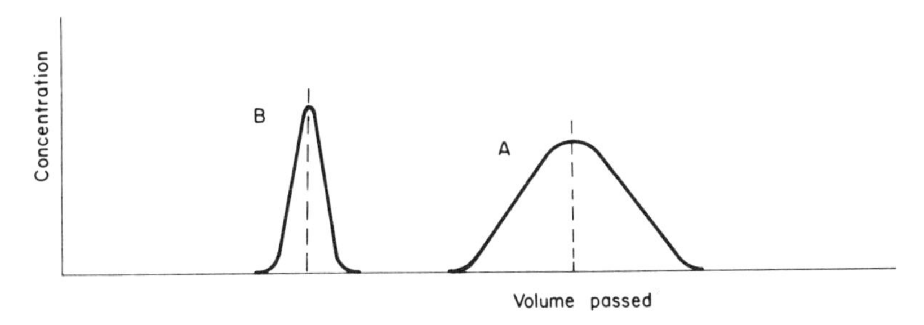

Fig. 9-3. Concentration of substrates A and B plotted against the volume passed. Concentrations are measured in the solution at the point where it flows out of the column.

smaller volume of solvent is needed to get it out, or, as we say, to elute it) and, since it has spent less time in the column than the slower-moving substrate, it has spread less; the zone or "peak" is sharper on emergence from the column than that of the other substrate.

Thus there are two variables in elution chromatography: zone migration and zone spreading. We shall consider migration first, and for simplicity we shall consider only one substrate.

II. ZONE MIGRATION

In their travel down the column, substrate molecules are slowed down, relative to the molecules of solvent or mobile phase, because they spend part of their time in the fixed phase or adsorbed on its surface. The more they are attracted to the fixed phase the slower they move. We can find a quantitative relationship if we consider that there is a constant partition ratio for the substrate between the two phases; that is, the ratio (concentration of substrate in fixed phase) : (concentration of substrate in moving phase) is a constant, regardless of what the concentrations are individually.

Let Fig. 9-4 represent a part of a chromatographic column. The column is shown horizontal for convenience in drawing, and the fixed phase and moving phase are shown as separate bands, whose thicknesses indicate the relative volumes of the two phases. (Normally, more than half

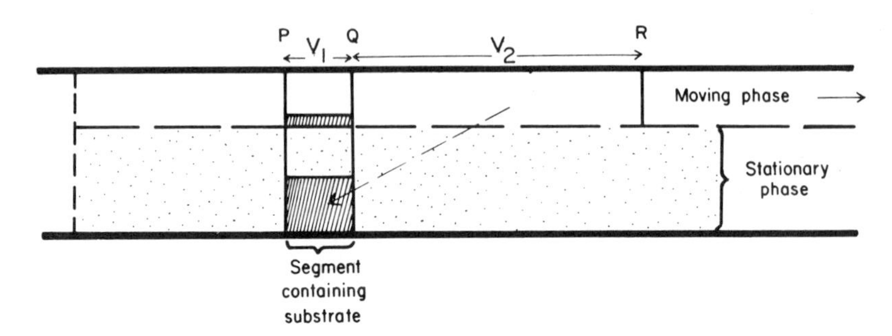

Fig. 9-4. Movement of substrate along a column; relation of volume of moving phase to the distribution ratio.

the space inside the column is occupied by the fixed phase.) The segment PQ represents a zone occupied by the moving substrate. The shaded areas in PQ represent the amounts of substrate in each of the two phases. We have shown four times as much substrate in the fixed phase as in the moving phase; that is, only one-fifth of the substrate is actually moving at a given moment. The moving part of the substrate is contained in the solvent volume V_1. Now, the four-fifths of the substrate that is now in the fixed phase was carried there by the solvent; it was in the moving phase originally, and it occupied a volume V_2, the volume of solvent (moving phase) between Q and R. Thus, in the time it takes for substrate molecules to travel from P to Q, the solvent (moving phase) molecules have traveled from P to R.

The ratio V_2/V_1 equals the ratio of substrate in the fixed phase to substrate in the moving phase in the segment PQ. Let us call this ratio the column distribution ratio, k. Then,

$$\frac{\text{solvent velocity}}{\text{substrate velocity}} = \frac{V_1 + V_2}{V_1} = k \qquad (9\text{-}1)$$

This is a very important relationship. It is important to note that k is not the same as the "partition coefficient" normally used, as, for example, in the Nernst distribution law for partition between two solvents. It is not the ratio of two concentrations, but the ratio of two amounts, the amounts of substrate in the fixed and moving phases in a given column segment. The value of k depends not only on the chemical characteristics of the substrate and the two phases, but on the degree of packing of the column, the fraction of the space in the column that is actually occupied by the stationary phase.

In column chromatography we measure the volumes of solvent that must be passed to elute (wash out of the column) the various substrates, or what amounts to the same thing, we measure the times that elapse between the injection of the substrate at one end of the column and its emergence at the other, when the moving phase is passed at a constant rate. The relation between solvent (or moving phase) velocity and substrate velocity is easily adapted to these measurements:

$$\text{(elution volume)} = \text{(void column volume)} \times (k + 1) \qquad (9\text{-}2)$$

where the "elution volume" is the volume of moving phase that enters or leaves the column between the injection of the substrate and its elution - the volume of solvent that is needed to wash the substrate out of the column - and the "void volume" is the volume of moving phase in the column, that is, the volume in between the particles of fixed phase. The term "1" in the factor $(k + 1)$ arises because one has to pass one void column volume simply to sweep out the solvent that is already in the column; this volume would have to be passed even if the substrate were not absorbed at all. Another way to express the relation is that the volume of moving phase that must pass by the substrate zone in order to move it from one end of the column to the other is k void column volumes.

Frequently the distribution ratio between the two phases is stated in this form:

$$\text{Distribution ratio, } D = \frac{\text{(quantity of substrate per gram of fixed phase)}}{\text{(quantity of substrate per milliliter of mobile phase)}}$$

This distribution ratio can be obtained from experiments in which a known weight of solid absorbent (fixed phase) is placed in a flask with a known volume of solvent (mobile phase), some substrate is added, and the mixture is shaken until equilibrium is reached and the substrate in distributed between the two phases. Many such measurements have been made, particularly for ion exchange distributions. We calculate elution volumes from values of D as follows:

Let the column contain w grams of solid absorbent and V_i milliliters of mobile phase. The symbol V_i stands for "interstitial volume." Then the column distribution ratio k is

$$k = \frac{\text{(substrate in } w \text{ grams of absorbent)}}{\text{(substrate in } V_i \text{ milliliters of solvent)}} = D\frac{w}{V_i}$$

The elution volume V_{el} is

$$V_{el} = V_i(k + 1) = Dw + V_i \qquad (9\text{-}3)$$

If D and w are known, and if V_{el} is much larger than V_i, it is not necessary to know V_i with much accuracy in order to calculate the elution volume V_{el}. This is fortunate, because it is sometimes difficult to evaluate the void volume V_i. A rough rule of thumb for a column packed with small spheres is that V_i is about 40 per cent of the total, or bulk volume of the column; the "bulk volume" is the volume that the absorbent would appear to occupy if it were poured into a graduated cylinder or buret.

In practical chromatography it is desirable to have the zones or bands of the different components well separated; that is, the distribution coefficients, D or k, should be as different as possible from one another. At the same time we do not want the distribution coefficients to be too large, or we shall have to pass inconveniently large amounts of solvent and wait an inconveniently long time. We can get good separations without large differences in distribution ratios if the zones are sharp. Zone spreading, therefore, is a factor in chromatography that is just as important as the rate of zone migration.

III. ZONE SPREADING

A small amount of substrate, placed at the top of a chromatographic column or injected into the flowing liquid or gas as it enters the column, forms at first a thin disk, a very sharp, narrow zone across the diameter of the column. As it moves along the column the zone widens and becomes more diffuse. There are basically three reasons why this happens. The first is called eddy diffusion. The molecules of substrate do not necessarily choose the most direct path between the granules of the fixed phase. When they meet a granule, some of them will go one way around it, some another. The situation is shown in Fig. 9-5. A few molecules will indeed take the shortest route, but this will be a matter of chance. Others will repeatedly take the long way around the granules, their path will be very tortuous, and they will be left behind the others as the molecules progress down the column. This, too, is a matter of chance. Taking all the molecules of substrate together, they will have a distribution of velocities, grouped around an average value, and the graph of concentration against volume of moving phase will be a bell-shaped curve, as shown in Fig. 9-6. Ideally the shape of this graph is that of the Gauss "normal curve of error."

The symmetrical shape can be predicted from the "random walk" model of zone spreading. In the middle of the zone, where the substrate concentration is maximum, the substrate is moving at the rate predicted from the distribution ratio in the way described in the last section. Superimposed

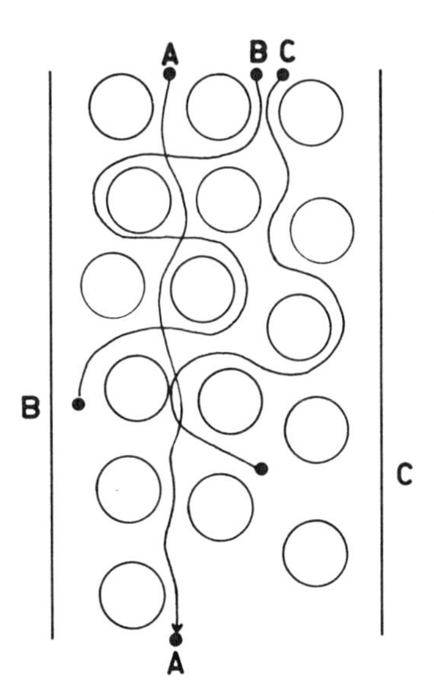

Fig. 9-5. "Eddy diffusion," one cause of band broadening. Molecule A moves ahead of the average, while molecule B lags behind.

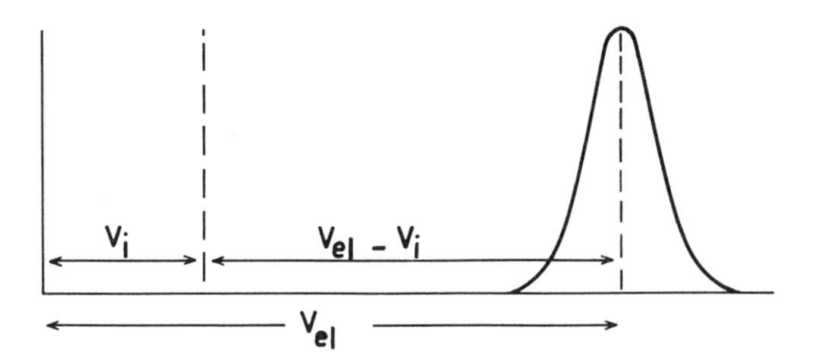

Fig. 9-6. Chromatographic peak. V_i is the interstitial, or void, volume; V_{e1} is the elution volume.

on the movement of the zone as a whole is the random motion that causes spreading. We take as our point of reference the center of the zone and consider the movements of molecules back and forth from this point of reference.

The idea of the one-dimensional "random walk" is this. Imagine that a man comes out of a door in a narrow street. He then takes a step either to the left or the right; he is just as likely to go one way as the other.

Then he takes another step. Again the chances of his going one way are the same as those of his going the other. After two steps, therefore, he may be two steps to the right, he may be two steps to the left, or he may be back where he started. (We assume, of course, that every step that he takes is of equal length.) He is twice as likely to be back where he started than he is to be two steps to the left or two steps to the right, for there are two ways in which he could get back to his starting point: left - right and right - left. Figure 9-7 indicates the idea.

After a large number of steps the probability is still greatest that the man will be back at his starting point (if the number of steps were even) or one step from it (if the number of steps were odd). There is a small but measurable probability that every step he took was in the same direction, either to the right or to the left, and there are intermediate probabilities for distances in between, as indicated by Fig. 9-8.

A way to visualize the probabilities of different events is to consider a large number of similar particles acting in different ways according to the laws of chance. Thus we could imagine a thousand men standing in single file along a straight line, then given the command to execute the "random walk" at right angles to the starting line. After each man has made many

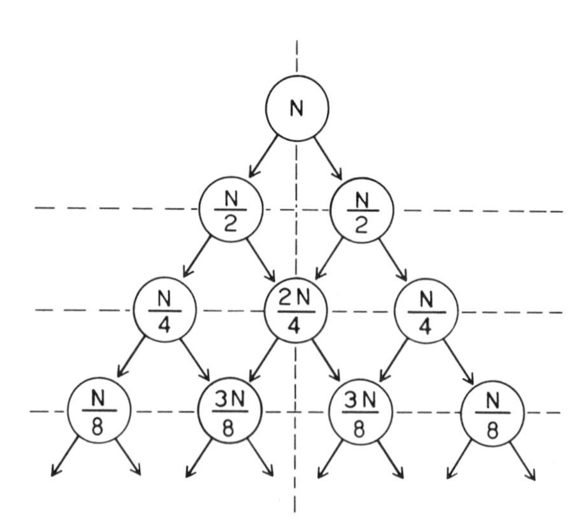

Fig. 9-7. The "random walk." Molecules start from the center line, represented at the top of the diagram; each horizontal line represents another step in the "walk." After four steps, molecules are in five different places; the numbers in each location are in the ratio 1:4:6:4:1.

steps we give the command to stop. Then we count the number of men at different numbers of steps from the starting line. Plotting the number of men against distance from the starting line we get a graph like that shown in Fig. 9-8, where the ordinate is the number of men per unit distance, and the abscissa is the distance x from the starting line. For an infinite number of men we would get a smooth symmetrical curve.

The mathematical treatment of the "random walk" is somewhat abstruse. It involves ideas that cannot be explained in a few words, and in this chapter it will suffice to give the result. If $p(x)$ is the probability of finding a particle at a distance x from its starting point,

$$P(x) = \frac{1}{\sigma} \frac{1}{\sqrt{2\pi}} \exp\left(-\frac{x^2}{2\sigma^2}\right) \tag{9-4}$$

$p(x)\,dx$ is the fraction of particles that will be found between the distances x and $x + dx$ from the central point. Since the particle has to be somewhere along the line of movement, the integral $\int_{-\infty}^{+\infty} p(x)\,dx$ must be 1, and so it is.

The quantity σ is called the <u>standard</u> <u>deviation</u>. This is a very important quantity in statistics. It is a measure of the degree of spreading, or the random error of a set of experimental results. In the "random walk," $\sigma^2 = M\ell^2$, where M is the number of steps and ℓ is the length of each step. The significance of σ in chromatography may be seen from Fig. 9-8. The shape of the ideal chromatographic elution curve is given by this modification of equation (9-4):

$$C = C_{max} \exp\left(-\frac{x^2}{2\sigma^2}\right) \tag{9-5}$$

where C is the concentration of solute (or substrate) at a volume x from the elution volume at the peak maximum, and C_{max} is the maximum concentration of solute in the effluent. By differentiating C with respect to x it will readily be seen that the concentration-volume curve has its maximum slope when $x = \pm\sigma$. At this point, $C = C_{max}\, e^{-1/2} = 0.606\, C_{max}$. A convenient way to find the standard deviation σ from a chromatographic curve is simply to draw tangents to the curve at its steepest points. This is easier than it sounds, because the steepest parts of the curve are almost straight over a considerable distance, and one simply places a ruler against the sides of the curve. The tangents intersect the base line ($C = O$) at distances -2σ and $+2\sigma$ from the center line. That is, the base of the triangle formed by the base line and the two tangents has length 4σ.

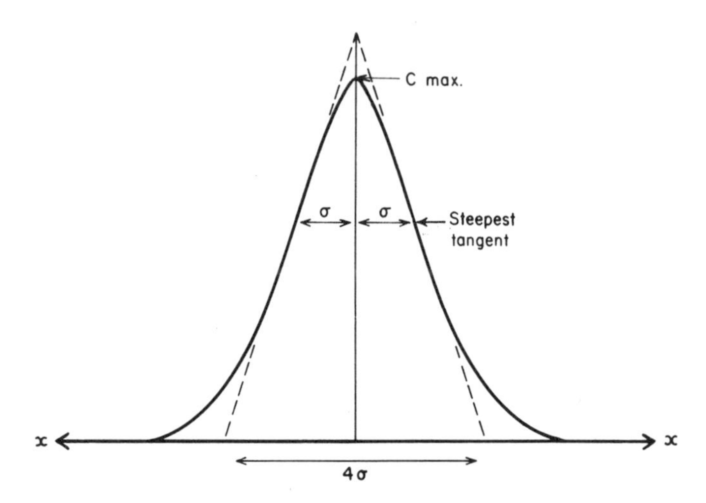

Fig. 9-8. The Gauss error curve, ideal shape of a chromatographic peak. Note that the base of the triangle formed by drawing tangents at the steepest points has a width of four standard deviations.

The maximum concentration C_{max} equals $m/\sqrt{2\pi\sigma^2}$, where m is the mass of substrate, or solute, that produces the peak. We see that in ideal, symmetrical, Gaussian chromatographic curves the height of the peak is proportional to the amount of material and can be used directly for its quantitative determination. However, we cannot compare the height of the peak of one substance with the height of the peak of another substance and say that, if the heights are equal, the amounts are equal. The peak height also depends on the standard deviation; the broader the peak, the less is its height; the standard deviations given by two different substances are in general not the same. The area under the concentration-volume curve is a better measure of the amount of solute. If the curves are not symmetrical one should always use the area, not the peak height, for quantitative measurement of the amount of solute.

IV. THE "THEORETICAL PLATE" CONCEPT: RESOLVING POWER OF CHROMATOGRAPHIC COLUMNS

The spreading of zones and the power of a column to resolve, or separate, mixtures of solutes is often expressed in terms of a concept called the "theoretical plate." The idea comes from fractional distillation, where the condensed liquid percolates back into the boiling flask by flowing over a series of disks or "plates" that permit the liquid to come to equilibrium with the vapor that rises up the column. The more "plates" and the better the contact between liquid and vapor, the better is the separation of mixtures of components. The same considerations apply to a chromatographic column. Of course a chromatographic column has no "plates" in it in a physical sense. The "theoretical plate" is a purely imaginary concept, and the expressions "plate height" and "plate number" are nothing more than parameters designed to express the resolving power of the column. The "theoretical plate" is an imaginary zone or segment of the column in which the composition of the mobile phase flowing out of the segment is in equilibrium with the average composition of the fixed phase within the segment. Note the word "average." In reality the composition of the fixed phase varies continuously along the length of the column. The moving phase is never in equilibrium with the fixed phase; its composition always lags behind the composition that would be in equilibrium with the fixed phase through which it is passing. However, the more closely it approaches equilibrium the shorter is the "theoretical plate" and the sharper are the chromatographic bands.

The number of theoretical plates in a column is commonly called N. It is related to the standard deviation, and hence to band width, as follows:

$$N = \left(\frac{V_{max}}{\sigma}\right)^2 \tag{9-6}$$

where V_{max} is the volume of effluent at the peak of the band corrected for the void volume, that is, $V_{max} = V_{el} - V_i$ (Fig. 9-6).

The plate height, or "height equivalent of the theoretical plate," is simply the column length divided by the number of plates, N. It is given the symbol H.

V. FACTORS INFLUENCING PLATE HEIGHT

We have already indicated one effect that causes zone spreading and thus affects the plate height; this is the "eddy diffusion" or random movement of solute around the particles, as shown in Fig. 9-5. Two other factors that influence plate height are these.

A. Longitudinal Diffusion

Within the moving phase the molecules of substrate diffuse back and forth along the axis of movement. This effect would cause the zone to spread, even if the granules of fixed phase were not there. Diffusion in a liquid is very slow, and if the moving phase is a liquid, the back-and-forth (or longitudinal) diffusion is so slow as to be insignificant unless the rate of liquid flow is extremely small. In gas chromatography, however, longitudinal diffusion in the carrier gas may be quite appreciable.

B. Mass Transport

This is the movement of substrate from one phase to another, from the moving to the fixed phase and back again. Some substrate molecules penetrate the granules of fixed phase for only a short distance and then come out again, while others diffuse inside the granules and stay there for a relatively long time. We have already noted that the slower the movement of substrate between one phase and another, the longer is the "theoretical plate."

The in-and-out movement of substrate molecules between the moving phase and the fixed phase is controlled by two rates of diffusion, one within the particles of the fixed phase themselves ("particle diffusion") and one across the layer of solution or carrier gas that is in contact with the fixed phase and to some extent immobilized by it ("film diffusion"). Sometimes one effect dominates and sometimes the other. In general, particle diffusion is more important at slower flow rates. When particle diffusion dominates and controls the plate height, the plate height is directly proportional to the square of the particle radius. In ion-exchange chromatography, where the whole volume of the particles of fixed phase takes part, it is advantageous to use small particles. In gas chromatography, where the active "fixed phase" is usually a film of liquid covering the surface of a solid granule, the radius of the granule is not very important, but the thickness of the liquid coating controls "particle diffusion" and should be as small as possible.

Eddy diffusion, longitudinal diffusion, and mass transport are affected in different ways by the rate of flow of the moving phase. The contribution of eddy diffusion to plate height is not affected by the flow rate. Longitudinal diffusion is less important the greater the flow rate, whereas mass transport is more important the faster the flow. These relations are summarized by the Van Deemter equation:

$$\text{Plate height,}\quad H = A + Bu + \frac{C}{u} \qquad (9\text{-}7)$$

Here, A, B, and C are constants for a particular column, moving phase, and substrate, and u is the flow rate. The "A" term shows the effect of eddy diffusion, the "B" term the effect of mass transport, and the "C" term the effect of longitudinal diffusion. The corresponding graph of H against the flow rate u is shown in Fig. 9-9. This relation is useful in gas chromatography, where all three factors influence plate height, and there is a certain rate of flow of the carrier gas that gives the sharpest bands. In liquid chromatography the "C" term is usually insignificant.

It remains to mention one more effect that limits the sharpness of chromatographic bands; this is the way in which the column is packed. It is quite difficult to pack an absorbent, or fixed phase, into a column in such an even manner that the moving phase does not flow down one side of the column a little more rapidly than down the other. Particularly this is true in liquid chromatography. If the substrates are colored ("chromatography" in the original sense of the word, used by Tswett) the zones will often be seen to be slightly irregular and deformed as they move down the column.

VI. RESOLUTION: SUMMARY

The resolution of a chromatographic column is its ability to separate or distinguish two or more substrates. It depends on two factors: first,

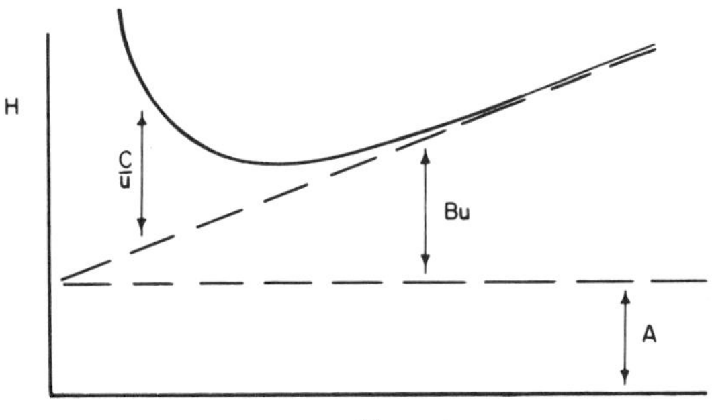

Fig. 9-9. The Van Deemter equation: plate height vs flow rate.

the separation between the two peaks, and second, the sharpness of the bands. Some workers express the resolution as follows:

$$\text{Resolution of two substrates} = \frac{\Delta V}{4\sigma} \qquad (9\text{-}8)$$

This definition is arbitrary and can have no exact meaning, because the standard deviation σ will be different for the two bands, and one must use an average value. But a glance at Fig. 9-8 will show that if two substrates come out of a column, one after the other, with peak volumes separated by 4σ or more, the two peaks will appear well separated with a pronounced "valley" between them. If the separation between peak volumes is less than 2σ it is hard to tell that one has two peaks; the graph of total concentration against volume has only one maximum.

The most obvious way to improve resolution is to use a longer column. Doubling the length of the column, other factors remaining equal, doubles ΔV. However, it increases σ also, multiplying it by the square root of two. To double the resolution, therefore, one must use a column four times as long. Other, more effective ways to increase the resolution are the following:

(1) Reduce the flow rate. This will be effective if the mass transport term determines the plate height.

(2) Use smaller particles of the fixed phase. This is specially effective in ion-exchange chromatography, where the entire volume of the fixed phase enters into the reaction. Smaller particles, however, impose more resistance to flow, and this effect is serious; according to the Poiseuille relation, the rate of flow through a tube of radius r depends inversely on the fourth power of r, for a given pressure gradient. Gravity flow cannot be used in liquid chromatography if the particles are very small, for the flow rate would be altogether too small. It is usual to use pumps generating pressures of tens of atmospheres.

(3) Raise the temperature. Raising the temperature of a liquid decreases its viscosity and allows diffusion to occur faster. However, temperature affects the distribution ratios and may lower ΔV [equation (9-8)] more than it lowers σ.

VII. UNSYMMETRICAL BANDS: TAILING

Throughout most of this discussion we have assumed that the concentration-volume graphs are symmetrical Gaussian curves, as shown in

Fig. 9-2, 9-3, 9-6, and 9-8. In practice they may be unsymmetrical, as shown in Fig. 9-10. The form (a) is by far the most common; the concentration of substrate rises rapidly but falls slowly; the last portions of the substrate are retained in the column and come out as a "tail." This phenomenon is called "tailing." Occasionally one observes the reverse of tailing, where some of the material moves ahead faster than the main band; this is indicated by curve (b).

Let us recall the conditions that are necessary to give a symmetrical, Gaussian curve. They are:

(1) The distribution ratio k [equation (9-1)] must be constant and independent of the loading, the amount of substrate per unit volume of column. For k to be constant the sample size and the "loading" must be small. The commonest cause of unsymmetrical peaks, especially in gas chromatography, is the overloading of the column by using too large a sample.

(2) When it is first introduced into the column, the substrate must occupy a very thin layer; ideally, it should form a disk of zero thickness. This condition means that the sample must be small, and also that it must be injected quickly, not gradually.

(3) Derivation of the Gauss equation, equation (9-4), requires the conversion of a factorial into an exponential function, using Stirling's approximation. This approximation is valid for large numbers. It is applied to the number of steps in the "random walk." It means that the corrected elution volume, $V_{el} - V_i$, must be large compared to V_i, but "large"

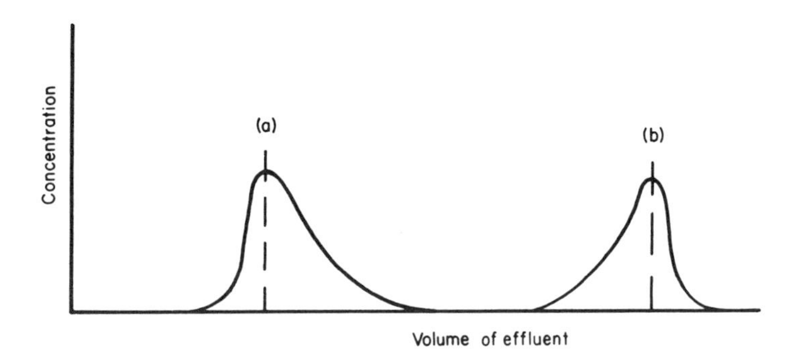

Fig. 9-10. Unsymmetrical peaks. (a) shows normal tailing; (b) has its steeper slope on the back of the band. The commonest cause of this effect is a distribution ratio k that increases with increasing concentration.

in this case means a factor of two or three; Stirling's approximation works quite well even for rather small numbers. But this restriction means that one cannot expect a band to be symmetrical if the retention volume V_{el} is only a little larger than the void volume.

Of these three conditions the hardest to meet is the first. In gas chromatography one often finds that the liquid used as the stationary phase is not a good solvent for the substrate vapors. The distribution ratio, (solute in fixed phase) to (solute in moving phase), falls rapidly as the concentration rises, and the peaks look like (a) in Fig. 9-10. The cure is to use a much smaller sample, or better, to use another column with a different stationary phase. Examples will be given in Chapter 11.

Tailing is also an indication that the absorbent or stationary phase is nonuniform and contains sites or regions that absorb the substrate more strongly than the rest and release it more slowly.

VIII. DISPLACEMENT CHROMATOGRAPHY

Relatively large quantities of substrates can be separated on chromatographic columns by using the technique called displacement. A mixture of substrates is poured into the column until half of the column, or more, is saturated with these materials. Already the less strongly adsorbed components will have moved further along the column, and the more strongly adsorbed components will have remained by the inlet. Now, a substance is passed that is more strongly adsorbed than any of the substrates. This will displace all the substrates along the column, pushing them ahead like a piston. As they move they become more and more separated and form bands with narrow overlapping zones. The graph of solute concentration against volume of solution flowing out of the column looks like Fig. 9-11. Over certain periods of time the solution leaving the column contains essentially only one component in pure form.

Separation is never complete, because there are always intermediate fractions of solution that contain two components together. Displacement chromatography is therefore little used in quantitative analysis, but it has a very important place in the large-scale separation of mixtures. The lanthanide elements, or rare earths, are separated on the commercial scale by displacement on ion exchange resin columns.

The important question in displacement chromatography, and in analyt-

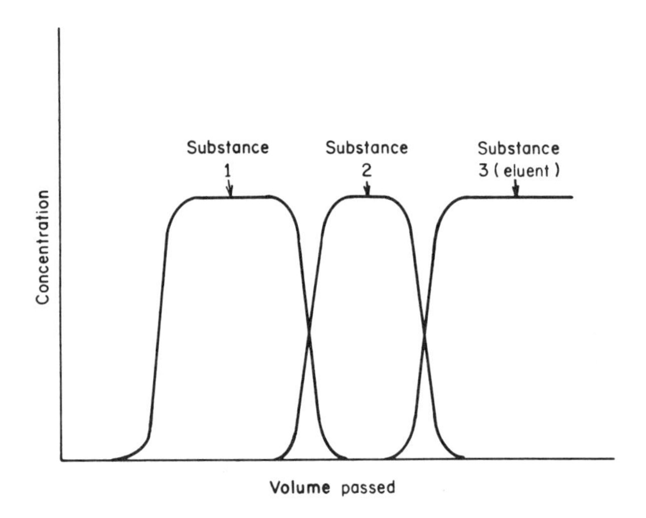

Fig. 9-11. Displacement chromatography.

ical column separations also, concerns the degree of sharpness of the "front," the region at the front of the zone or band where the substrate concentration is rising, or the region at the rear of the band where the concentration is falling away to zero. The sharpness of the front or rear of the band is determined by the same factors that determine the width of the bands in elution chromatography; they depend on the "theoretical plate height," discussed in Section IV. There is one more factor, namely, the curvature of the adsorption isotherm, the graph relating the concentration of substrate in the stationary phase to the concentration in the moving phase. For small amounts of substrate we can consider this graph to be a straight line, and we did this when we derived equation (9-3) and even, though the connection is less obvious, in deriving the Gauss equation (9-5). With large concentrations of substrate we cannot consider the graph to be straight. If the distribution ratio falls with rising concentration (Fig. 9-12a) the front is self-sharpening; if it rises with rising concentration (Fig. 9-12b), the front tends to become more diffuse as it moves along the column. (We shall not prove this statement here, however.) Another way to express the relation, which is not as precise but is more understandable, is that the front is self-sharpening when a substrate that is more strongly adsorbed displaces another substrate that is less strongly adsorbed.

Displacement chromatography is important in ion exchange separations, in cases where one component is absorbed on a column and is later desorbed in preparation for its quantitative determination.

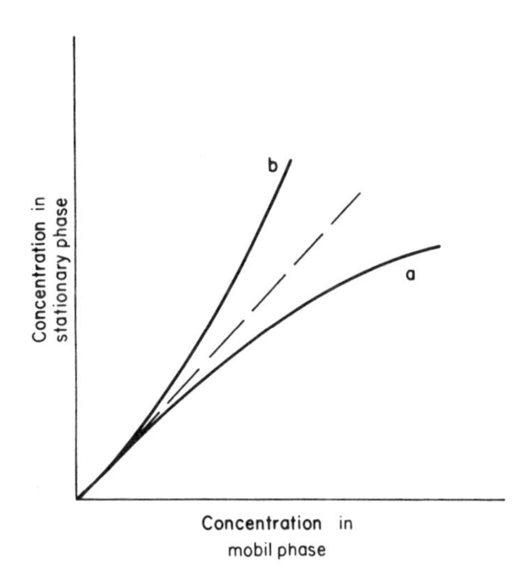

Fig. 9-12. Absorption isotherms.

QUESTIONS

1. A solid absorbent is shaken with a solution of a colored substance until equilibrium is reached. The absorbance of the solution was originally 1.30, before it came into contact with the solid, and 0.45 after shaking. The volume of solution was 25.0 ml, the weight of the absorbent 2.00 g. A column is now prepared that contains 10.0 g of the absorbent and 4.0 ml (interstitial volume) of the solvent. A little of the solution of the colored substance is added at the top of the column, then solvent is passed until the colored substance comes out of the bottom of the column. What volume of solvent must be passed?

2. How is the separation of two substances on a chromatographic column affected by (a) column length, (b) particle size of the stationary phase, (c) flow rate, (d) temperature?

3. Suppose you have a solute that associates with itself to form a dimer, for example, acetic acid, that is in equilibrium with $(CH_3COOH)_2$. In liquid or gas chromatography, would this substance give one peak or two? Explain your answer.

4. In the chromatography of plant pigments performed by Tswett, why did the addition of alcohol make the bands move faster?

5. A well-known experiment in absorption chromatography is the separation of cis- and trans-azobenzene, $C_6H_5N{=}NC_6H_5$. With alumina as the stationary phase it is found that the trans isomer travels faster. Can you explain this?

Chapter 10

ION-EXCHANGE CHROMATOGRAPHY

I. INTRODUCTION: ION EXCHANGERS

Ion-exchange chromatography is the kind of chromatography most used in inorganic analysis. Ion exchange is the reversible exchange of ions between a solution and a solid insoluble body in contact with it. This body, called the ion exchanger, must have certain characteristics. First, it must have ions of its own. Second, it must have a molecular structure that is sufficiently porous or open to allow the ions to diffuse in and out.

Ion exchangers may be inorganic or organic, natural or synthetic. The most familiar are organic polymers called "ion-exchange resins." The resins most commonly used are based on crosslinked polystyrene, and have a molecular structure like that shown in Fig. 10-1. They are made by mixing the two liquids, styrene ($C_6H_5CH{=}CH_2$) and divinylbenzene ($H_2C{=}CH \cdot C_6H_4 \cdot CH{=}CH_2$), in a ratio that is usually about 12:1 but may be varied according to the needs of the synthesis, adding a small amount of a free-radical catalyst like benzoyl peroxide, and immediately stirring the liquid mixture into water with a small amount of a surface-active agent in it. Stirring continues, and keeps the liquid reactants suspended as little drops, which may be smaller or larger according to the speed of stirring. In a short time they become solid spherical beads, as the styrene and divinylbenzene molecules join together to form very long chains and networks. Next, the ionic groups (sulfonic acid groups are shown in Fig. 10-1) are introduced by chemical treatment, such as reaction with fuming sulfuric acid. Finally the resin is washed and classified according to particle size.

Ion-exchange resins have characteristics that the user should understand when he selects a certain resin for a certain application. First, the

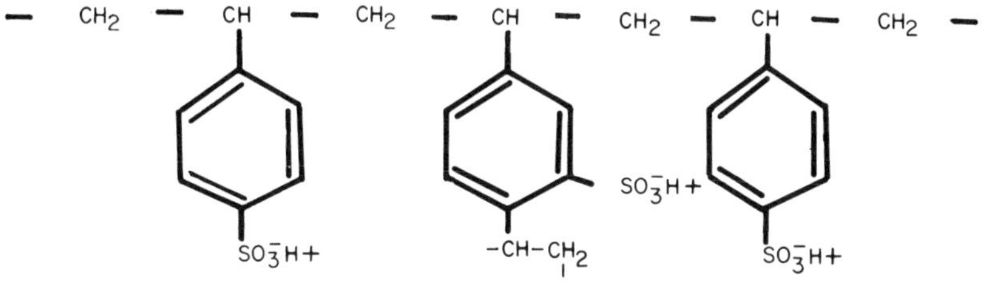

Fig. 10-1. Chemical structure of sulfonated polystyrene cation-exchange resin.

polymer type, or "backbone." In the example described this is the styrene-divinylbenzene copolymer, but it could be something else, like an acrylic acid polymer or a natural material like cellulose. Second, the functional groups or fixed ionic groups. In the example described these are sulfonate ions, but they could be carboxylate ions, or positive ions like quaternary ammonium ions. A resin to be used in the experiments of this chapter has the functional groups $-CH_2-N(CH_3)_3^+$. Thus the fixed ions may be negatively charged or positively charged. If they are negatively charged, they hold small positive ions or cations; if they are positively charged, they hold small negative ions or anions. It is these small ions, or counter-ions, that are exchanged. An exchanger with fixed negative ions is a cation exchanger; one with fixed positive ions is an anion exchanger. One speaks of strong-acid and weak-acid cation exchangers according to the nature of the functional groups and their degree of ionization. Likewise one speaks of strong-base and weak-base anion exchangers.

The degree of crosslinking tells us the tightness of the polymer network. The greater the crosslinking, the more dense is the material and the more ions there are per cubic centimeter. However, high crosslinking makes it difficult for ions, and especially large ions, to diffuse in and out. The common commercial resins of the polystyrene type have 8 per cent crosslinking, that is, the styrene-divinylbenzene mixture from which they were made contained 8 per cent of divinylbenzene.

Finally the particle size is important. One should choose the desired particle size when the resin is purchased. The resin beads are extraordinarily strong and difficult to grind, and one should not buy a large bead size and expect to make smaller sizes later by grinding this up. Particle sizes are commonly expressed on the U. S. standard sieve scale as the number of openings per inch, the "mesh size." Resins used for industrial water purification must permit fast flow and are of large mesh size, usually 20-30 mesh. Those used for exacting chromatographic analysis have much finer particles, like 200-400 mesh. The price one pays for fine particle size is a very great resistance to flow. Using resins of such fineness one needs high-pressure systems to pump the solutions; they will not flow under gravity at a satisfactory rate. For small chromatographic columns operated under gravity flow, such as we shall use in this chapter, a mesh size of 50-100 or perhaps 100-200 is best.

When ordering resins for the laboratory, it is helpful to know some of the commercial names that appear in catalogues. These names will be commonly met:

Dowex-50 or Dowex-50W. This is a strong-acid cation-exchange resin based on sulfonated, crosslinked polystyrene, that is, the type shown in Fig. 10-1. It is sold in the "hydrogen form," that is, with hydrogen ions as the replaceable cations or counter-ions, and in the "sodium form," with sodium ions as the replaceable cations.

Dowex-1 and Dowex-2. Both of these resins are strong-base anion exchangers with polystyrene "backbone" and the functional groups $-CH_2N(CH_3)_3^+$ and $-CH_2N(CH_3)_2C_2H_4OH^+$, respectively. The hydroxide form of Dowex-2 is a somewhat weaker base than that of Dowex-1. These resins are sold in the chloride form, with chloride ion as the replaceable anion. The hydroxide form is somewhat unstable and decomposes on long storage.

Dowex A-1. This is a "chelating resin" with polystyrene "backbone" and the functional group $-CH_2N(C_2H_4COOH)_2$, the same group that appears in the reagent EDTA. It is used for concentrating small traces of metal ions from large volumes of solution.

Amberlite IRC-50. This is a weak-acid cation-exchange resin whose structural unit is $-C(CH_3)COOH-CH_2-$. It is sold in the hydrogen form and also in the sodium form.

The crosslinking and mesh size of the resins are usually indicated as follows: Dowex-50Wx4, 200-400 mesh. This means the resin was made with 4 per cent of divinylbenzene.

II. EQUILIBRIUM AND SELECTIVITY

As we noted in our definition of the term, ion exchange is reversible. A typical cation-exchange process is the replacement of some of the sodium ions in a sodium-loaded exchanger by potassium ions in an external solution:

$$Res^-Na^+ + K^+ = Res^-K^+ + Na^+$$

If we stir a quantity of sodium-form exchanger with a solution of potassium chloride, we get an equilibrium distribution with some sodium and some potassium ions in the exchanger, some sodium and some potassium ions in the solution. The distribution obeys the mass-action law and can be described by an equilibrium constant Q:

$$Q = \frac{(Res^-K^+)\ (Na^+)}{(Res^-Na^+)\ (K^+)}$$

For Dowex-50x8 the value of Q is about 2. Potassium ions are held more strongly than sodium ions. We say that the "selectivity" for potassium ions is greater than that for sodium ions.

Some typical selectivity data for ion-exchange resins are given in Table 10-1. The actual numerical values depend on the characteristics of the resin, but in general, selectivity for cation exchange increases in the order Li(least)-Na-K-Rb-Cs for the alkali metals and Mg(least)-Ca-Sr-Ba-Ra for the alkaline earths. Ions of "heavy metals," like Cu and Pb, are held more strongly than Mg and Ca. For anion exchange, the order of halide ions is F(least)-Cl-Br-I.

TABLE 10-1

Ion-Exchange Selectivities; Values of Q[a]

Univalent cations (against Li)		Divalent cations (against Ca)		Univalent anions (against Cl)	
H	1.3	Mg	0.61	F	0.08
Na	2.0	Ca	1.00	Cl	1.00
K	2.9	Sr	1.25	Br	3.5
Rb	3.1	Ba	2.2	I	13.5
Cs	3.2	Zn	0.7	NO_3	3.0
Ag	8.5	Cd	0.8	ClO_4	23.0
Tl	12.5	Pb	1.9	OH	0.5
		Cu	0.75		
		Ni	0.75		

[a]These data apply to polystyrene-based resins, the cation exchangers with functional sulfonic groups, the anion exchangers with quaternary ammonium groups; crosslinking is 8%. It must be emphasized that these numbers vary from one batch of resin to another.

In expressing exchanges between ions of different charge, for example the replacement of Ca^{2+} by $2Na^+$, we must be careful of the units. The numerical value of the equilibrium constant depends on the units used to express concentrations. We must also note the very important effect that the higher the total concentration, the more the equilibrium is displaced in favor of putting the more highly charged ion into solution. Conversely, the more dilute the solution, the more of the highly charged ion there is in the exchanger. This follows from the law of mass action, which expresses the equilibrium constant as

$$Q = \frac{(Res_2Ca)(Na^+)^2}{(Res\ Na)^2(Ca^{2+})}$$

This effect is put to good use in the softening of water.

The fact that ion-exchange resins prefer one kind of ion to another is what makes ion-exchange chromatography possible. As we shall see, ion exchange is very often combined with complex formation to obtain additional selectivity.

III. ION-EXCHANGE COLUMNS

Nearly always, ion exchangers are used in columns, in the manner discussed in Chapter 9. Columns permit the complete replacement of one kind of ion by another even though the reaction is reversible. It is easily possible to replace one ion by another even though this goes against the selectivity order. Thus, we could pass a solution of sodium chloride through a column of potassium-loaded resin, and, for a time at least, obtain pure potassium chloride solution at the column outlet. The solution that passes through the column - sodium chloride in this example - continually meets fresh, unreacted resin, in this case potassium-loaded resin, and the excess of this reactant constantly displaces the equilibrium. Column operation is a very good way to make a reversible reaction go to completion, and we can make it go to completion in either direction.

Because this is so, we can use an ion-exchange column over and over again. When the column becomes exhausted, that is, when the ions that should be retained by the column start breaking through, we simply "regenerate" the column by passing more of the ions that were in the column at the start. In Experiment 10-1 we use a cation-exchange resin to absorb copper ions from a dilute solution; in Experiment 10-2 we take calcium ions out of a solution that also contains phosphate. After the experiment is over we do not throw the resin away; we pass a solution of hydrochloric acid to remove the calcium and other metal cations from the resin, then wash the column with water. Now the column is "regenerated" and ready to be used again.

This chapter is headed "Ion-Exchange Chromatography," but ion-exchange columns are often used for chemical separations or substitutions where no chromatography (that is, differential migration) is involved. Examples of nonchromatographic uses are: separation of traces, separation of interfering ions, and determination of total salt content of solutions. These are illustrated by the first three experiments of this chapter.

Experiment 10-1 illustrates the concentration of a trace constituent. A solution containing a low concentration of copper ions is passed through a column of cation-exchange resin which absorbs the copper ions and, so to speak, filters them out of the solution. Later the copper ions are removed from the column by passing a relatively small amount of dilute hydrochloric acid. Now all the copper that was originally in a large volume of solution is present in a small volume of relatively concentrated solution, in which it can be determined by titration.

In concentrating traces by ion exchange we must never forget that ion exchange is a competitive process. Other ions that are present can compete for places in the resin with the trace metal ions that are being absorbed. We could not collect traces of copper from sea water by the procedure of Experiment 10-1; the high concentration of sodium ions would prevent the absorption of the copper. We would need a resin with a specially high selectivity for copper ions, like the chelating resin mentioned in Section I.

IV. SEPARATING INTERFERING IONS

It often happens that one ion interferes with the analytical determination of another. For example, phosphate ions interfere with the titration of calcium or magnesium ions by EDTA, and calcium ions interfere with the acid-base titration of phosphate ions. Calcium ions and phosphate ions have opposite charges, however, and it is easy to separate them by ion exchange. If a solution containing both calcium and phosphate ions, slightly acidified to prevent precipitation, is passed through a column of a cation-exchange resin loaded with hydrogen ions, the calcium ions are retained by the resin while phosphate ions pass on. The resin does not "separate" calcium and phosphate in the sense of pulling them apart; electrical attractions between oppositely charged ions are far too strong for this to be possible. What the resin does is to give the ions new partners. Thus, we start with calcium phosphate and finish with phosphoric acid and calcium chloride (if hydrochloric acid is used to strip the calcium ions off the resin).

A similar method is used to bring into solution certain sparingly soluble salts, for example calcium sulfate. We could dissolve a sample of gypsum mineral (mainly $CaSO_4 \cdot 2H_2O$) by stirring the powdered mineral with an excess of hydrogen-form cation-exchange resin in a beaker, then filtering the solution through a short column containing more of the same resin. The solution thus prepared contains all the sulfate in the original mineral and none of the calcium or other metallic ions. The sulfate can be determined by titration with barium perchlorate in 50 per cent isopropanol at pH 2.5, using the indicator "thorin," the disodium salt of 2-hydroxy-3,6-disulfo-1-naphthylazobenzene-arsonic acid.

V. DETERMINING TOTAL IONIC CONCENTRATIONS

The solution is passed through a column of strong-acid hydrogen-form cation exchanger, which converts all the salts to their corresponding acids, which are then titrated with standard base. This method determines the concentration of the anions of <u>strong</u> acids, not weak ones like carbonic acid.

VI. SEPARATION OF METALS BY ANION EXCHANGE OF CHLORIDE COMPLEXES

This is truly an illustration of ion-exchange chromatography. It is somewhat paradoxical that metals, which we normally consider to form cations, are most effectively separated by an <u>anion</u>-exchange resin. The point is that most metals, other than the alkalis and alkaline earths, form negatively charged complexes with chloride ions. These complexes differ greatly from one another in their stability and also in their strength of attachment to anion-exchange resins.

The case of cobalt will serve as a good example, because the ions of cobalt are colored and shifts in the equilibrium can be seen in the color of the solutions. In a dilute solution of a cobalt(II) salt in water the principal ionic species is the hydrated cation, $Co(OH_2)_6^{2+}$, which is pink. If concentrated hydrochloric acid is now stirred into the pink solution the color changes, turning through various shades of purple and blue. When the hydrochloric acid is 4 molar the solution is more blue than pink; at 6 molar the color is pure blue. The color change corresponds to the shifting of this equilibrium:

$$\underset{\text{(cation, pink)}}{Co(OH_2)_6^{2+}} + 4Cl^- \rightleftharpoons \underset{\text{(anion, blue)}}{CoCl_4^{2-}} + 6H_2O$$

The anions are absorbed on an anion-exchange resin; the cations are not. If we were to add hydrochloric acid gradually to a vessel containing a dilute cobalt chloride solution and some beads of an anion-exchange resin, we would see the cobalt going into the resin and turning it blue as we raised the hydrochloric acid concentration above 4 molar. The distribution ratio D (see Chapter 9), which expresses the tendency of cobalt to enter the resin, rises rapidly with the concentration of hydrochloric acid, starting at about 3 molar and rising to a maximum near 9 molar. It rises as the proportion of cobalt present as $CoCl_4^{2-}$ rises. Beyond 9 molar the value of D falls somewhat, because the complex is now fully stabilized and additional chloride ions merely compete with $CoCl_4^{2-}$ for places in the resin.

The tendency is shown in Fig. 10-2, where the logarithm of D is plotted against the molar concentration of hydrochloric acid for zinc, cobalt(II), and iron(III). Every element has its own characteristic curve. If these curves are known, it is possible to devise methods of chromatographic separation of half the elements in the periodic table, using a strong-base anion-exchange resin as the fixed phase and hydrochloric acid solutions of different concentrations as the mobile phases.

Experiment 10-4 illustrates such a separation. Before we discuss this experiment let us note the metals whose ions are <u>not</u> absorbed from hydrochloric acid of any concentration. They are: the alkali metals, alkaline-earth metals except for Be, aluminum, yttrium, lanthanum, and the rare earths, thorium and nickel. Except for nickel ion, the ions in this list all have "inert gas" electronic structures and coordinate with

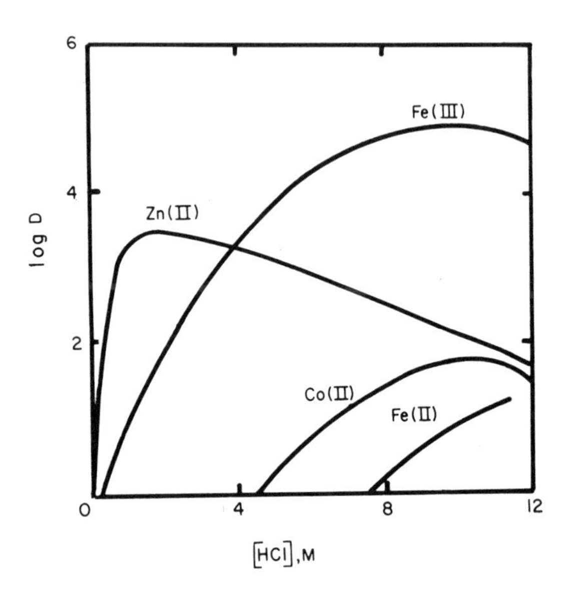

Fig. 10-2. Absorption of metals from hydrochloric acid solutions. D is the distribution ratio in milliliters per gram.

chloride ions weakly or not at all. They show little tendency for covalent bonding.

In Experiment 10-4 the ions of iron, cobalt, and nickel are separated by anion exchange, using Fig. 10-2 as a guide. They are placed on a column of strong-base anion-exchange resin in the form of a solution in 9 molar hydrochloric acid. Nickel ions are not absorbed, but iron(III) and cobalt(II) are both held by the resin. After washing nickel out of the column with 9 molar hydrochloric acid we pass a solution of hydrochloric acid that is 3 to 4 molar. This washes cobalt out of the column, leaving behind the iron, which, however, starts to spread down the column because its distribution ratio has fallen to about 100. After removing the cobalt the iron is washed out with 1 molar hydrochloric acid. Schematically the process can be shown like this:

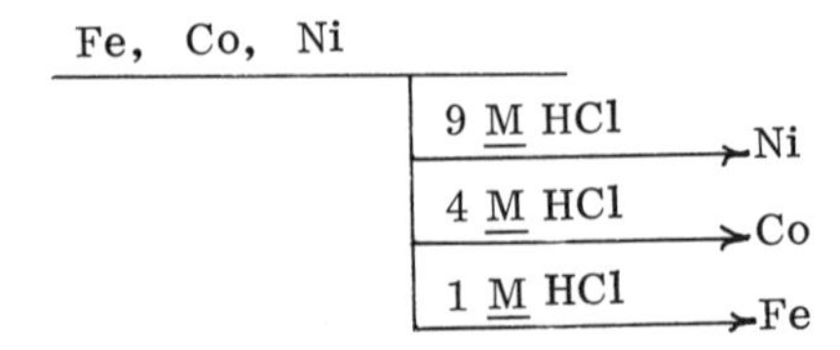

Next, the same separation is made using another solvent. The metal ions are dissolved in 6 molar hydrochloric acid; to this solution is added four times its volume of acetone. The mixed solution contains 80 per cent of acetone by volume and is 1.2 molar in hydrochloric acid. It is poured into a resin column that has previously been washed with this solvent mixture (4 volumes acetone to 1 volume 6 molar aqueous HCl). Now the iron passes through the column without being absorbed; nickel is absorbed as well as cobalt. However, nickel is held relatively weakly, and is washed out first, leaving the cobalt in the column, by diluting the solvent mixture with half its volume of water. Finally the cobalt is removed by passing aqueous hydrochloric acid of concentration 1 molar or less, or it can be washed out of the column by simply passing water. We note that diluting the solvent with water serves the same function as reducing the hydrochloric acid concentration. The separation can be represented thus:

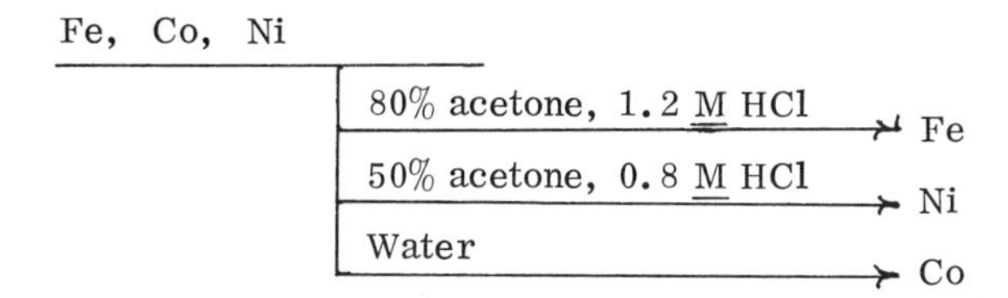

This experiment shows the very striking changes that can be made by changing the solvent. An interpretation of the strange behavior of iron is that the low dielectric constant of 80 per cent acetone favors the production of tightly associated ion pairs, $H_3O^+FeCl_4^-$ perhaps, that are electrically neutral and therefore not bound by the ion-exchange resin.

VII. SEPARATION OF ORGANIC COMPOUNDS BY ION EXCHANGE

The applications of ion exchange are not limited to inorganic chemistry. Organic compounds that form ions, either positive or negative, can be separated by ion-exchange chromatography; ionic impurities can be separated from nonionic compounds, and vice versa; and under certain circumstances, ion-exchange resins act as solid absorbents or solvents for nonionic materials.

An example of a simple separation that can be accomplished by ion exchange is that of a well-known pharmaceutical product, tablets for the relief of headaches that contain caffeine, aspirin, and phenacetin. Caffeine is a weak base, and in acid solutions it forms a cation; aspirin or acetylsalicylic acid is a weak acid, and in basic solution it forms an anion; phenacetin is a neutral substance that does not form an anion or a cation, yet is absorbed by the "solvent" action of the resin. The formulas are:

Caffeine:

Acetylsalicylic acid:
(Aspirin)

Phenacetin:

A plan of separation by ion exchange is the following:

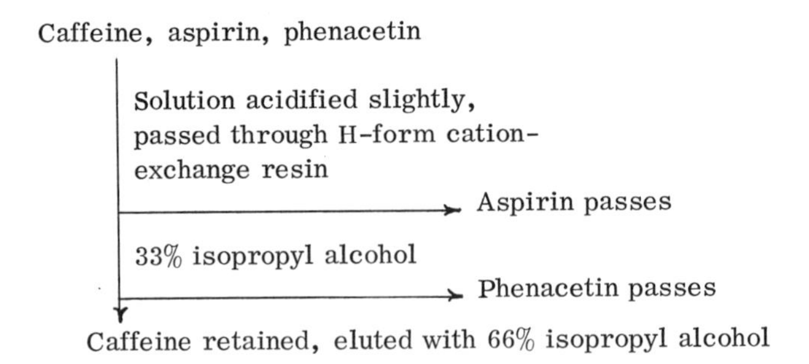

The compounds may be determined by their absorption of ultraviolet light (see Chapter 6). A parallel separation may be made by solvent extraction. Experiment 10-5 shows the separation of caffeine and aspirin only, without phenacetin; some pharmaceutical products contain only these two substances.

Undoubtedly the most important application of ion exchange to the analysis of organic compounds is the analysis of mixtures of amino acids.

Amino acids are the "building blocks" of proteins and are liberated from proteins when these are heated under pressure with dilute hydrochloric acid. Proteins normally contain some fifteen to twenty different amino acids combined in different proportions. All these acids contain the group

$$-\underset{}{\overset{NH_2}{\overset{|}{C}}}H - COOH \ .$$

In acidic solutions this group forms a cation by protonating the $-NH_2$; in basic solutions it forms an anion by losing a proton from the -COOH. In acidic solutions, therefore, one can absorb amino acids on a cation-exchange resin. They are bound with differing strengths, some strongly and some weakly, and they can be separated by chromatography. One takes a long column of cation-exchange resin in the hydrogen or ammonium form and adds a small amount of the amino acid mixture to the top, or inlet, of the column. Then one could separate the acids by passing dilute hydrochloric acid or ammonium chloride; the most weakly absorbed amino acid would travel fastest, the most strongly absorbed one slowest, by the reasoning explained in Chapter 9. In practice one uses buffer solutions of citric acid with sodium or ammonium citrate, and one uses the technique called "gradient elution." One starts with a buffer of low pH, say 3.0 or below. At this pH almost all the amino acids form cations and are absorbed fairly strongly. Only the most weakly absorbed acids move ahead at any considerable speed; these are those in which the acid character outweighs the basic character, for example aspartic acid, $HOOC \cdot CH_2 \cdot$-$CHNH_2 \cdot COOH$, and glutamic acid, $HOOC \cdot CH_2 \cdot CH_2 \cdot CHNH_2 \cdot COOH$, which have two -COOH groups and only one $-NH_2$. After these have passed down the column the pH of the moving phase is raised by cautiously adding sodium hydroxide or (better) sodium citrate. In "gradient elution" the sodium citrate is added continuously to the reservoir of solvent, so that the pH of the solution rises steadily as the liquid flows. In this way the movement of the amino acids along the column is regulated so that they emerge one after another at more or less regular intervals. In the middle stages of the elution come the "neutral" amino acids containing one -COOH and one $-NH_2$, such as glycine, $CH_2NH_2 \cdot COOH$, alanine, $CH_3CHNH_2 \cdot COOH$, and tyrosine, $HO \cdot C_6H_4 \cdot CH_2 \cdot CHNH_2 \cdot COOH$. Last to emerge are the acids having two basic groups and only one acidic group, for example, lysine, $H_2N \cdot CH_2CH_2CH_2CHNH_2 \cdot COOH$, and arginine, $HN{=}C(NH_2)NHCH_2CH_2CH_2CHNH_2 \cdot COOH$.

The acids are detected continuously and automatically as they pass out of the column. A common way is to mix the flowing solution with ninhydrin reagent (Chapter 11) and measure the absorption of light by the red products that are formed.

A problem rather similar to the analysis of amino acids and nearly as important is the analysis of the hydrolysis products of nucleic acids, particularly deoxyribonucleic acid or DNA. There is a great variety of these products, ranging in complexity from the four bases, adenine, guanine, thymine, and cytosine, to nucleosides (base-sugar combinations) and nucleotides (base-sugar-phosphate combinations). Cation and anion exchange are both used to analyze these mixtures by chromatography.

The analysis of mixtures of carboxylic acids is important in plant biochemistry and in the examination of fruit juices. Acids occurring in fruits include oxalic, lactic, tartaric, malic, citric, malonic, and many more. They can be separated by chromatography on a column of anion-exchange resin, using a solution of sodium nitrate as the moving phase.

Sugars can be separated on ion-exchange resin columns, even though sugars carry no ionic charge. So can polyhydroxy compounds related to sugars, like mannitol, sorbitol, and pentaerythritol. There are two ways to do this. One is to add sodium borate (borax) which forms anions with _cis_-dihydroxy and polyhydroxy compounds, thus:

```
  |                          |
HC-OH                      HC-O\  (-)  /OH
  |     +   BO2-   <==>      |     B
HC-OH                      HC-O/       \OH
  |                          |
```

These anions are absorbed on a column of anion-exchange resin in the borate form. The second way to absorb sugars on a resin is simply to use the resin as a solid organic solvent. A strong-base anion-exchange resin in the sulfate form absorbs sugars, sugar alcohols, and other compounds from 85-90 per cent ethanol. They are displaced by passing aqueous alcohol, starting with 85 per cent alcohol and adding more water as needed; the more water is added, the more weakly are the compounds held.

VIII. OTHER APPLICATIONS

There are many applications of ion exchange to analytical chemistry and chemical separations, besides those we have described. One of the earliest applications of ion-exchange resins to chemical separations was the separation of the lanthanides, or rare-earth elements, by cation exchange. These elements all form triply charged ions whose properties are very similar. Traditionally they have been very difficult to separate. The ion-exchange column, however, makes it possible to use small differences in absorbing power to produce efficient separations. The lanthanide ions are all absorbed with about the same force, but they differ in the stabilities of their complex ions. The smaller the radius of the lanthanide ion, the more stable is its complex ion with citrate or with other ligands:

$$M^{3+} + Cit^{3-} \rightleftharpoons MCit \text{ (uncharged)}$$

Since the complex ion is uncharged, it is not held by a cation-exchange resin. The distribution ratio of a lanthanide element between a cation-exchange resin and a citrate solution is smaller the greater the stability of the complex. A citrate solution can therefore be used as the moving phase, or eluent, for the chromatographic separation of the lanthanides. The element lutecium, which has the smallest ionic radius and the most stable citrate complex, comes out of the column first, while lanthanum, with the largest ionic radius and least stable citrate complex, comes out last.

The concentration of actual triply charged citrate ions in a citrate solution depends on the pH, and the higher the pH the more rapidly are the metal ions stripped from the resin. In separating the lanthanides it is customary to use gradient elution, as is done with the amino acids. One starts with low pH and low complexing power, then, as elution proceeds and the metals become harder and harder to remove from the resin, the pH is raised continuously to raise the concentration of triply charged citrate ions.

This process is used on a large scale to separate the lanthanides (rare-earths) in commercial quantities. They are produced in high purity for special uses. The technique of displacement is used (Chapter 9) since this is much better than elution chromatography for preparing large quantities.

Ion-exchange chromatography has many uses in radiochemical analysis, in atomic power, and in the separation of the artificially produced transuranium elements. One such element, mendelevium (atomic number 101), was identified after collecting only four atoms of it in an ion-exchange column.

EXPERIMENTS

Experiment 10-1

Separation and Concentration of Traces of Copper

Materials. Copper sulfate and sodium thiosulfate solutions, each 0.1 M; sulfuric acid, 2 M; solid potassium iodide and potassium (or ammonium) thiocyanate; starch solution, made by stirring 1 g of starch with a few milliliters of cold water and pouring the suspension into 100 ml of boiling water; buret, pipet (10-ml), and volumetric flask (25-ml or 50-ml); ion-exchange column (see Fig. 10-3); cation-exchange resin, sulfonated polystyrene type (such as Dowex-50), 50-100 mesh.

Object of Experiment. To observe the selective absorption of copper by the resin and use this to remove copper ions from water at low concentrations.

First prepare the solutions of copper sulfate and sodium thiosulfate. The copper sulfate solution may be made by weighing crystals of $CuSO_4 \cdot 5H_2O$, formula weight 249.6; an accuracy of 1% is sufficient. Thus one may weigh about 6 g of copper sulfate crystals to ± 0.05 g and dissolve in water, transfer to a 250-ml volumetric flask, and make up to the mark, after adding about 1 ml of 2 M sulfuric acid to prevent hydrolysis. The thiosulfate solution is made by weighing sodium thiosulfate, $Na_2S_2O_3 \cdot 5H_2O$, formula weight 248.2, and dissolving it in water, adding 1 g of sodium bicarbonate or borax as a preservative if the solution is to be kept for more than a week, then making up to a known volume. For this experiment it will suffice to take the copper sulfate solution as standard and use it to standardize the sodium thiosulfate, as follows:

Pipet 10 ml of the 0.1 M copper sulfate solution into an Erlenmeyer flask; add about 10 ml of water. Add 1 - 2 g of solid potassium iodide and about 1 g of solid potassium or ammonium thiocyanate; swirl the flask until these are dissolved. A white precipitate of cuprous iodide is formed together with a brown solution of iodine in excess iodide:

$$2Cu^{2+} + 4I^- = 2CuI + I_2 \; (+I^- = I_3^-)$$

Now titrate with sodium thiosulfate. The brown color of the iodine will fade to a yellow; before the yellow disappears, add 1 or 2 ml of starch

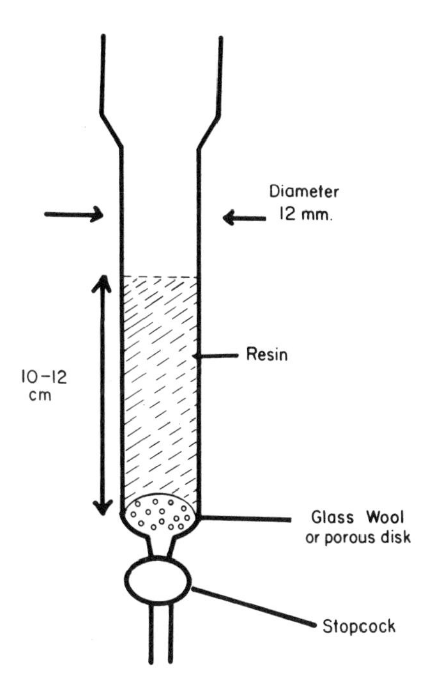

Fig. 10-3. Ion-exchange column for general laboratory use.

solution. This gives an intensely blue color with the iodine. Continue to titrate until the blue color just disappears. The reaction is

$$I_2 + 2S_2O_3^{2-} = 2I^- + S_4O_6^{2-}$$

Thus 1 mol of cupric ions is equivalent to 1 mol of thiosulfate. (The thiocyanate is added to prevent adsorption of brown triiodide ions on the surface of the white cuprous iodide.) Do two or more titrations until consistent results are obtained.

Now set up the ion-exchange column, as illustrated in Fig. 10-3. The part occupied by the resin should be about 10 - 12 cm long, and the tube should be about 10 - 12 mm inside diameter. The tube should be about 25 cm long. It is closed at the bottom with a plug of glass wool to hold the resin in place, and a short piece of rubber tubing is attached with a screw clamp to control the rate of flow of the solution. To add the resin to the column, first put the resin you need into a small beaker, then cover it with water containing a few milliliters of 2 M sulfuric acid, stir, and pour the resin suspension into the column. Wash the resin in the column by passing some 50 ml dilute (0.2-0.5 M) sulfuric acid followed by 50 ml distilled water. The column is now ready for use. (Note that the resin should be in the hydrogen form, that is, with hydrogen ions as the replaceable cations, before use. If it is not, you may have to pass more dilute sulfuric acid to remove the other cations it contains.)

Now take about 500 ml of tap water in a flask or beaker and add to it, with a pipet, 5.00 ml of the standard 0.1 M copper sulfate. Add a little dilute sulfuric acid, enough to make the water just acid to test paper; the amount will depend on the bicarbonate content (alkalinity) of the tap water. Mix; then slowly pour the solution through the resin column, so that it takes about 1 hr for it all to flow through. (You could put the dilute copper solution into a large separating funnel and let it drip slowly into the column.) If all is well, you will see the copper ions accumulating as a blue-green zone at the top of the resin which spreads downward as more copper is absorbed. The liquid flowing out of the column should be colorless.

When you have passed all the dilute copper solution, rinse the resin by pouring into the column about 10 ml of distilled water. Discard the water that comes out of the column, and also the 500 ml of solution from which the copper has been removed. Now pour into the column about 15 ml 2 N (1 M) sulfuric acid and let it flow slowly, about 1 drop/sec. The first few milliliters of solution to flow out of the column will be colorless; do not collect this. But, as soon as the first drop of blue solution appears at the column outlet, place an empty, clean small beaker under the outlet and collect all the liquid. Cupric ions are blue, and you will see them being pushed off the resin by the acid. After all the acid has entered the column, wash it down with 5 ml of water. By this time, all the copper should be out of the column. Transfer the solution in the beaker, which contains the copper, to a 25-ml volumetric flask, make to the mark with distilled water and mix.

Take 10 ml of this solution with a pipet, and titrate it with thiosulfate as you did the standard. Repeat with another 10 ml. And, if there is any doubt whether all the copper has been removed from the resin, pass a few milliliters of 1 M sulfuric acid through the resin and then a few milliliters of water, and titrate the entire solution that comes out of the column. Calculate the total amount of copper removed from the column, and see how it compares with the known amount of copper added to the tap water at the start.

Keep the column of resin; it is ready for use by the next student.

The quantities, sizes of flasks, etc., can of course be changed according to the needs of the laboratory. The idea is to get experience in ion-exchange techniques, using ions that are colored and can be seen. One can handle lower concentrations of copper by using atomic absorption (Chap-(Chapter 8) or colorimetry (Chapter 6) to determine the copper.

Experiment 10-2

Determination of Calcium and Phosphate in Baking Powder or Phosphate Rock

This experiment is essentially the same as Experiment 1-2, and is repeated here for convenience. In Experiment 1-2, only the phosphate in the baking powder was determined. Here the calcium is determined too. The same method, basically, will serve to determine phosphate, calcium, and iron, as well as other metals, in fertilizers or phosphate rock. The details of operating an ion-exchange resin column were noted in Experiment 1-2 and again in Experiment 10-1. Read these if you wish to refresh your memory.

(a) Calcium in Baking Powder Containing Little or No Iron. Follow the instructions for Experiment 1-2 until you come to the sentence "Wash the column with hydrochloric acid to displace the accumulated calcium ions." Use about 25 ml 2 M hydrochloric acid, and collect the acid and the washings as they come out of the column. Pour these solutions into a 100-ml volumetric flask, rinsing the column and the beaker used to receive the effluent from the column, to make sure to transfer all the calcium into the volumetric flask. Make up to the mark and mix well.

The solution in the flask probably contains between 5 and 10 mequiv of calcium ions, if 0.8 - 1.0 g of baking powder was taken as advised in Experiment 1-2. Take 10.0 ml of the solution with a pipet, place in an Erlenmeyer flask, neutralize carefully with 5 M sodium hydroxide (use test paper), and add an additional 10 drops of 5 M sodium hydroxide. Add enough murexide indicator to give a pale rose-red, and titrate with EDTA (the disodium salt of ethylene diamine tetraacetic acid), 0.025 M, until the solution turns purple-violet. Repeat until consistent results are obtained. Calculate the proportion of calcium in the baking powder, using the units (i) milliequivalents per gram, (ii) per cent Ca by weight.

Details of the preparation and standardization of EDTA and of the murexide indicator will be given in Experiment 10-4.

(b) Calcium and Iron in Phosphate Rock or Fertilizer. Use a finely ground phosphate rock or a fertilizer high in calcium phosphate. Weigh about 0.5 g of phosphate rock or 1 - 2 g of fertilizer (the best sample size obviously depends on the composition of the fertilizer), add about 20 ml 5 M nitric acid to the sample in a small beaker, and evaporate cautiously to near dryness, leaving about 2 ml of liquid. If it appears that not all the sample has disintegrated, add more nitric acid and evaporate again. There will probably be a residue of silica, which may be gelatinous.

If the evaporation is carried too far, iron will form brown-red ferric oxide which will be difficult to redissolve. One should stop just short of producing ferric oxide; but, if too much nitric acid is left behind, the hydrogen ions will prevent the iron and calcium ions from being taken up by the resin.

Dilute with about 50 ml of water, filter through fast filter paper to remove any precipitated silica, and pass the filtered solution slowly through a column of hydrogen-form cation-exchange resin, as described in Experiment 1-2 or 10-1. For the fertilizer it may be advantageous to use a longer column containing more resin than the one described. As a guide, note that 1 ml of resin (bulk volume, in place in the column) holds about 2 mequiv of cations.

Collect the solution flowing from the column; wash the column with about 25 ml of distilled water. Combine the washings with the main effluent and determine the phosphate by titration with standard base, as described in Experiment 1-2. The colorimetric vanadomolybdate method, Experiment 6-3, may be used instead. Here, it will be unnecessary to extract the iron with ethyl acetate, since the iron has already been removed.

To determine the iron and calcium, pass about 20 ml 2 M hydrochloric acid (more if necessary; use the size of the column as a guide, with the information given above, that the resin holds 2 mequiv cations per milliliter) to remove the cations from the resin. Rinse the column with 1-2 bed volumes of water - say, 20 ml of water if the column has bulk volume 10 ml - and add the rinse water to the main effluent. Transfer the solution to a 100-ml volumetric flask and make up to the mark.

The iron in this solution is best determined spectrophotometrically, as in Experiment 6-2. Note that a high phosphate concentration might have interfered with the determination of iron in fertilizer as described in Experiment 6-2; iron(III) phosphate is very insoluble at pH 4, and iron(II) phosphate is also sparingly soluble.

To determine the calcium one titrates with EDTA, using the procedure described in Experiment 10-1; however, iron oxide precipitates when sodium hydroxide is added and must be filtered before the titration is done.

Experiment 10-3

Total Ionic Concentration of Tap Water

Materials. Ion-exchange resin column, as in Experiment 10-1, with cation-exchange resin in hydrogen form, 50-100 mesh; standard hydrochloric acid and sodium hydroxide, each about 0.02 - 0.05 N, of known concentrations; methyl orange and phenolphthalein indicators; burets, glassware.

Principle of the Method. The use of ion exchange to determine total ionic concentrations was mentioned briefly on page 266. The method is very useful for natural water (ground water, well water, river water) and city water supplies. The total ionic concentration may be anywhere from nearly zero up to 20 mequiv/l. (0.02 N). Waters with salt concentrations higher than this are not good to drink.

The cations normally present in potable water are sodium, calcium, and magnesium; the anions are bicarbonate, chloride, and sulfate. Sometimes other ions are present, and a measurement of the total ionic concentration, coupled with determinations of individual ions, tells us whether or not other ions are present.

The bicarbonate ion is the anion of a weak acid, and it is, therefore, a weak base. As such, it can be titrated with standard acid. This experiment, therefore, has two parts: one sample of the water is titrated with hydrochloric acid to find its bicarbonate content, while another sample is passed through the ion exchanger and titrated with sodium hydroxide to determine the concentration of the other ions.

Titration with Hydrochloric Acid. Pipet 25 ml or 50 ml of the water sample (depending on the ionic concentration) into an Erlenmeyer flask, add 2 or 3 drops of methyl orange, and titrate with hydrochloric acid to the first change in color of the indicator. It changes from yellow to a distinct orange, corresponding to pH 4 to 4.5. A red color means that the titration has gone too far. If you have not done this titration before, you should practice by titrating three or four samples and keeping the solutions, after they have turned color, as a guide to the first color change, which is the correct one. A practised operator can locate the end point to within 1 or 2 drops.

Knowing the normality of the hydrochloric acid, calculate the normality of the bicarbonate in the sample, and the number of milliequivalents per liter, which is one thousand times the normality.

Ion-Exchange Procedure. First, make sure that the resin column has been well rinsed with distilled water; the rinse water, after passing through the resin, should not need more than 1 or 2 drops of sodium hydroxide to turn phenolphthalein pink (at least, after any carbin dioxide has been boiled out; see below).

Take a measured sample of the water, which may be 25, 50, or 100 ml, depending on its salt content and on the concentration of your standard base; pass it through the resin column at about 1 - 2 drops/sec; rinse the column with some 20 ml distilled water, and add the rinse water to the main effluent. Now boil this solution to remove carbon dioxide. Little bubbles of carbon dioxide will appear when the solution is first heated, for the resin converts bicarbonate ions into carbonic acid. When the water starts boiling with big bubbles and "bumping," it will be free of carbon dioxide. Now cool until the flask can be comfortably held in the hand (the contents will now be below 40°C), add phenolphthalein indicator, and titrate with standard sodium hydroxide until the indicator just turns pink. From your titration, calculate the number of milliequivalents of acid formed from 1 l. of the original water sample. This value will normally be the sum of the concentrations of chloride and sulfate ions; if the water contained nitrate, which is sometimes the case, the nitrate ions will be included too.

By adding the concentrations of strong-acid anions (obtained from the sodium hydroxide titration) to the concentration of bicarbonate (obtained from the hydrochloric acid titration), calculate the total ionic concentration in milliequivalents per liter.

Experiment 10-4

Separation of Iron, Cobalt, and Nickel by Anion Exchange

Materials. Strong-base anion-exchange resin, Dowex-1 or Dowex-2 or equivalent, in chloride form, 50-100 mesh or 100-200 mesh; tube for resin column, about 10-12 mm inside diameter and 25-30 cm long, as shown in Fig. 10-3; hydrated ferric chloride, cobalt chloride, and nickel chloride (other salts may be substituted); hydrochloric acid, acetone, EDTA, ammonium thiocyanate, isoamyl alcohol, separating funnel; spectrophotometer, Spectronic-20 or equivalent; volumetric glassware; Eriochrome Black T and murexide indicators.

Object of Experiment. To illustrate the separation of metals by anion exchange in hydrochloric acid (page 266). This section of the text should be read carefully before starting the experiment. The quantities of iron, cobalt, and nickel will be determined after these elements have been separated; these determinations will illustrate the use of EDTA, as well as colorimetric (spectrophotometric) analysis.

First, prepare a mixed solution containing all three of the elements to be separated, as follows: Weigh approximately 5 g of $FeCl_3 \cdot 6H_2O$, 5 g of $NiCl_2 \cdot 6H_2O$, and 3 g of $CoCl_2 \cdot 6H_2O$; place all three salts together in a beaker or flask; add water and stir until all are dissolved; and make to a total volume of 100 ml. This stock solution will be used for both parts of the experiment, the separations in aqueous solution and in aqueous acetone.

You will also need a standard EDTA solution, 0.025 M. Make this by dissolving about 9.3 g of the disodium salt of ethylenediamine tetraacetic acid, $Na_2H_2C_{10}H_{12}O_8N_2 \cdot 2H_2O$, formula weight 372.3, in water and making to 1 l. The reagent-grade salt is pure enough that a standard solution can be made by direct weighing; however, in very dry or very moist climates there may be some doubt about the exact quantity of water of crystallization, and it is best to standardize the solution. A good standard is pure cadmium or zinc metal. A suitable quantity of the metal is weighed accurately, placed in an Erlenmeyer flask, and dissolved in dilute hydrochloric acid, warming if necessary to speed the reaction. Cadmium dissolves very slowly. Finally the solution is made to the mark in a volumetric flask. You, the student, will know how to calculate the weights and volumes to be used. To titrate the zinc or cadmium solution, proceed as follows: Take a measured volume, add concentrated ammonia drop by drop to neutralize the excess of acid, then add five times as much ammonia as you have already used. This will make the pH close to 10. Add Eriochrome Black T indicator. Solutions of this indicator are unstable, and the best way to add the indicator is in the form of a powder made by grinding one part by weight of the pure solid indicator with ten parts of sodium chloride. (They should be ground to a fine powder and mixed well.) Take a little of the mixture on the end of a small spatula, enough to color the solution a definite purple-red. The correct amount of indicator is found by practice. A common mistake is to add too much. Now, titrate with EDTA until the color changes to a pure, clear blue.

Prepare the resin column by mixing some resin with 9 M hydrochloric acid in a small beaker. (Concentrated hydrochloric acid is 12 M.) Pour the suspension into the glass tube, as was done in Experiment 10-1, and let the acid drain out of the bottom until the level is the same as that of the resin; do not let air bubbles get into the resin. The height of the resin in the tube should be about 10 cm.

Pipet 5.0 ml of the Fe-Co-Ni stock solution into a small beaker and add 15 ml concentrated hydrochloric acid, mix, and record the change in color that you observe. Now pour this solution through the resin column, letting it pass at about 1 drop/sec, and collect the effluent. Rinse the column with 10 ml 9 M hydrochloric acid. Set aside the combined rinse and effluent; this solution contains the nickel.

Now pass 4 M hydrochloric acid. You will see the dark blue band of cobalt move down the column, away from the brown band of iron. The upper part of the blue band turns pink as the diluted acid meets it. The 4 M acid pushes the cobalt out of the tube; you will be able to see when the cobalt is all out of the column. Collect the cobalt solution. Record carefully in your notebook all that you see.

Now pass 1 M hydrochloric acid. This will wash out the iron. Again the color of the solution (yellow) shows clearly when the iron is all removed from the column. Collect the iron solution.

You now have three solutions containing nickel, cobalt, and iron. Measure the amount of each element as follows:

Nickel. First evaporate the solution in the hood to get rid of most of the hydrochloric acid. Do not let it go completely to dryness, but stop while there still remain a few drops of hydrochloric acid. Dilute with water, transfer to a 100-ml volumetric flask, and make up to the mark. Use 25-ml portions for titration. To each portion add ammonia (concentrated or 6 M) until the solution is neutral to test paper, then add enough ammonia - five times as much as you have already added - to make the pH near 10. Add murexide indicator (1 mg of the pure powder, or 10 mg of a mixture made by grinding 1 part of murexide with 10 parts of sodium chloride) to color the solution light yellow; titrate with 0.025 M EDTA until the color becomes purple-blue.

Iron. Dilute the solution to 100 ml in a volumetric flask; take 25-ml portions for titration. Neutralize the excess acid by adding 6 M ammonia carefully until the solution starts to turn brown; this should be about pH 3. Add 0.5 g of solid salicylic (or sulfosalicylic) acid, and again add ammonia to pH 3 (use test paper). Warm to 40°C and titrate with EDTA. The end point is marked by a change from brown-red to yellow-green. If the end point is not sharp, the most likely reason is faulty pH adjustment. The pH should be between 3 and 4. Adding 1 g of glycollic acid will provide good buffering.

Cobalt. Cobalt could be titrated with EDTA, as was nickel, but it is most important to keep the concentration of ammonium salts to a minimum in this titration. An easier way is the spectrophotometric method with thiocyanate.

The reagents needed for this method are ammonium thiocyanate (40 g per 100 ml), isoamyl alcohol, 6 M ammonia for pH adjustment, and a standard cobalt solution containing 1.0 mmol/l. Prepare the standard from pure $CoCl_2 \cdot 6H_2O$, formula weight 237.9.

Take your cobalt chloride solution from the column and dilute to 100 ml in a volumetric flask. Take 5.0 ml for the measurement. (Or, dilute the solution to 250 ml and take 10.0 ml for the measurement. You need a sample containing about 0.02 mmol of cobalt.) Dilute the sample with water to about 25 ml, and add ammonia to pH 3 - 4. Add 25 ml 40% ammonium thiocyanate, and transfer to a separating funnel. Add 25 ml isoamyl alcohol (approximately), shake well, and separate the isoamyl alcohol layer (which should be blue). Extract the aqueous solution with two more 10-ml portions of isoamyl alcohol to make sure you extract all the cobalt. Mix the three isoamyl alcohol extracts, transfer them to a 50-ml volumetric flask, and make to the mark with more isoamyl alcohol. Read the absorbance at 620 nm in a 1-cm cell or test tube, using a suitable spectrophotometer. Refer to Chapter 6 for details of spectrophotometry.

The proportions given should give you a solution whose absorbance is between 0.5 and 1.0. If the absorbance is outside this range, repeat with a larger or smaller portion of your cobalt chloride solution.

Prepare two standards of known cobalt content from your standard cobalt chloride solution of 1.0 mmol/l., and calculate the number of millimoles of cobalt in the entire solution you recovered from the ion-exchange column.

Separation in Aqueous Acetone. Start by washing the column with a mixture of 4 volumes of acetone and 1 volume of 6 M aqueous hydrochloric acid. Take 5.0 ml of the same Fe-Co-Ni sample solution that you used before; add 5 ml of concentrated hydrochloric acid and 40 ml acetone. Pour this solution through the column at about 2 drops/sec. Rinse the column with about 20 ml of the mixture of 4 volumes acetone and 1 volume 6 M HCl. The effluent and rinse together will contain all the iron in the sample. Note that the solution, instead of being yellow like the solution of iron(III) in aqueous hydrochloric acid, is nearly colorless.

Take the nickel and cobalt off the column, one at a time, following the outline on page 269. The colors of the solutions that pass out of the column are not as distinct as they would be in aqueous solution, and chemical tests should be made, if necessary, to verify that the removal of iron (first) and then nickel (second) is complete.

Iron, cobalt, and nickel may be determined individually in the same way as was described above, first evaporating to remove most of the acetone. Or, at the discretion of the instructor, the student may merely test the purity of each of the effluent fractions, using qualitative tests. One advantage of the acetone solvent is the low viscosity of the solutions; the flow rates can be speeded considerably if one wishes, and the entire separation can be run in 10 to 15 min. However, the separations are not complete; there is some cross-contamination. A good student exercise is to compare the degree of contamination at two or three different flow rates.

Experiment 10-5

Separation and Determination of Caffeine in Pharmaceutical Preparations

Materials. Column of cation-exchange resin, sulfonated polystyrene type, as used for Experiment 10-1; spectrophotometer and silica cells for ultraviolet absorption; standard glassware.

An object of this experiment is to gain experience in designing one's own procedure. This is not the standard method for determining caffeine, yet it works, provided one chooses the right range of conditions. You, the student, are expected to decide the sample size, what size of volumetric flasks to use, what dilutions to make, and what standards to run.

You will need this information about caffeine: Formula weight 194; maximum absorption in ultraviolet at 272 nm in acid solution, with molar absorptivity 10,500 mol^{-1} l. cm^{-1}; ionization constant as a base, about 3×10^{-14}; solubility in water, 2 g/100 ml at 25°C. As a base, caffeine is extremely weak, yet it is quantitatively absorbed on a short column of cation-exchange resin and desorbed by dilute hydrochloric acid, about 1 M. It is likely that the relatively strong binding (more than would be expected from its weakness as a base) is due to interaction of its pi-electrons with those of the aromatic rings of the styrene-divinylbenzene polymer.

For an "unknown" use caffeine-aspirin tablets, sold for headache relief. They normally contain about 10% caffeine and have some starch added as a binder. Take one tablet or perhaps half a tablet, weigh it, place in a small beaker with enough cold water to dissolve all the caffeine, break it up with a glass rod and stir for a few minutes. Most of the aspirin will remain undissolved, as flaky white crystals. Filter through coarse paper; this removes the undissolved aspirin and suspended starch. Wash the material on the filter with a little water. Note: do not boil; boiling disperses the starch and forms a gel that is impossible to filter.

Mix the filtrate and washings, whose total volume should not exceed 50-100 ml, and pass slowly through the column of ion-exchange resin. Wash the column with distilled water. Discard the effluent and washings.

Now pass through the resin 2 or 3 bulk column volumes of 1 M hydrochloric acid to strip the caffeine from the resin. Collect the effluent and washings; make up to a convenient volume in a volumetric flask. Perform another quantitative dilution, as needed, to bring the absorbance of the solution into the range 0.2 - 1.0. Measure the absorbance at 272 nm, using 1-cm silica or quartz cells. Compare the absorbance with that of standard solutions prepared from pure caffeine.

It is important to separate the caffeine effectively from aspirin, for aspirin has a maximum absorbance at 273 nm, and the spectra are very similar. However, the molar absorptivity of aspirin at 273 nm is only 800.

Make any tests you can devise to test the completeness of removal of caffeine from the resin and the degree of contamination of the final solution by aspirin. And, because this is in the nature of a research problem, be particularly careful in writing up the experiment to give full experimental details and to assess possible experimental errors.

It is interesting to apply this method to other products containing caffeine, for example, powdered coffee.

QUESTIONS

1. A column of hydrogen-form ion-exchange resin has bulk volume 15 ml and contains 35 mequiv of exchangeable hydrogen ions. Its bulk volume is 6 ml. A little KCl is placed at the top of the column, and 0.75 M HCl is passed. The maximum potassium ion concentration ap-

pears at 150 ml. Calculate (a) the distribution ratio D (Chapter 9), (b) the equilibrium constant Q for the ion exchange reaction (page 263).

2. You have a bottle of strong-base anion-exchange resin in the chloride form. You weigh out 1.00 g and heat it to 110°C to dry it and determine the moisture content; the resin after drying weighs 0.680 g. Then you take another 1.00-g sample and stir it with potassium nitrate solution, then titrate the liberated chloride ions; the total liberated chloride ions are 2.60 mequiv. Finally you weigh out 8.50 g of resin from the bottle, stir it with water, and pour it into a tube to make a chromatographic column. The volume occupied by the resin is 30.0 ml.

   (a) What is the capacity of the column, in mequiv of exchangeable anions?

   (b) What is the equivalent weight of the dry resin, and how does this value compare with that expected from the formula of this resin, given on page 261?

   (c) How is it that 8.5 g of resin can fill a volume as large as 30 ml? Why does the resin not float?

3. For the anion-exchange reaction $ResCl + Br^- = ResBr + Cl^-$ the equilibrium constant is 2.8 with the resin described in question 2. If you place a small amount of bromide at the top of the column of question 2, and elute it with 0.50 <u>M</u> KCl, what volume is needed to elute the peak concentration of Br^-? (See Chapter 9. It is easiest to calculate D first. Take the void fraction as 40%.)

4. In Experiment 10-1, where copper is absorbed by the resin from a dilute solution, the solution is first made slightly acidic. What would happen if (a) no acid, (b) too much acid were added?

5. The cation exchange resin used in Experiments 10-1, 10-2, and 10-3 has an ionic capacity of about 2.0 mequiv/ml of bed volume. Suppose the resin in the column has a bulk volume of 12 ml and will be used, according to Experiment 10-2, to determine the phosphate in phosphate rock. We plan to regenerate the column after we have exchanged <u>half</u> of the hydrogen ions in the column. What is the maximum weight of hydroxyapatite, $Ca_5(PO_4)_3OH$, that can be treated by the column before it must be regenerated?

6. A solution containing Fe(III), Co(II), and Zn(II) is to be analyzed by anion exchange in aqueous hydrochloric acid solutions. Using Fig. 10-2 as a guide, suggest a procedure.

7. In titrating nickel or cobalt with EDTA, as in Experiment 10-4, the concentration of ammonium salts must be carefully controlled; it must not be too high, nor must it be very low. Explain the reasons for this control.

Chapter 11

PAPER AND THIN-LAYER CHROMATOGRAPHY

I. INTRODUCTION

Paper and thin-layer chromatography do not use columns. The fixed phase is in one case a sheet of filter paper, and in the other case a layer of finely divided absorbent spread on a plate of glass or plastic. The substrate is placed near one edge of the paper or absorbent surface as a small spot. A drop of solution of the sample is applied and allowed to dry. Then this edge of the paper or plate is dipped in the "developing solvent" which is the mobile phase. The solvent rises up the paper or plate by capillary action, taking the substrate with it. The flow of solvent is stopped before it reaches the upper end of the paper or plate. The paper or plate is removed from the developing chamber and the position of the solvent front is noted or marked with a pencil. The substrate will normally not have moved as fast as the solvent; the more strongly it was bound by the paper or the solid absorbent, the more it is held back. The position of the spot of substrate is noted. If the spot is colored its location is easy; otherwise, it must be made visible by one of the techniques to be discussed below.

The ratio

$$\frac{\text{(distance traveled by substrate)}}{\text{(distance traveled by solvent)}} = R_F$$

The symbol R_F, which is universally used to denote this ratio, stands for "ratio of fronts." From the considerations of Chapter 9,

$$\frac{1 - R_F}{R_F} = k, \text{ the "column distribution ratio"}$$

Here, k is the ratio of amounts of substrate in the fixed and moving phases per unit area of the plate or paper.

If two substrates move at different rates, a spot that originally contains them both will split into two spots as it moves; see Fig. 11-1. The spots or zones stay fairly sharp as they move; the "effective plate height" is small, resolution is good, and complex mixtures give a large number of spots. At the same time the sensitivity is high; small amounts of material can be analyzed. Paper and thin-layer chromatography are used extensively in biochemistry for the analysis of complex mixtures like mixtures of amino acids or sugars. They can also be applied to inorganic analysis and have been used in geochemical prospecting.

Paper chromatography is the older of the two techniques. It was invented by Martin and Synge in 1941 and earned them the Nobel Prize in 1952; see Chapter 12. It soon became a standard technique in biochemical laboratories and made analyses possible that could be done in no other way. It played a key part in Calvin's investigations of photosynthesis, the process by which green plants convert carbon dioxide and water into sugars. Thin-layer chromatography in a sense is older, having been proposed in 1938 by Izmailov and Shraiber in Russia, but it did not come into general use until 1960. Today it is more popular than paper chromatography, because it is faster, gives better resolution, and is more versa-

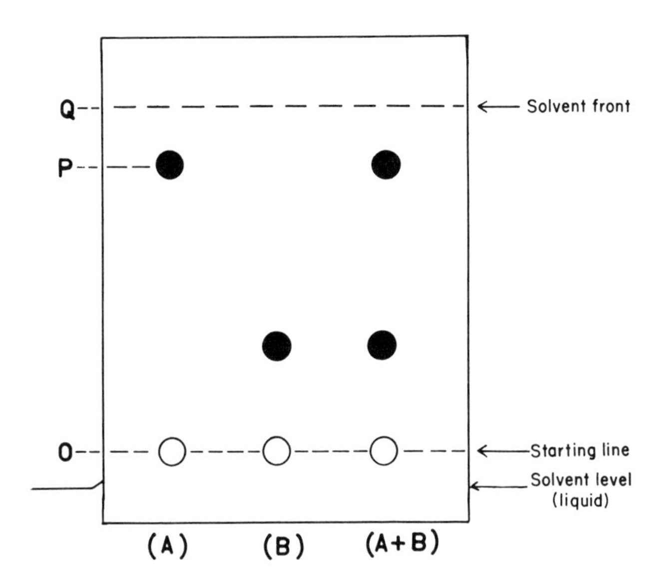

Fig. 11-1. Spots in paper or thin-layer chromatography.

tile. Different solid absorbents can be used and the thickness of the layer is easily varied.

The two techniques have different mechanisms. In paper chromatography the moving liquid flows between the fibers of cellulose, but these are not the true stationary phase; the true stationary phase is a very thin film of liquid, usually water, adhering to the surface of the fibers. The paper appears dry. If it looks wet, this means that the spaces between the fibers are filled with water and the developing solvent (which is usually not water, but another liquid) cannot flow. However, if the fibers are really dry the substrates will not move. Thus, paper chromatography will not work in dry climates unless the paper is exposed to air of high humidity before being placed in the developing solvent.

Paper chromatography, therefore, is a kind of _partition chromatography_ in which the substrates are distributed between two liquids, the stationary liquid (water) that is held on the fibers of the paper, and the moving liquid or developing solvent.

In thin-layer chromatography, on the other hand, the active fixed phase is the finely divided solid itself, and plates for thin-layer chromatography must often be dried in the oven to remove adsorbed water before they are used.

Of course there are many variations possible with each technique. Thin-layer plates can be coated with powdered cellulose, giving an effect like paper chromatography, with better resolution because of random orientation of short fibers. Paper can be made to hold a surface film of nonpolar liquid, instead of water, by first treating it with acetyl chloride to convert the polar hydroxyl groups of cellulose into less polar acetyl groups, then dipping it into the nonpolar liquid. Sometimes the nonpolar liquid will stick to the paper without this treatment. The paper is then "dried" to remove the excess of nonpolar liquid and to leave only a very thin film coating the cellulose fibers. Now the paper can be used for chromatography with a _polar_ liquid like water as the moving phase. This kind of chromatography is called _reversed-phase partition chromatography_.

Paper can also be treated with chemical reagents to convert the hydroxyl groups into ionic groups of various kinds, like carboxyl, -COOH, and diethylaminoethyl, $-C_2H_4N(C_2H_5)_2$. These make the paper become an ion exchanger. Special papers are also made that have finely ground ion-exchange resins incorporated between the cellulose fibers. Plates for thin-layer chromatography may be coated with ion exchangers, either resinous or inorganic.

II. DETECTION (VISUALIZATION)

In both thin-layer and paper chromatography one must know where the zones or spots of substrate are. If they are colored and can be seen, this is no problem. If they are invisible by ordinary light, perhaps they are fluorescent under ultraviolet light. A useful piece of equipment for chromatographic analysis is an ultraviolet lamp. Two types of lamp are available, the "long-wave" ultraviolet lamp giving radiation of 380 nanometers, and the "short-wave" lamp giving 254 nanometers. Both wavelengths are emitted by mercury vapor; the wavelength desired is selected by a filter. The short-wavelength lamps are made of silica, because glass does not transmit this wavelength. In using the short wavelength one must avoid looking directly at the lamp, because the radiation is harmful to the eyes.

Substances that absorb ultraviolet light can be made visible on a thin-layer plate by spraying the plate lightly with a fluorescent reagent like fluorescein or Rhodamine B. Under ultraviolet radiation the spots appear as shadows against a bright background. One must, of course, use the ultraviolet wavelength that is most strongly absorbed, and this is usually the short wavelength, 254 nanometers. Thin-layer plates are sold that have incorporated into them a fluorescent substance, usually zinc silicate. Then it is not necessary to spray the plate, but simply observe it, after drying, with ultraviolet light.

In Chapter 6 there is a short discussion of how organic substances absorb ultraviolet light. Aromatic compounds and compounds containing conjugated double bonds absorb significantly, and could, in principle, be detected on thin-layer plates by the method just described. However, the intensity of absorption varies greatly from one compound to another. Aromatic compounds that contain only a phenyl group not conjugated to another double bond or electron-donor atom, for example, benzyl alcohol or phenethylamine, show relatively weak absorption, with molar absorptivity about 250-300 at 254 namometers, and low concentrations are hard to detect by their ultraviolet "shadows." Compounds like benzaldehyde or cinnamic acid, or polynuclear aromatics like naphthalene, absorb much more strongly, with molar absorptivities of 10,000 or more, and are much easier to detect.

The commonest way to make spots visible is to spray the plate or paper, after drying, with a reagent that forms a colored product with the material in the spots. Many such reagents are used, and it would be impossible to mention more than a few. One important spray reagent is ninhydrin, triketohydrindene:

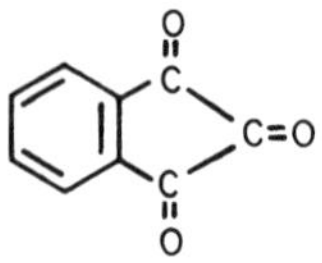

This compound, sprayed as an 0.5 per cent solution in acetone, combines with amino acids and many primary amines to form red, purple, or yellow products. The colors develop after heating for a few minutes at 100°C in an oven. Another all-purpose reagent that can be used with thin-layer plates, but not with paper, is a dilute solution of potassium dichromate in concentrated sulfuric acid. Dichromate (yellow) is reduced to chromic sulfate (green) by most organic compounds, and it is particularly useful for sugars. Paper, of course, is oxidized and charred by this reagent. The vapor of sulfur trioxide, produced on warming "fuming sulfuric acid," chars organic compounds and makes them visible as dark spots on thin-layer plates.

Apparatus for spraying is shown in Fig. 11-2. The push-button kind that uses a can of propellant (usually a fluorinated hydrocarbon) under pressure is very convenient but hard to control; often it delivers too much reagent. In drug stores one can sometimes buy small spray bottles that are operated by squeezing a rubber bulb. They have steel tubes and nozzles and cannot be used with corrosive reagents.

A detection method used in biochemical research employs radioactive tracers. Spots that are radioactive, say from carbon-14, will darken a photographic plate laid next to them.

III. ARRANGEMENTS FOR PAPER CHROMATOGRAPHY

The kind of paper used is the open-textured paper suitable for filtering coarse analytical precipitates. "Whatman No. 1" is a grade that is much used. This paper can be bought in rectangular sheets, but if these are not available, strips can be cut out of large-sized filter-paper circles. A convenient size of strip is 2 centimeters wide by 15 - 25 centimeters long. A strip of paper like this will take only one sample; if several samples are to be run on one piece of paper, a rectangle 20 - 25 centimeters wide is needed. For running several samples side by side it is convenient to use the slotted paper shown in Fig. 11-3. This can be bought commer-

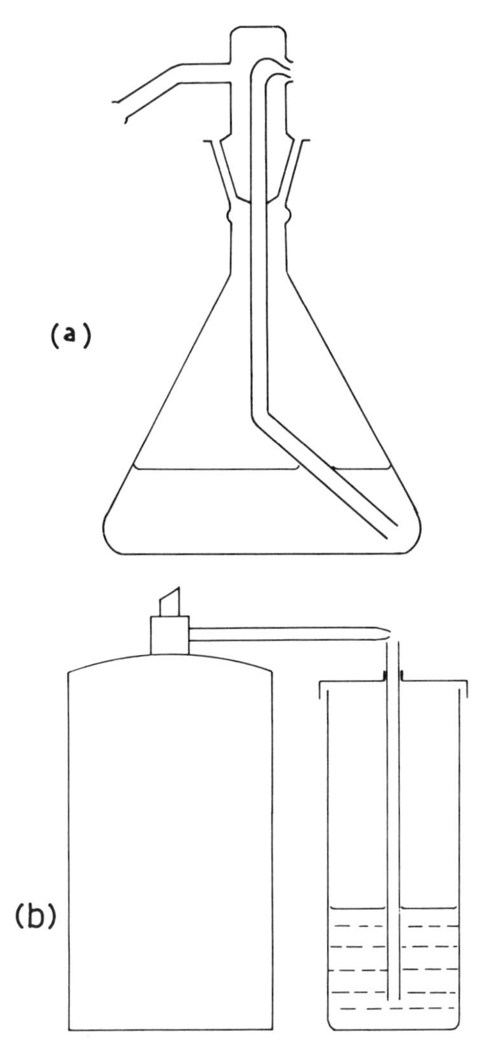

Fig. 11-2. Spray bottles. (a) is an all-glass bottle operated with compressed air; spraying is controlled by placing the finger on the hole shown just below the spray nozzle. (b) shows an aerosol can and glass bottle that contains the reagent; the plastic holder is omitted for clarity.

cially, or the slots can be cut out of ordinary rectangular sheets with a razor blade. The slotted paper (overall dimensions 12 by 25 centimeters) was designed for geochemical prospecting (see Experiment 11-2) where one needs a simple, rapid test without great resolution.

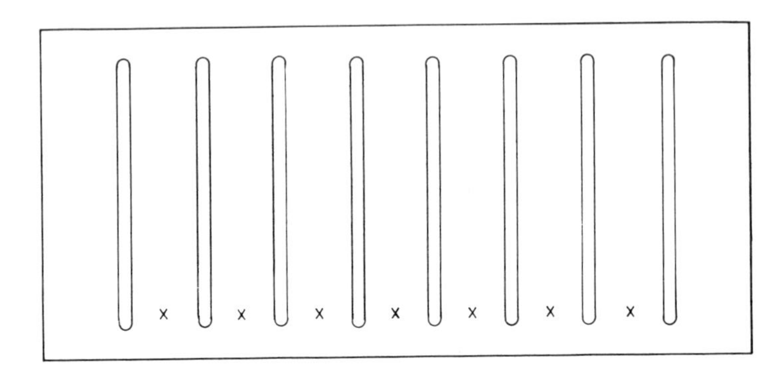

Fig. 11-3. Slotted chromatographic paper. Samples are placed at the points marked "x."

Samples for chromatography are placed about 2 centimeters from one end of the paper, as shown in various figures of this chapter, in the form of spots or narrow horizontal lines. Generally they are applied as solutions and allowed to dry. Then the paper is "developed" by allowing the solvent, the mobile phase, to rise up the length of the paper. This must be done in an enclosed vessel that is saturated, if possible, with the vapor of the solvent. Figure 11-4 shows typical arrangements. The simplest, used with paper strips, is simply a wide test tube or a glass cylinder closed at the top with a large cork or rubber stopper, on the under side of which is a paper clip to hold the strip. Larger pieces of paper need larger tanks and more elaborate supports; also, more care is needed to saturate the air inside the tank with solvent vapor. Large sheets or circles of filter paper, soaked in solvent and supported on the sides of the developing tank, help to fill the space with solvent vapor.

Rectangles of paper, like the slotted papers described, may be rolled into a cylinder, fixed with a paper clip, and stood upright in a large beaker or cylindrical jar, which is then closed with a watch glass, a flat piece of glass, or a plastic sheet; see Fig. 11-4(c). A practical detail is this: where the two edges of the paper are clipped together to form the cylinder, the solvent will run rapidly up between them unless the two edges are separated. Separation is easily done by making a little fold in one of the edges; see the figure.

In the arrangements shown, the solvent moves upward through the paper; these arrangements are called "ascending paper chromatography." There is another arrangement called "descending paper chromatography" in which the solvent is contained in a small trough at the top of the developing tank. The paper hangs over a horizontal glass rod with one end in the

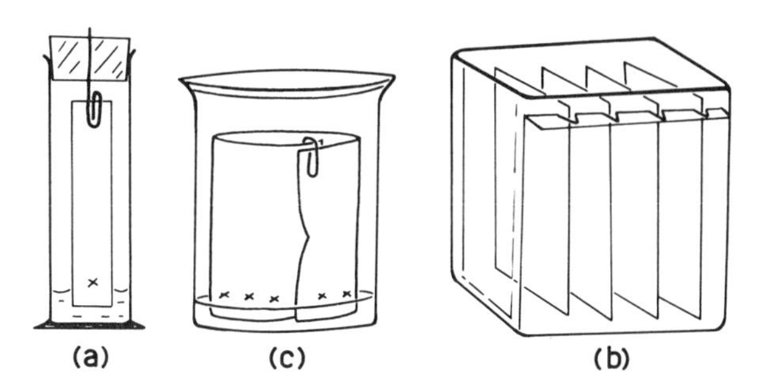

Fig. 11-4. Vessels for developing paper and thin-layer plates.

solvent and the other end hanging down in the developing tank. The solvent rises over the glass rod, then flows downward through the paper. This arrangement is more cumbersome than ascending chromatography. The flow is a little faster, but it is hardly worth the trouble.

IV. THIN-LAYER CHROMATOGRAPHY: PLATE PREPARATION

The arrangements for developing thin-layer plates are even simpler than those for paper chromatography, because the upper end of the plate does not need to be supported; it rests against the side of the developing vessel. Two sizes of plates are commonly used; 5 by 20 centimeters and 20 by 20 centimeters. The narrower plates will take two samples at a time, or three; the larger, square plates will take more samples and can be used for two-dimensional chromatography (see Section VI).

Plates can be bought commercially with different kinds of coatings, or they can be made in one's own laboratory. The two materials most commonly used for coatings are silica gel and aluminum oxide. They are ground to a very fine powder, with particle sizes of 1 - 5 microns, and to make them stick on to the glass plates, they are mixed with a binder that is usually Plaster of Paris, $2CaSO_4 \cdot H_2O$, but may be starch. Absorbents for thin-layer chromatography can be bought with the binder already added; "Silica Gel G" contains about 12 per cent of Plaster of Paris. A silica gel powder is also available that sticks to the glass plate without the need for a binder.

To coat a plate with "Silica Gel G" or alumina containing calcium sulfate binder, the absorbent is applied as a thin suspension which is prepared by mixing 1 gram of powdered absorbent to 2 milliliters of water. This must be spread on the plate within a minute of mixing, because the plaster sets rapidly. Various devices are used to spread the suspensions. The simplest way is to fix plastic tape along two parallel edges of the plate to be coated (see Fig. 11-5), using two or more thicknesses if necessary to give a total depth of 0.25 millimeters, then pour the suspension on the plate and spread it with a glass rod or a plastic ruler. A more uniform coating is obtained, and several plates can be coated at one time, by using an applicator such as that shown in Fig. 11-6. Whatever system is used, it is essential that the glass plates be really clean and free from grease before they are coated.

After coating, the plates are allowed to dry in air; then, before use, they are dried in an oven.

Coatings that are quite satisfactory for many types of work can be made by dipping small glass plates into the suspension of the absorbent, withdrawing them slowly (about 3 - 5 seconds), and placing them horizontally to dry. Microscope slides can be used, but plates about 15 centimeters long are better. Suspensions of absorbents in chloroform-methanol mixtures have been used for preparing plates by dipping, for these solvents evaporate faster than water.

V. SOLVENTS FOR PAPER AND THIN-LAYER CHROMATOGRAPHY

The selection of a solvent for a particular analysis seems difficult, because sc many solvents and solvent mixtures are in use. It is hard to

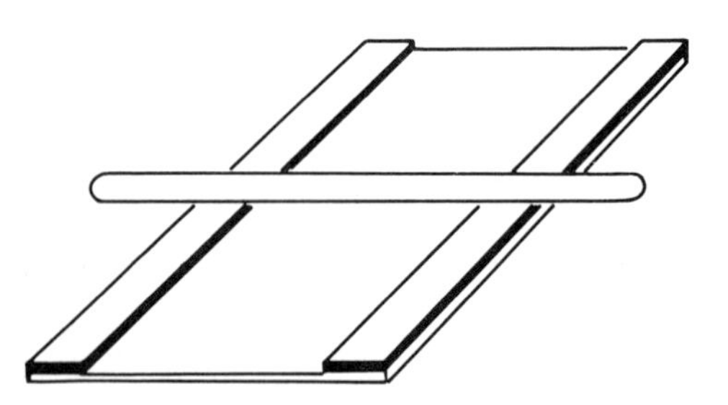

Fig. 11-5. A simple way to spread thin-layer plates. Strips of several layers of plastic tape are placed along the edges of the plate. A glass rod is used to spread absorbent between the strips. The tape is removed after the absorbent has dried.

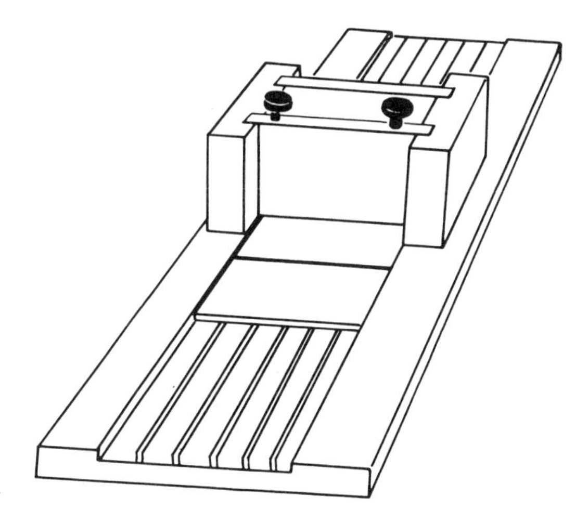

Fig. 11-6. Spreader for thin-layer plates. The "Camag" spreader is illustrated. The suspension of absorbent is placed in the reservoir, and the plates are pushed from back to front. The height of the front "gate" can be adjusted to give the film thickness desired.

predict the solvent that will give the best results; one must proceed by trial and error. The most important characteristic of a solvent is its polarity, and after this, its acidic or basic character. Usually, solvent mixtures are used; sometimes single, pure solvents are best. Table 11-1

TABLE 11-1

Pure Solvents Used in Chromatography

Water	*n*-Butanol
Formamide	Ethyl acetate
Methanol	Ether
Acetic acid	*n*-Butyl acetate
Ethanol	Chloroform
Isopropanol	Benzene
Acetone	Toluene
n-Propanol	Cyclohexane
tert-Butanol	Petroleum ether
Phenol	

lists pure solvents in order of decreasing polarity and decreasing ability to form hydrogen bonds. Table 11-2 lists a few of the mixed solvents commonly used in paper and thin-layer chromatography.

TABLE 11-2

Solvent Mixtures for Paper and Thin-Layer Chromatography

Solvent mixture
Isopropanol-water-conc. ammonia, 9:1:2
n-Butanol-water-acetic acid, 4:5:1
Phenol saturated with water
Benzene-methanol, 4:1 or other proportion, depending on polarity desired
Benzene-acetone, 1:1
Formamide-chloroform; formamide-benzene
Dimethylformamide-cyclohexane

VI. TWO-DIMENSIONAL CHROMATOGRAPHY

When analyzing a complex mixture it frequently happens that no one solvent or solvent mixture will separate all the components. The analysis of amino acid mixtures is a good example. In such cases one can use two solvents in succession, in directions at right angles to each other. One takes a square paper or thin-layer plate and places the sample spot near one corner, say 2 centimeters from each edge. Then one develops the spot with one solvent, for example the n-butanol-water-acetic acid mixture given in Table 11-2. This resolves the mixture into a line of spots, some overlapping, along a line parallel to one edge of the paper or plate. Without spraying the spots, one dries them, then puts the plate or paper back in the developing tank, with a second solvent, for example phenol-water (3:1 by weight), with the line of spots at the bottom of the tank. The second solvent moves the spots upward at right angles to the original direction of flow and gives a two-dimensional pattern, as shown in Fig. 11-7. Then the plate or paper is sprayed with a reagent to visualize the spots.

Spots are identified, as with one-dimensional chromatograms, by comparing their positions with the positions of spots from known compounds

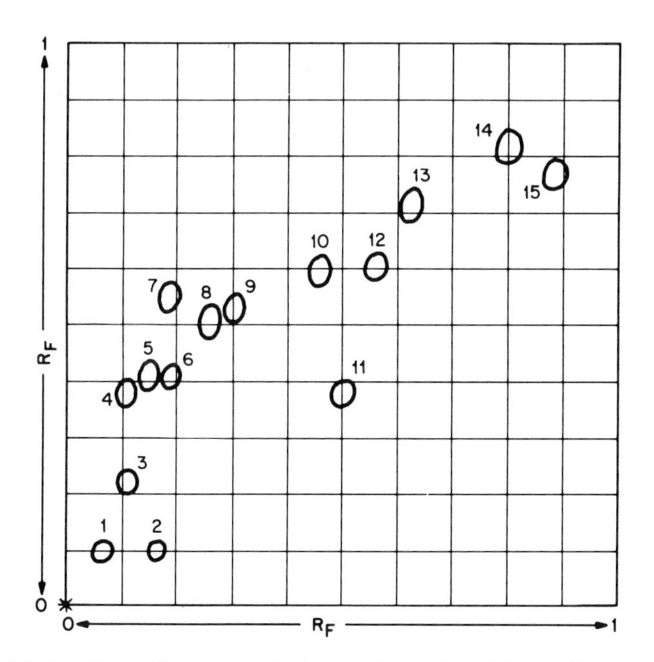

Fig. 11-7. Two-dimensional chromatography of 15 amino acids. See A. R. Fahmy et al., Helv. Chim. Acta 44, 2022 (1961).

obtained under strictly similar conditions. If the quantity of sample was sufficient - and samples of 0.5 to 1 milligram can be used in thin-layer chromatography if necessary - one can identify constituents by scraping off the plate the absorbent containing the "unknown" spot (provided it has not been changed by combining it with a spray reagent), dissolving the material in a solvent, and studying its ultraviolet and infrared spectra.

EXPERIMENTS

Note: Some of these experiments use paper, others thin-layer plates. With minor modifications the paper chromatography experiments can be done with thin layers and *vice versa*.

Experiment 11-1

Examination of Indicators and Inks by Paper Chromatography

Equipment, Materials. Paper, sheets 25 x 25 cm more or less; cylindrical developing tank, as shown in Fig. 11-4; glass capillaries 1 x 50 mm, of the type used to take blood samples; isopropanol, n-butanol, glacial acetic acid, concentrated ammonia (ammonium hydroxide); graduated cylinder; indicators; ink samples.

Object. To demonstrate the principle of paper chromatography and examine commercial products without the need for spraying and visualization of the spots.

(a) Indicators. These will be run in their alkaline forms. Prepare 0.1% solutions of Congo red, phenol red, bromthymol blue, bromphenol blue, bromcresol purple, methyl red, phenolphthalein. (Not all these need be used, and other indicators may be substituted if they are available.) Take a sheet of chromatographic paper of the size mentioned above (or smaller if desired), rule a line with a lead pencil about 2 or 3 cm from one edge, and place spots of the different indicators, about 3 cm apart, along this line. One or two spots may be made of mixtures, a good mixture being Congo red, phenol red, and bromphenol blue; these should have R_F values of 0.0, 0.2, and 0.6, roughly, in the solvent to be described. Place small spots on the paper with the aid of the capillary tubes.

Prepare the solvent by mixing 45 ml of isopropanol or n-butanol, 10 ml of water, and 5 ml of concentrated aqueous ammonia. Pour this mixture into the tank, then roll the paper into the form of a cylinder and fix it with a paper clip as shown in Fig. 11-4. Hold the cylinder in the vapor of the solvent for a few seconds until the ammonia vapor has converted the indicators into their alkaline forms. Then place the paper in the solvent, with the spots downward; close the top of the cylinder with a glass plate or watch glass or with a plastic film. Let the solvent rise at least 15 cm up the paper. Then remove the paper from the developing tank, note the position of the solvent front by marking lightly with a lead pencil, let the solvent evaporate; measure the positions of the colored spots and their distances from the starting line, and calculate the R_F values. Compare the R_F values found in the mixtures with those found by running single components; they should be the same.

If you have time, study the effect of the solvent composition by running one test with isopropanol and another with n-butanol. The differences will show the effect of solvent polarity; butanol is less polar than propanol. Other indicators may, of course, be studied.

(b) Inks. The samples for analysis are colored inks of different kinds. They should include brown and black inks, for one will find that the black ink may be a mixture of colored components, and that the brown ink may not contain a brown substance. Inks from ball-point pens may be used.

The best solvent for this experiment is the n-butanol-water-acetic acid mixture, 4:5:1 by volume, listed in Table 11-2. Shake the liquids together vigorously when you mix them; the mixture will not be homogeneous, but may be used none the less; the phase rich in butanol will run up the paper. Spot the inks on the paper as you did in the last experiment. If some of the samples are not resolved clearly into their various components, if the zones are elongated or overlap, it will be interesting to try two-dimensional paper chromatography, using butanol-water-acetic acid as one solvent and butanol-water-ammonia [see part (a)] as the other.

Experiment 11-2

Inorganic Paper Chromatography

Equipment, Materials. The best paper to use for this experiment is Whatman No. 1 Slotted Paper, Type CRL/1, made especially for chromatography. The sheets measure 11 x 21 cm. If this is not available, rectangular sheets of the same size or larger may be used. In addition one needs: Beaker, 600 ml; watch glass to cover same; glass desiccator; spray bottle; graduated pipet, size 0.1 ml; uranyl nitrate; chlorides of nickel, cobalt, and copper; aluminum nitrate hydrate, $Al(NO_3)_3 \cdot 9H_2O$; hydrated calcium or magnesium nitrate (this may not be needed; see below); nitric acid, hydrochloric acid, ethyl acetate, methyl ethyl ketone, potassium ferrocyanide, rubeanic acid (dithiooxamide); volumetric glassware.

Object. The tests to be described are used in geochemical exploration to locate ore deposits. They are simple and sensitive but not very accurate. They illustrate the use of paper chromatography in qualitative and semiquantitative inorganic analysis.

(a) Uranium. The method depends on the selective extraction of uranyl nitrate by ethyl acetate and the "salting-out" of uranyl nitrate by saturated aluminum nitrate.

First prepare a standard solution containing 1 g of uranium per liter, or 1000 ppm, as follows: Dissolve 0.531 g of uranyl nitrate, $UO_2(NO_3)_2 \cdot 6H_2O$, in 65 ml concentrated nitric acid, add 80 ml water and 150 g aluminum nitrate hydrate. Shake and warm, if necessary, to dissolve the aluminum nitrate; the object is to saturate the solution with this salt. Transfer to a 250-ml volumetric flask and make up to the mark with water.

Also prepare a nitric acid - aluminum nitrate solution by mixing concentrated nitric acid and water in the ratio 1 to 3 by volume and adding enough aluminum nitrate to saturate the solution, that is, nearly 2 g to every milliliter of diluted acid. Make about 100-200 ml of this solution. Use it to make diluted uranium standards containing 100 and 500 ppm of uranium. For example, pipet 5.0 ml of the 1000-ppm standard into a 50-ml volumetric flask and add the nitric acid - aluminum nitrate solution to make up to the mark; this solution will contain 100 ppm, or, stated more precisely, 100 mg of uranium per liter of solution.

To test the chromatographic technique, place spots of solution at one end of the paper strips between the slots of the CRL/1 paper, or place them at intervals of 2.5 or 3 cm along a line ruled parallel to the long edge of a rectangular sheet. At the top of the sheet, where the solvent will not reach it, write lightly with a lead pencil the number of micrograms of uranium. These quantities are suggested:

0.5 μg; 0.005 ml of 100-ppm standard
1.0 μg; 0.010 ml " " " "
2.0 μg; 0.020 ml " " " "
5.0 μg; 0.010 ml of 500-ppm standard
10.0 μg; 0.020 ml " " " "

In addition, place two or three spots to which you add 0.005 ml of roughly 10% ferric nitrate. Do not try to place more than 0.025 ml of solution in any one spot; this is as much as the CRL/1 paper will hold. (A thicker paper will hold more.)

Now let the spots dry in the air. After they have dried the water content of the paper must be adjusted to an atmosphere of about 50% relative humidity. If this is the normal humidity of your laboratory air, nothing more need be done. If, however, the humidity is low, you must place the paper in a desiccator whose lower part contains moist calcium nitrate or magnesium nitrate crystals. These maintain the humidity constant at 52%. Leave the paper in the desiccator for 1/2 hr.

Now prepare the developing solution, as follows: Into a dry 600-ml beaker pour 6 ml concentrated nitric acid and 1.0 ml water, mix, then add 30 ml ethyl acetate and mix again. (This solution must be used within 3 hr after mixing.) Remove the chromatographic paper from the desiccator, bend it into a cylinder, and fix with a paper clip as shown in Fig. 11-4, making sure that one edge has a fold in it to prevent the solution running up between the two edges under the paper clip. Place the paper in the developing solution with the sample spots down. Cover the beaker with a watch glass, and leave it until the solvent has risen to within 1-2 cm of the top. With the slotted CRL/1 paper this takes about 30 min. Other papers may take longer.

Then remove the paper from the beaker and wave it in the air for 1 or 2 min until the solvent has evaporated. Now spray the paper with a 5% solution of potassium ferrocyanide, holding the paper against a larger sheet of paper to avoid spraying ferrocyanide all over the laboratory. The uranium will become visible as brown stains just below the solvent front ($R_F = 1$). Iron, if it was present, remains close to the starting line and is visible as a dark blue area.

Compare the intensities of the brown uranium ferrocyanide stains. What is the smallest quantity of uranium that can be detected? Do the intensities of the stains from the samples that contained iron agree with the intensities from samples that contained only uranium?

If "unknowns" are available, test them qualitatively and quantitatively by this method. Choose the sample size to give between 1 and 10 μg of uranium, and run at least three standards in parallel with the unknowns. Rock samples may be tested for uranium by taking a weighed quantity of finely ground material and heating in a test tube with a measured volume of the nitric acid - aluminum nitrate solution that you prepared. As a guide to what to expect, ordinary granite may contain 20-30 ppm (μg/g) of uranium, and some red sandstones in uranium-bearing areas contain up to 100 ppm of uranium.

The paper chromatographic technique gives a rough estimate, say to within a factor of 2, of the amount of uranium present. If more accuracy is required one can cut out the section of paper that has the uranium in it, burn the paper in a platinum crucible, and convert the uranium to soluble form and determine it by fluorimetry.

(b) Copper, Cobalt, Nickel. For these tests you need standard solutions of each element. Weigh 0.120 g cobalt chloride, $CoCl_2 \cdot 6H_2O$, transfer to a 250-ml volumetric flask, add 100 ml concentrated hydrochloric acid and 10 ml concentrated nitric acid, and make to the mark with water. This solution contains 200 ppm (mg/l.) of cobalt. Prepare similar solutions, each 200 ppm in its metal ion, with 0.120 g $NiCl_2 \cdot 6H_2O$ and 0.085 g $CuCl_2 \cdot 2H_2O$. Prepare a fourth solution by mixing equal volumes of these three solutions, say 10 ml of each.

Place spots on the chromatographic paper, as in part (a), using 0.010 and 0.020 ml of each of the four solutions. Mark the upper edge of the paper as before, to note which spot was which. Place the paper, as before, in a desiccator with moist calcium or magnesium nitrate to adjust its water content to a humidity of 52%, unless, of course, the laboratory humidity is high.

Into a dry 600-ml beaker place 5 ml concentrated hydrochloric acid, 3 ml water, and 25 ml methyl ethyl ketone (2-butanone); mix. Bend the chromatographic paper into a cylinder, fix with a paper clip, and place in the beaker with the developing solvent, as in part (a). Cover the beaker with a watch glass and leave until the solvent has risen to within 2 cm of the top of the paper. Then remove, and let the paper dry in the air. Now expose it to ammonia vapor by hanging it in a large beaker containing some concentrated ammonia solution; leave it in the ammonia vapor until fuming has stopped and the free acid in the paper has been neutralized. Now spray the paper with a 0.1% solution of rubeanic acid, or dithiooxamide, in 60% ethyl alcohol. Three colored bands appear: nickel gives a blue band close to the starting line; cobalt gives a yellow band, and copper gives a greenish band with the largest R_F.

Measure the R_F values of the three metals and include these values in your report. If you have time, repeat the experiment with a higher proportion of water, say 5 ml instead of 3 ml, in the developing solvent and observe the effect that this change has on the R_F values. (The separation between copper and cobalt is sharply reduced.) For an "unknown" you can determine the copper content of a soil (compare Experiment 6-5); dry the soil in the oven, and bring the copper into soluble form by fusing the soil with potassium bisulfate.

Experiment 11-3

Ion-Exchange Paper Chromatography

Materials. Strong-base anion-exchange resin paper, Amberlite SB-2; various reagents.

Object. To explore the anion-exchange behavior of metals in hydrochloric acid, using the discussion of Chapter 10 as a guide; to investigate the use of this technique in qualitative analysis.

The paper to be used in this experiment contains a strong-base quaternary ammonium type resin in the chloride form, the same resin that is used in Experiment 10-4, but very finely ground and mixed with cellulose fibers to form the paper. The paper contains about 50% of resin by weight. A typical way to use it is as follows:

Cut a strip of the paper, about 1 cm wide and 20-25 cm long. Set up an arrangement like Fig. 11-4a, so that the strip of paper can be hung by a paper clip or a hook with the lower end dipping into the developing solution, without its touching the walls of the vessel. Make a light pencil mark on the paper about 3 cm from the lower end, but do not place the sample spot right away; wait until the solvent front has moved past the level where the spot will be placed, then place the spot of sample on the wet part of the paper, behind the solvent front.

For a first test, place a mixed solution containing the chlorides of iron(III), cobalt, and nickel, dissolved in 6 M hydrochloric acid, on the paper. Use 6 M hydrochloric acid as the developing solvent. Probably the spots of iron, cobalt, and nickel will be seen as they travel; if the quantities are too small, the ions can be visualized by drying the paper, exposing it to ammonia gas, then spraying with rubeanic acid, as in Experiment 11-3.

Other tests can be devised from a knowledge of the anion-exchange behavior of metals in hydrochloric acid solutions. It is always important to place the sample spot behind the advancing solvent front, to ensure that the resin has come into equilibrium with the hydrochloric acid. If this is not done, the developing spots show bad streaking.

In your written account, note the R_F values, remembering that R_F is the ratio of the rates of movement of substrate and solvent, and that in this experiment the solvent and substrate do not start from the same point. Calculate also relative values of k, the "column distribution ratio."

Experiment 11-4

Thin-Layer Chromatography of Amino Acids

Materials. Thin-layer plates coated with Silica Gel G, or the equipment for coating such plates; developing tank, spray bottle, capillary tubes (Experiment 11-1); oven at 100°C; ninhydrin, acetone, acetic acid; ethanol, *n*-butanol; samples of pure amino acids.

Object. The analysis of amino acid mixtures is probably the most important application of thin-layer chromatography and one of the easiest. Thin-layer chromatography is several times faster than paper chromatography for the purpose and several times as sensitive. Conditions can be varied widely; a range of solvent mixtures can be used, and two-dimensional chromatography is desirable for complex mixtures. We shall describe the one-dimensional procedure for simplicity.

The plates to be used in this experiment have a silica gel coating with calcium sulfate binder. It is easiest to use the prepared plates, 5 x 25 cm, but you can coat your own plates, and they can be of any convenient size. Before use, dry the plates in the oven at 105°C for 15-30 min, and let cool thoroughly in the air for another 30 min before applying the samples. Rule a "starting line" 2 - 3 cm from the end of the plate and apply spots of amino acids about 1.5 cm apart; thus, 3 samples can be run at once on a plate 5 cm wide. Apply the amino acids as 0.25% solutions in water or aqueous alcohol. The choice of amino acids depends on what one has in stock; some of the commoner amino acids, and their approximate R_F values in the solvents to be used, are given in Table 11-3.

Place in the developing tank enough solvent to give a layer 1 cm deep, and before inserting the plate, shake the solvent around in order to create an atmosphere of solvent vapor in the tank. One of these solvents may be used:

Solvent A: *n*-Butanol-glacial acetic acid-water, 3:1:1 by volume

Solvent B: *n*-Propanol-water, 7:3 by volume

TABLE 11-3

R_F Values for Amino Acids in Various Solvents

Acid	Butanol-water-acetic acid, 3:1:1	*n*-Propanol-water, 7:3	Phenol-water, 3:1 by wt.
Alanine	0.27	0.37	0.29
beta-Alanine	0.27	0.26	0.30
Arginine (in HCl)	0.08	0.02	0.10
Aspartic acid	0.21	0.33	0.09
Cystine (in HCl)	0.16	0.32	0.27
Glutamic acid	0.27	0.35	0.14
Glycine	0.22	0.32	0.24
Leucine	0.47	0.55	0.48
Lysine (in HCl)	0.05	0.02	0.09
Methionine	0.40	0.51	0.49
Phenylalanine	0.49	0.58	0.55
Tryptophane	0.56	0.62	0.63
Tyrosine	0.47	0.55	0.47
Valine	0.35	0.45	0.40

Place the plate in the tank and let the solvent rise to within 5 cm of the top of the plate. Then remove the plate, mark the position of the solvent front, let the plate dry in the air, and spray it with a 0.2% solution of ninhydrin in acetone to which a few drops of pyridine have been added. Put the plate in an oven at 105 - 110°C and leave until the colored spots appear, which takes about 3 min. Measure the R_F values.

For an "unknown," take a mixture of pure amino acids, and run a spot of the unknown sample in parallel with two or three known, pure amino acids on the same plate. The R_F values may vary somewhat from one plate to another, especially if the plates are "home-made," and small changes in the solvent composition affect R_F, but the ratio of R_F values for different acids remains fairly constant as long as the solvent is not radically changed. Thus, if you have a table of observed R_F values and run only two or three known acids for reference, you will be able to identify perhaps ten or more acids in an unknown mixture.

For another "unknown" use a protein hydrolyzate. Take 100 mg of a protein, for example, gelatin or casein (from skim milk powder), place in a glass tube about 8 mm inside diameter and sealed at one end, place in the tube some 2 ml 6 M hydrochloric acid, and seal the tube with a blowtorch flame; see Fig. 11-8. Wrap the tube in a cloth, to guard against breakage, place it in a small beaker, and put it in an oven at 105°C and leave overnight. Next morning remove the tube and let it cool. Open the tube carefully by scratching it with a file and touching a red-hot glass rod to the file scratch. Wash the contents (which should be almost colorless, not brown) into a small beaker. Evaporate nearly to dryness to remove the excess hydrochloric acid, then add a few milliliters of water and evaporate again. Now dissolve the solid amino acids in about 5 ml of water and apply a drop of the solution to a thin-layer plate. Run several pure amino acids in parallel and identify as many as possible of the amino acids from the protein, estimating their relative abundances. If the protein hydrolyzate gives many bands that merge continuously into one another, try again with a smaller sample of the solution. The method is very sensitive, and a small drop is all that is needed.

It is interesting to run the hydrolyzate of gelatin side by side with that of a protein like casein. Gelatin is nutritionally incomplete because it lacks certain essential amino acids, notably tyrosine.

Fruit juices may be examined for their amino acid content. Take about 5 ml of freshly squeezed orange juice and another 5 ml of fresh lemon juice, filter each if necessary, then add 15 ml of alcohol to each. This precipitates proteins and salts. Filter again or centrifuge. Apply 1 drop of each solution to the thin-layer plate. Also run spots in which more than

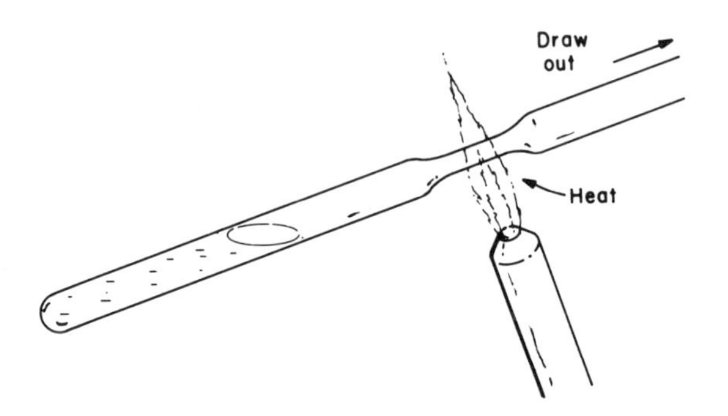

Fig. 11-8. Sealing a glass tube.

1 drop, say 4 or 8 drops, are added; to avoid the spots spreading unduly, add 1 or 2 drops, then evaporate and add more solution in the same place. Again, run samples of pure amino acids in parallel on the same plates.

In all work with amino acids be very careful to avoid touching the surface of the plate or paper with the fingers. Perspiration contains amino acids. In fact, one could run a chromatogram of the amino acids of one's fingers by rubbing the fingers with a few drops of approximately 0.1 M acetic acid and placing a spot of the solution on a thin-layer plate.

Amino acids may appropriately be separated and identified by two-dimensional thin-layer or paper chromatography.

QUESTIONS

1. A key concept in thin-layer chromatography and in column chromatography is that of polarity. How can polarity be defined? Which is the more polar absorbent, alumina or silica gel? Silica gel or powdered cellulose? Which is the more polar solvent, chloroform or benzene? Ethyl alcohol or diethyl ether?

2. If, in thin-layer chromatography, a particular substance has $R_F = 0.80$, how is this substance distributed between the stationary and mobile phases? What is the distribution ratio k (see Chapter 9)?

3. In the method for paper chromatography of uranium, Experiment 11-2, why is it so important to saturate the standard and sample solutions with aluminum nitrate?

4. In Experiment 11-4, what changes occur when the solvent meets the ion-exchange resin in the paper, and why is it important to let the solvent wet the resin before applying the sample?

5. In paper or thin-layer chromatography of amino acids the sample must be freed as far as possible from dissolved salts and inorganic acids and bases before applying to the paper or thin-layer plate. Why is this necessary? How may one remove electrolytes from amino acid solutions?

Chapter 12

GAS CHROMATOGRAPHY

I. INTRODUCTION

More analyses are performed by gas chromatography, probably, than by any other method of analysis. Gas chromatography is used for organic compounds, and the manufacture of organic compounds is the major part of chemical industry. The petrochemical industry, in particular, uses this technique on a very large scale. Yet gas chromatography was unknown in 1940, and has seen its greatest development since 1955. Its discoverers were the British chemists A. J. P. Martin and R. L. M. Synge, who won the Nobel Prize in 1952. Their collaborator A. T. James was a leader in the early development of gas chromatography.

The moving phase in gas chromatography is an unreactive gas like helium or nitrogen. The stationary phase may be a solid absorbent like silica gel or a "molecular sieve," but usually it is a film of high-boiling liquid held on the surface of a solid, granular support like diatomaceous earth or ground-up brick. Hence the name "gas-liquid chromatography," or GLC. A fixed phase that is actually a porous solid, but acts like an organic solvent, is a styrene-divinylbenzene copolymer like that from which ion-exchange resins are made (see Chapter 10).

The great advantage of gas chromatography is its speed. Gases are much less viscous than liquids and can move much faster through packed columns. Diffusion through gases is thousands of times faster than diffusion through liquids, and the rate of mass transfer, on which the resolving power of a chromatographic column depends (see Chapter 9, Section V, page 252), is correspondingly greater. To separate ten components by liquid chromatography used to take five or six hours, though the new high-pressure techniques are much faster. A similar separation by gas chromatography may take five or six minutes.

A disadvantage to gas chromatography, or at any rate a limitation, is the fact that the samples must be gases or volatile liquids. They must exist as gases at the temperature of the column. This means that inorganic ions, for example, cannot be analyzed by gas chromatography unless they can first be converted into volatile compounds, like the chelated compounds with acetylacetone. Many organic compounds have to be converted into volatile derivatives before they can be analyzed by gas chromatography. Examples are the amino acids, which decompose on heating and do not vaporize. By treatment with an alkyl trifluoroacetate they are converted to esters of the formula

$$\begin{array}{l} R\cdot CH\cdot COOR' \\ \quad\;| \\ \quad NH\cdot COCF_3 \end{array}$$

(the original acid is $R\cdot CHNH_2\cdot COOH$) and are then separated by gas chromatography. Carbohydrates can be analyzed by gas chromatography if they are first converted into esters. To form derivatives takes time and opens the possibility of decomposition and isomerization of the compounds that are being analyzed. Nevertheless, the speed and sensitivity of gas chromatography make this technique attractive.

II. BASIC INSTRUMENTATION

Gas chromatography requires a certain minimum amount of instrumentation. Some gas chromatographs are very elaborate and expensive; the more elaborate equipment allows the user to take advantage of the great versability that gas chromatography offers. All gas chromatographs, however, have the same basic components. They are illustrated in Fig. 12-1. They are:

(1) A cylinder of "carrier gas," the inert gas that is to be used as the moving phase. As was noted, this is generally helium or nitrogen. Attached to the cylinder is a pressure-regulating valve and a gauge to show the pressure. Close control of carrier gas pressure and flow rate are important, and there will generally be a second regulating valve and gauge in the gas chromatograph itself.

(2) A flow meter to measure the rate of flow of the carrier gas. This is usually of the "Rotameter" type, a vertical glass tube with a V-shaped

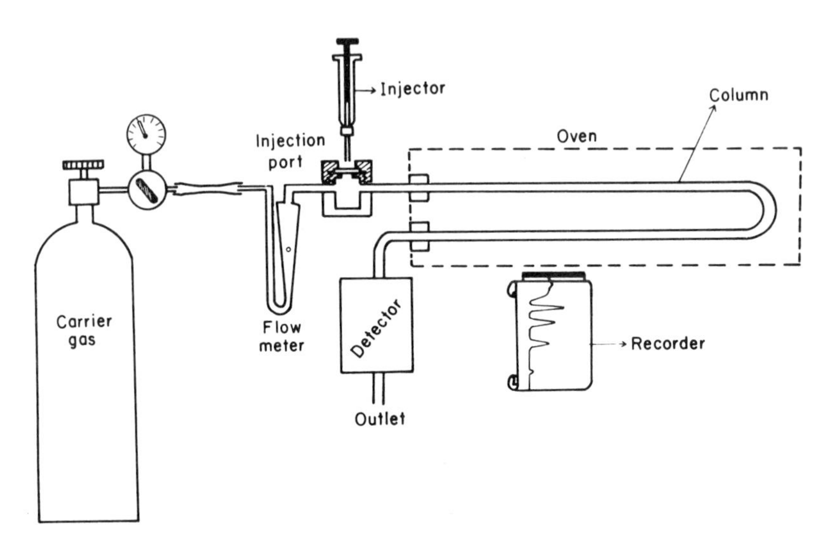

Fig. 12-1. General plan of a gas chromatograph.

section that is wider at the top than at the bottom. Inside this tube is a small metal ball or a cone-shaped piece of metal that is lifted by the gas stream and rises higher in the tube, the faster the gas is flowing. The level at which the ball "floats" shows the speed of the gas.

(3) The "injection port," the place where the sample is introduced into the carrier gas stream. Usually the samples are liquids at room temperature and are injected by a special syringe, like the hypodermic syringes used in medicine but smaller. The needle is introduced through a septum of silicone rubber which closes again after the needle is withdrawn. It is important for the sample to be vaporized completely as soon as it is introduced. Therefore, the injection port is embedded in a heavy block of metal that can be heated electrically. The temperature is adjustable, and a gauge on the front of the instrument indicates the temperature. The injection port should be at a higher temperature than that of the column.

(4) The column. This is the heart of the chromatograph. It consists of a tube of copper or stainless steel, or sometimes of glass, that is at least 1 meter long and sometimes 2 meters or more; the inside diameter is variable, depending on the type of detector used. With thermal conductivity detectors the diameter is about 4 millimeters; for flame ionization detectors it may be half this. Gas chromatographs used for preparative purposes have much wider columns.

The column is enclosed in an oven, and to save space the column is bent in the form of a "U," as shown in Fig. 12-1, or into a coil. The temperature of the oven is controlled by thermostates, and air is circulated inside the oven by a fan, to make the temperature uniform. In most instruments the oven temperature can be adjusted continuously to any desired value. Some low-cost instruments have only certain temperature settings. More expensive instruments have "temperature programming," whereby the temperature of the oven can be continuously raised in a controlled way as the analysis proceeds. In any case, a dial or an indicator system allows the operator to read the temperature of the oven. The importance of the column temperature will be seen from the experiments.

The columns are packed with solid granules that carry a film of nonvolatile liquid that is the real stationary phase. The nature of the stationary phase is very important; different types of sample need different stationary phases. We shall discuss this point later. Users of gas chromatographs should keep two or three different columns on hand, filled with different kinds of stationary phase, so that the columns can be changed in the instrument as needed. (Note that once a column has been packed and bent into a coil or U-shape for use in the chromatograph, it is virtually impossible to empty it and repack it. It is difficult to pack a column well, and most users of gas chromatographs buy their columns ready packed from the instrument manufacturer.)

(5) The detector. As the carrier gas passes out of the column it takes with it the various substrates which emerge at different times, depending on how strongly they were held by the fixed phase. A graph of substrate concentration against volume of carrier gas, or time, looks like Fig. 9-2 or, better, Fig. 12-2. To obtain such a graph it is necessary to have a way to continuously measure the composition of the emerging carrier gas. One needs a nonselective method that will detect any component that is different from the carrier gas itself.

The most popular method, until recently, has been the measurement of thermal conductivity. The heavier a molecule is, the more slowly it moves, and the more poorly it conducts heat across a gas. (There are other factors, like the number of ways in which the molecule can rotate

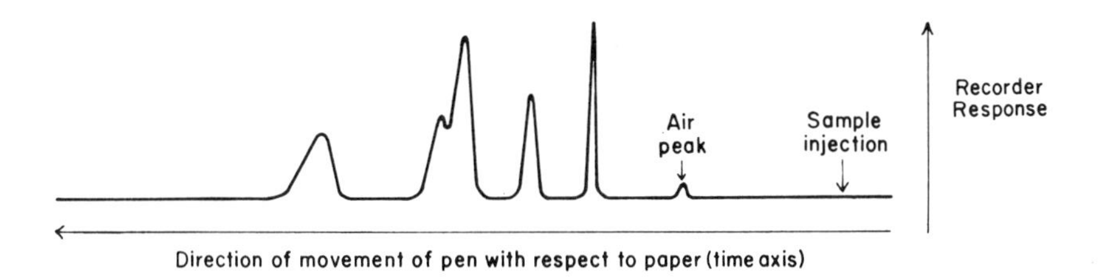

Fig. 12-2. Typical recorder trace of a gas chromatograph.

and vibrate, but the main factor in the ability of a gas to conduct heat is the speed at which its molecules can diffuse.) The carrier gas has small, rapidly moving molecules and conducts heat well. Helium is the best choice of carrier gas if thermal conductivity is used. (Hydrogen conducts even better than helium, but it is flammable and dangerous to use, and moreover it reacts with many organic compounds.) The molecules of organic compounds, on the other hand, are relatively heavy, and lower the heat conductivity if they are present.

Another kind of detector that has become very popular is the flame ionization detector. The principle of this will be discussed in Section IV-B. This detector is much more sensitive than the thermal conductivity detector, and it can be used with nitrogen as the carrier gas, which is a distinct advantage in locations where helium is hard to get.

(6) The recorder. The recorder traces a curve like Fig. 12-2 automatically and makes the analysis visible. It is connected to the detector and receives a signal from the detector in the form of a very small electric current that rises or falls when substrate passes through the detector. The current flows through a high resistance and sets up an electromotive force between one end of the resistance and the other, and it is this electromotive force that controls the movement of the pen of the recorder across the paper. The line traced by the recorder is therefore a graph of electromotive force against time. By appropriate calibration it may be interpreted as a graph of substrate concentration against volume of carrier gas.

The position of a particular peak on the time axis tells us the identity of the substance causing the peak; the area under the peak, between the curve traced by the pen and the base line, tells us the amount of the substance.

III. TYPES OF COLUMN PACKINGS

As we have noted, the effective stationary phase in gas-liquid chromatography is the film of liquid that covers the surface of the solid granules of the "support." The nature of the support is not specially important as long as it holds the film of liquid well. The support is normally a type of firebrick ground to a diameter of 0.2 millimeters or less. Diatomaceous earth (kieselguhr) is also used. The porous granules have an irregular surface and will absorb a lot of liquid without appearing noticeably wet. For chromatography they are loaded with about 10 to 15 per cent their weight of liquid.

To coat a support with the high-boiling liquid that is to be used as the stationary phase one first dissolves the high-boiling liquid in a volatile solvent like benzene. This solution is poured over the support, which is placed in a large dish, and the two components are stirred well together. The quantities should be calculated beforehand to give the desired ratio of high-boiling liquid to support. The dish is now heated on a steam bath and the contents stirred until most of the solvent has evaporated. The solid granules appear dry or slightly sticky, but they still contain a rather large proportion of the volatile solvent. Most of the remaining solvent can be removed by vacuum evaporation; the granules are transferred to a round-bottom standard-taper flask which fits the evaporator. The flask is rotated while the vacuum is applied.

Now the granules are ready to be placed in the column, and as we noted, the proper filling of the column is difficult and needs practice. The tube must be straight and vertical at the time it is filled, and it must be tapped continuously to ensure that the granules pack regularly. After it is filled, the column tube is bent into the desired shape, the end fittings are attached, and it is ready to be placed in the instrument.

Columns purchased commercially are prepared as we have described. The granules still contain some volatile solvent, which must be removed before the column can be used. The manufacturer's instructions tell the user to mount the new column in the chromatograph and disconnect the detector, then heat the oven to the temperature at which the column will be used and pass carrier gas for several hours to sweep out the last traces of the volatile solvent. Finally the outlet to the column is connected to the detector (note that the column connections must be tight!) and the chromatograph is ready to be used.

The stationary phase is a liquid of high-boiling point. We have called it an "involatile" liquid, but any liquid will vaporize or decompose if the temperature is high enough. Some stationary phases vaporize at a lower temperature than others. Manufacturers specify a maximum operating temperature for their columns, and one must be careful not to exceed this temperature. If the column is made too hot, the stationary phase vaporizes and the column becomes useless.

Liquids used for stationary phases are classified according to <u>polarity</u>. Compounds containing hydroxyl groups, ether, or ester groups are called <u>polar</u>; hydrocarbons, compounds containing only carbon and hydrogen, are called nonpolar. Among hydrocarbons we make a distinction between aromatic and aliphatic compounds. The point is that the stationary phase in gas chromatography must be a good solvent for the substrates, the compounds that are to be separated. It must dissolve them well enough to retain them in the column long enough to discriminate between one substrate and another, and also the distribution ratio k (Chapter 9) should as far as possible be independent of the concentration of the substrate. That is, solutions of the volatile substrate in the involatile stationary phase should obey Henry's law (see textbooks of physical chemistry). Only if this is so will symmetrical elution curves like those shown in Figs. 9-6 and 9-8 be obtained. Now, the rule of solubility is "like dissolves like." Nonpolar liquids are good solvents for nonpolar substrates, aromatic hydrocarbons are the best solvents for other aromatic hydrocarbons, polar liquids are the best solvents for other polar substances. In gas chromatography the stationary phase should be chosen such that it is a good solvent for the substances being analyzed. To analyze a mixture of alcohols or esters one chooses a polar stationary phase; to analyze a mixture of hydrocarbons one chooses a nonpolar stationary phase. This is why the operator of a gas chromatograph should have different kinds of columns available. Table 12-1 lists a few of the liquids commonly used for stationary phases. Very many stationary phases are described in the literature.

In addition to the stationary phases listed in Table 12-1 we should mention the product "Poropak." This is the styrene-divinylbenzene copolymer mentioned on page 260. The polymer beads are prepared in such a way as to be porous and to admit substrate molecules easily. Thus the whole particle serves as the nonvolatile solvent; no liquid coating need be added. The material is, of course, best for hydrocarbons and specially aromatics.

IV. DETECTORS

A. The Thermal Conductivity Detector

This detector, in its simplest form, is shown in Fig. 12-3. There are two compartments, hollow cylinders in a block of highly conducting metal such as aluminum, and along the axis of each is mounted a wire in the form of a spiral. The inlet leads for the wires are insulated from the metal block. An electric current is passed through each wire and raises its

TABLE 12-1

Stationary Phase Liquids for Gas Chromatography

Type	Name	Formula or chemical type	Maximum operating temperature, °C
Nonpolar	Squalane	$-CH(CH_3)-(CH_2)_3-$	150
	Apiezon L oil	Hydrocarbon, better for aromatics than squalane	250
Moderately polar	Silicone oil DC-200	$-Si(CH_3)_2-O-$	200
Polar	Dinonyl phthalate	$C_6H_4(CO\text{-}OC_9H_{19})_2$ (benzene ring with two adjacent CO-OC$_9$H$_{19}$ groups)	130
	DEGS (diethylene glycol succinate)	$CH_2-CO-O-(CH_2)_2-O-(CH_2)_2-O-$ / $CH_2-CO-O-(CH_2)_2-O-(CH_2)_2-O-$	200
	Carbowax (polyethylene glycol)	$-CH_2-CH_2-O-$	150

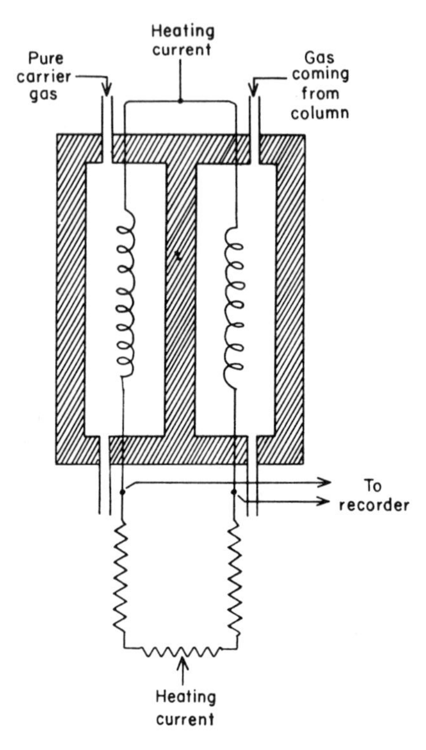

Fig. 12-3. Thermal conductivity detector. In instruments of high precision, the wires are of platinum and are straight. A spring at one end keeps them under tension while expanding or contracting. The cheaper instruments use thermistors.

temperature. Through one compartment pure carrier gas is passed; through the other, the gas flowing out of the column. A common arrangement is to pass the carrier gas through one compartment before it enters the column and before the sample is injected, then to lead it through the injection port and the column, then back through the second compartment. If the wires on the two sides of the cell are similar and heated by equal currents, and if everything else is the same, the two will be heated to the same temperature and their resistances will be the same. If, however, the vapor of an organic compound coming out of the chromatograph enters one compartment and lowers the conductivity of the gas inside it, the wire will not be able to lose heat as fast as it did before, and its temperature will rise. As its temperature rises, so does its electrical resistance.

The electrical resistances of the wires in the two cell compartments are compared by means of the Wheatstone bridge circuit shown in Fig.

12-3 (see also Chapter 2). With nothing but carrier gas passing through the two sides of the cell, the sliding rheostat shown at the bottom of the diagram is set so that the recorder pen reads zero or is positioned at a convenient place near one end of the scale of the chart paper. As long as there is no change in the gases passing through the cell the recorder pen will trace a straight line parallel to the side of the chart paper (Fig. 12-2), unless, of course, the recorder itself is unstable or drifting. When a substrate enters the cell the balance of the Wheatstone bridge will be disturbed and the recorder pen will move, tracing the line shown in Fig. 12-2. By proper circuitry it can be arranged that the movement of the pen is proportional to the concentration of substrate. However, we note that two different substrates having the same concentration will not necessarily give the same deflection; this depends on the thermal conductivities of the two substrates.

All detectors have in their electrical circuit a switch to vary the range, that is, the signal coming out of the detector cell that will cause a full-scale deflection of the recorder pen. For low concentrations of substrate a high sensitivity is used, for high concentrations a low sensitivity. With thermal conductivity detectors it is also possible to regulate the heating current that passes through the two filaments. The higher the current, the greater the sensitivity.

There is one precaution that must be observed with thermal conductivity detectors: never turn on the heating current unless the carrier gas is flowing. If there is no movement of gas through the cell the wires will get very hot and will melt or burn out. Some detectors use thermistors, which are more robust and do not easily burn out, but these have disadvantages; their response is not proportional to the substrate concentration.

Thermal conductivity detectors work best with helium as carrier gas, but they can also be used with nitrogen. Nitrogen has a much lower heat conductivity than helium, however, and the sensitivity with nitrogen is only a tenth to a fifth of what it would be with helium. When using nitrogen one must also be careful to use a smaller heating current, since the heat of the wires is not conducted away as fast as it would be with helium. Instructions are given in the manuals supplied by the manufacturers.

B. The Flame Ionization Detector

This detector is shown in Fig. 12-4. The principle is this: A flame of burning hydrogen has a small electrical conductivity due to the forma-

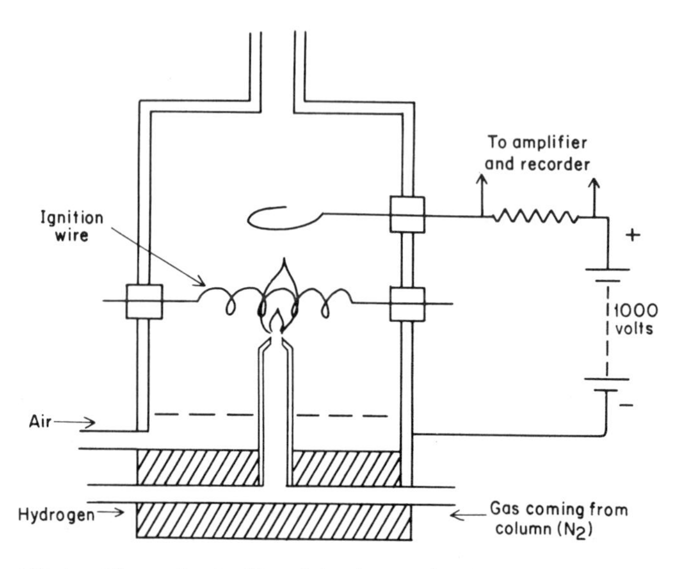

Fig. 12-4. Flame ionization detector. The ignition wire is placed to one side of the flame.

tion of gaseous ions such as HO_2^+. A very small amount of a carbon compound introduced into the flame causes very many more ions to be formed and greatly increases the electric current that passes through the flame in the arrangement shown in Fig. 12-4. The current is nevertheless very small, and a sensitive amplifier is necessary to magnify it to the point where it will drive the recorder pen. Recently, amplifiers have been developed that are very sensitive and at the same time are robust and reliable.

Nitrogen is used as the carrier gas. In addition, compressed air and a supply of hydrogen are needed. The hydrogen may be obtained from the electrolysis of water; commercial units are available that make it unnecessary to have a hydrogen cylinder.

The hydrogen flame ionization detector has several advantages. First is its extreme sensitivity. It will respond to as little as 1 nanogram (10^{-9} gram) of organic compound. The high sensitivity is in a sense an inconvenience. The samples introduced must be very small, 1 microliter or less; it is difficult to measure them with any degree of accuracy, and one worries whether evaporation from the tip of the syringe needle might not change the composition of the sample before it enters the column. A way to avoid these difficulties is to use a sample-splitting device, which allows only a small fraction of the injected sample to enter the column while rejecting the rest; see Fig. 12-5.

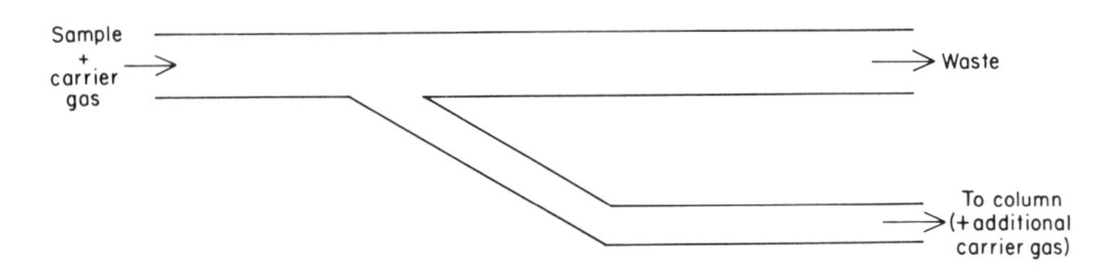

Fig. 12-5. Sample splitter. The tube that goes to "waste" has a valve to control the fraction of gas that goes to the detector.

The ionization current is proportional to the amount of material in the flame over a very large range, over a factor of one million. There are not many analytical methods in which the signal is directly proportional to the concentration over a range as great as this. Another feature of the flame ionization detector is its selectivity. It does not respond to carbon dioxide, and carbon atoms linked to oxygen, as in the carbonyl group :C=O, do not produce a current either. Water does not affect the detector, which is a great advantage in analyzing aqueous solutions. The thermal conductivity detector, on the other hand, responds to every substance in the sample.

The high sensitivity of this detector makes it possible to use narrower columns than with the thermal conductivity detector. It also makes possible the use of capillary columns, in which there is no solid packing material; the liquid stationary phase is a coating on the walls of a capillary a few tenths of a millimeter in diameter and a hundred or more meters long. The resolution of such columns is fantastically great, but the samples introduced must be very small.

C. The Electron Capture Detector

This detector is shown in Fig. 12-6. As the name implies, compounds entering the detector capture electrons and lower the current that passes across it. The effect is similar to the absorption of light quanta by a colored substance and follows an equation like Beer's law:

$$(\text{current}) = (\text{base line current}) \times e^{-kc}$$

where c is the concentration of the absorbing (capturing) species and k is a constant that depends on the nature of the substance and on the voltage across the cell.

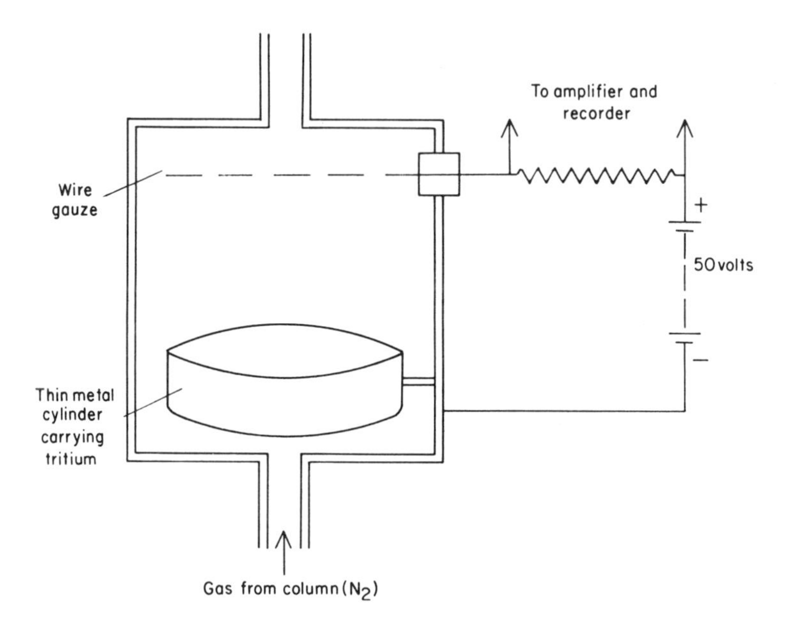

Fig. 12-6. Electron capture detector.

The electrons are supplied by tritium, radioactive hydrogen of mass 3, whose half-life is 12 years. It is held as a solid hydride in a metal foil. The carrier gas is nitrogen. Electrons from the tritium pass a short distance into the nitrogen and produce ions that carry the current. The potential across the cell is variable and of the order of 50 volts.

The electron capture detector is very sensitive to some compounds and insensitive to others. It is most sensitive to chlorine compounds, which makes it valuable for studying insecticides. It is sensitive to compounds containing sulfur and nitrogen, to metal alkyls like tetraethyl lead, and to oxygenated compounds like ketones and alcohols, and is relatively insensitive to hydrocarbons.

Other types of detector are available; thus there is a special detector sensitive to phosphorus compounds. By using two or more detectors simultaneously in the same chromatograph one can analyze complex mixtures in a simple way, and the more elaborate instruments have facilities for doing this.

V. PEAK IDENTIFICATION AND QUANTITATIVE ANALYSIS

The chromatogram of a complex mixture, like gasoline, may show twenty or thirty peaks, perhaps more. How is one to tell which peak corresponds to what compound? The only simple way is to compare the unknown chromatogram with chromatograms obtained with known, pure compounds under exactly the same conditions of column, temperature, and flow rate. In other words, one identifies a compound by its retention time. In simple mixtures, with, say, less than ten components whose nature is known (a group of alcohols, for example, or isomers of pentane) this kind of identification is satisfactory. For more complex mixtures it is not. One must then collect the substrate causing a particular peak as it comes out of the column and examine it. Now, the amount of material in a chromatographic peak is so small that most methods of analysis are far too crude and insensitive to give any information. About the only method that has sufficient sensitivity is mass spectrometry. Combination gas chromatograph - mass spectrometers are used in big industrial laboratories to give positive identification of the chromatograph peaks. Needless to say, this equipment is very expensive. Preparative-scale gas chromatographs, even the smaller ones, may give sufficient sample that it can be trapped in dry ice or liquid nitrogen and examined by infrared spectroscopy.

The more material is present, the higher is the peak on the chromatograph. Gas chromatography can obviously be used for quantitative as well as qualitative analysis. One must be able to correlate the quantity of material with the recorder response. Correlation is easiest when the movement of the recorder pen (which shows the electromotive force applied to the recorder) is directly proportional to the substrate concentration. This is true over a very large range for flame ionization detectors, and over a satisfactory range for thermal conductivity detectors. For a particular substrate, therefore, the area under the peak traced on the recorder chart is proportional to the quantity of substrate that has passed through the detector; see Fig. 12-7. If the peak is perfectly symmetrical and Gaussian (Fig. 9-8) the height is proportional to the area and hence to the amount of substance.

The correlation between peak area and quantity is different for different substances, and ideally one should prepare calibration graphs for each substance, injecting measured amounts of the substance and measuring the peak area for each amount. This is easier said than done, especially for flame ionization detectors, where the quantity of sample is so small that it is practically impossible to inject the same amount of sample twice in succession. If the characteristics of the detector are known, one has a basis for comparing the peak areas of two different substances. With the flame ionization detector, for example, all hydrocarbons give the same ionization current per carbon atom, while oxygenated compounds give smaller responses that nevertheless are known and reproducible. With the thermal conductivity detector the response depends on the difference

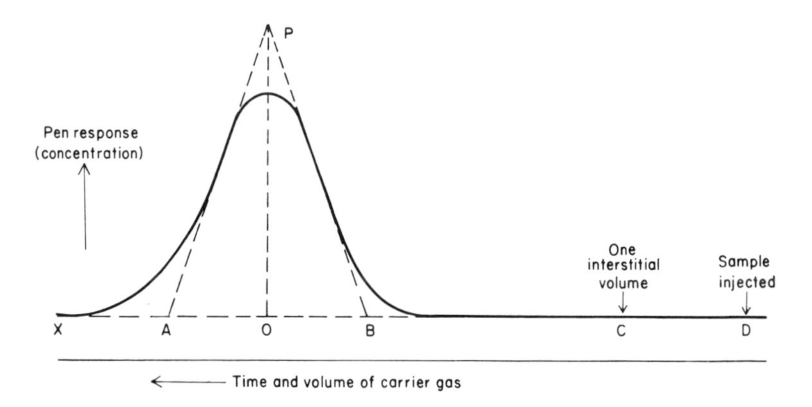

Fig. 12-7. A typical gas chromatographic peak.

in conductivity between the substrate and the carrier gas. With nitrogen as the carrier gas these differences, and hence the detector responses, vary greatly from one compound to another; with helium the differences are all more or less the same, and the detector response is roughly the same for all compounds on a mole basis (see Table 12-2).

TABLE 12-2

Thermal Conductivities of Gases at 50°C[a]

Hydrogen	471.1
Helium	376.1
Nitrogen	65.7
C_2H_6	58.3
n-C_3H_8	48.4
H_2O	46.7
CO_2	43.8
n-C_4H_{10}	43.4
C_2H_4	42.1

[a]Units: calories per second per square centimeter for 1°C per centimeter temperature gradient, x 10^6.

Very often, in chemical analysis, one is interested not so much in the percentage of a compound in the sample as in the ratio of the concentrations of two or more substances. Gas chromatography is used to follow the progress of chemical reactions. One may want to know the ratio of starting material to final product, or the ratio of two products that are forming at the same time - for example, secondary and tertiary butyl alcohol from the addition of water to isobutylene. Then one simply compares the areas of two peaks, multiplying the ratio of the areas by whatever factor is necessary to allow for differences in detector response. (In the case cited it would be safe to assume that the ratio of the concentrations was the same as the ratio of the peak areas.) It may even be sufficient to take the ratio of the two peak heights, if the peaks are symmetrical. The size of the sample is of no consequence, as long as it is not so great as to overload the column.

The same principle is used to measure the concentration of a single component of a sample using an internal standard. This method is illustrated in Experiment 12-4. One adds to the sample a known proportion of a substance that one knows is not present in the original, then compares the peak area of this substance with that of the "unknown" substance. Methyl ethyl ketone is added as the internal standard to measure the proportion of ethyl alcohol in an aqueous solution. This compound is not normally present in alcoholic beverages or industrial alcohol (it is very easy to verify this fact with the chromatograph), yet it is a polar compound like ethanol and its retention volume on a polar column is not greatly different from that of ethanol. We add a known proportion of methyl ethyl ketone to our sample, then run the chromatograph and measure the ratio of the peak areas of methyl ethyl ketone and ethanol. This ratio is compared with the peak ratio in a synthetic mixture that contains known amounts of both the alcohol and the ketone. Only the ratios are of interest; the size of sample injected into the chromatograph is unimportant.

The "internal standard" principle is used in other kinds of analysis, such as emission spectroscopy.

The peak area is conveniently found by drawing a triangle along the steepest tangents to the curve and along the base line and by taking half the product of base and height (Fig. 12-7). It may also be computed automatically by an attachment to the recorder.

VI. RETENTION VOLUMES OF HOMOLOGOUS SERIES

Within a homologous series of compounds, like the normal paraffins or the straight-chain primary alcohols, there is a relation between the number of carbon atoms in the compound and its retention volume that is useful to know. If one plots the logarithm of the corrected retention volume - that is, the quantity $V_{el} - V_i$ in Fig. 9-6, the retention volume minus the void or interstitial volume of the column - against the number of carbon atoms in the molecule one gets a straight line. This is true provided the temperature is constant and provided that a stationary phase is used that is a good solvent for the substrates; hydrocarbons should be compared with a nonpolar stationary phase, alcohols with a polar one. There is a sound thermodynamic basis for this relationship. We have seen (Chapter 9) that the corrected elution (or retention) volume is directly proportional to the distribution ratio of the substrate between the fixed and mobile phases. This ratio depends on the vapor pressure of the substrate in the stationary phase. Now the logarithm of the vapor pressure of a liquid is proportional to the standard free energy of vaporization of the liquid. The standard free energy of vaporization, in turn, goes up linearly with the number of carbon atoms in a homologous series; every carbon atom that is added adds the same number of calories to the vaporization energy.

If the logarithm of the corrected retention volume increases linearly with the number of carbon atoms, this means that the corrected retention volume itself is multiplied by a constant factor on going from one member of a homologous series to the next. Thus one might have the following set of values:

Number of carbon atoms:	1	2	3	4	5
Log(corrected retention volume):	0.5	0.8	1.1	1.4	1.7
Corrected retention volume:	3.2	6.4	12.8	25	50

In our imaginary series the volume is multiplied by 2 every time a carbon atom is added. A relation like this is useful for identifying unknown peaks and predicting retention volumes from a minimum of data, and one can see what it does to the appearance of the curve from Fig. 12-8. The later peaks take longer and longer to come and are flatter and flatter.

VII. TEMPERATURE CONTROL AND PROGRAMMING

A peak can be made to come out sooner by raising the temperature of the column. Figure 12-8 shows our imaginary homologous series run at

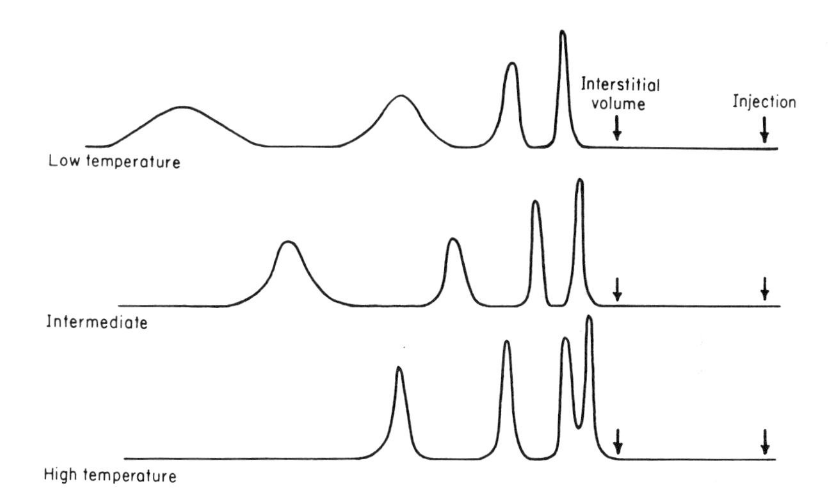

Fig. 12-8. Elution curves of a homologous series at three temperatures.

three different temperatures, increasing as we go down the figure. At the lowest temperature the graph is inconveniently extended; at the highest temperature the first two peaks are very close and cannot be distinguished properly. The intermediate temperature would be best for analyzing this mixture. There is another solution, however, which is a necessity in analyzing a mixture like gasoline where the components have a large range of volatilities. This is to raise the temperature while the components are passing through the column. One could do this by hand, raising the temperature in steps of 5°C every 5 minutes, for example, but a more convenient and more reproducible method is to raise the column temperature automatically at a predetermined rate. This feature is called temperature programming. The more expensive instruments have it, but, as we said, one can sit by the instrument and raise the temperature manually if necessary. The effect of temperature programming on the elution of a mixture is to save time and space the peaks more evenly.

Thermodynamics helps us understand the effect of temperature on retention volume. The corrected retention volume, OC in Fig. 12-7, is inversely proportional to the vapor pressure that the substrate exerts at a given concentration in the stationary phase. If the substrate forms ideal solutions with the liquid of the stationary phase, this vapor pressure is the vapor pressure of the pure substrate times its mole fraction: $p = p_0 N$. Now the vapor pressure of a liquid varies with temperature according to the Clausius-Clapeyron equation:

$$\frac{d \ln p}{dT} = \frac{\Delta H}{RT^2}$$

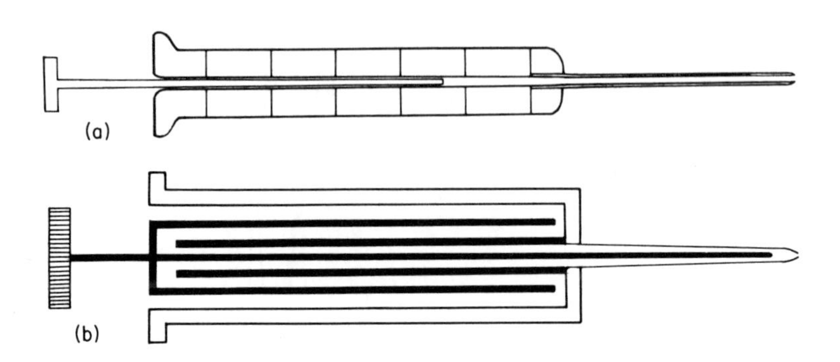

Fig. 12-9. Gas chromatography syringes.

which, when integrated, becomes

$$\log p - \log p_o = \frac{\Delta H}{2.3R}\left(\frac{1}{T_o} - \frac{1}{T}\right)$$

where p is the vapor pressure at temperature T and p_o the vapor pressure at temperature T_o. The net result is that if we plot the logarithm of the corrected retention volume against the reciprocal of the absolute temperature we get a straight line whose slope is proportional to the latent heat of evaporation, ΔH.

A strict application of thermodynamics to gas chromatography requires a knowledge of the activity coefficients of the substrates dissolved in the stationary phase, but one can make reasonable assumptions about these coefficients and deduce a surprising amount of thermodynamic data from gas chromatography. The reason for doing this is that gas chromatographic data are so easy to obtain, much easier than vapor pressure measurements, for example.

VIII. MEASUREMENT OF VOID VOLUME

To use chromatograms to derive thermodynamic data, and to estimate numbers of theoretical plates in the manner discussed in Chapter 9, one has to know the void volume, or at least the distance on the chart paper that corresponds to one void volume. (We note that the recorder is like a clock; the divisions that run across the chart paper at right angles to its length mark periods of time, and these can be converted into volumes by multiplying by the flow rate.) One way to find this distance is to inject some air into the carrier gas and observe the "air peak" (see Fig. 12-2). This works well with thermal conductivity detectors, because the conductivity of air is different from that of the carrier gas. With a flame ionization detector this will not work. In this case one may inject a substance that is retained very little, such as methanol or acetone, and note its retention time for different temperatures. As the temperature rises the retention time of a highly volatile substance will approach more and more closely the time that corresponds to one void volume. Figure 12-8 shows the idea.

Usually one does not need to know the actual volumes, but only their ratio, and the ratio of volumes is the same as the ratio of times. The floating ball type of flowmeter, illustrated in Fig. 12-1, does not give actual flow rates unless it is calibrated carefully. It is sensitive to small variations, however, and does tell one whether the flow rate is constant. If one should need to know the actual flow rate in cubic centimeters per minute, one can measure it by the simple device described in Experiment 12-3.

EXPERIMENTS

Experiment 12-1

Gaining Experience with the Gas Chromatograph

Equipment, Materials. Gas chromatograph, syringes; various pure liquids, as noted below: 5-ml or 10-ml beakers and a 10-ml graduated cylinder.

Purpose of Experiment. The object of this first experiment is to show what the gas chromatograph will do and to see the effects of column temperature and stationary phase.

General. (1) The chromatograph. Many models are available commercially, and the details of operation must be learned by reading the instruction manual. Low-priced instruments with flame ionization detectors include the Perkin-Elmer Model F-11, Varian Aerograph Model 200, and the Carle Basic Chromatograph Model 9000. Instruments with thermal conductivity detectors include the popular Perkin-Elmer Vapor Fractometer and the Carle Basic Chromatograph Model 6500. There are small differences in operation of the two types of instrument, as was noted in Section IV. In both types the first thing to do, before turning on the electronics or the detector circuits, is to start passing the carrier gas. Next, turn on the heaters for the column, the injection port, and the detector.

(2) The syringe. The size of syringe depends on the type of detector; for thermal conductivity instruments, a 10-μl syringe is used; for flame ionization detectors, the samples injected will normally be 1 μl or less and it is common to use a 1-μl syringe. One μl equals 10^{-6} l., or one-thousandth of a milliliter. These volumes are very small, and the syringes are delicate objects that must be handled with care. Two types of syringe are illustrated in Fig. 12-9. Type (a) has a fine wire plunger that is easily bent and broken. Type (b) has a more complicated plunger; the part that comes into contact with the liquid sample is a fine wire that goes inside the syringe needle; the upper part is more robust and serves as a guide. This type of syringe is more expensive, and when using it one must NEVER pull the plunger out of the glass barrel. Once it is pulled out, it is very difficult indeed to replace.

Using a syringe takes practice. The usual technique is to hold the barrel of the syringe between the fingers of the right hand, holding the index finger of the left hand gently against the plunger, and then push the syringe with the right hand through the rubber septum for injection. When the syringe needle has pierced the septum, and not before, one pushes the plunger gently. One must practice making injections until one can do it neatly and quickly without bending the plunger.

The septum can be punctured many times and still close again when the needle is withdrawn. Eventually, however, the holes will not close, and a new septum must be placed in the injection port. A supply of septa should always be kept on hand.

Experiments with a Polar Column. Using a column with a polar stationary phase, like Dow-Corning silicone oil DC-200 or DEGS (diethylene glycol succinate), inject a number of alcohols, esters, and ketones, one at a time, to find their retention times, then in mixtures. When changing from one sample to another, be sure to rinse the syringe with three or four portions of the new sample before making the injection. Include in your samples members of a homologous series, for example methyl, ethyl, and normal propyl alcohols, and look for a quantitative relation between the corrected retention times. Determine the retention time corresponding to the void column volume in one of the ways described above, page 328.

Run chromatographs at different temperatures. Start at about 75°C and then try lower and higher temperatures, at intervals of 20°C or so. Determine what temperature is best for the analysis of a mixture of alcohols, esters, and ketones. (The exact composition of the mixture can be chosen according to the chemicals you have on hand.) Measure the plate number (see Fig. 12-7) for two different substances at two or more temperatures.

Inject samples of one or two nonpolar substances, such as benzene or cyclohexane, and compare the symmetry of the peaks with those of polar substances.

Examine an unknown mixture supplied by the instructor to determine its composition. The "unknown" can be a solvent sold as a paint or lacquer thinner or a nail-polish diluent. These are mixtures of esters, alcohols and ethers. You will find some peaks whose retention times do not coincide with those of the pure substances you have run. Use the relation between retention times of members of a homologous series (discussed earlier in this chapter) to guess at the substances producing these peaks.

Commercial products containing nonvolatile ingredients, like the correction fluid used in typing stencils, can be run by gas chromatography if a glass liner is inserted into the injection port to catch the nonvolatile residue and prevent it contaminating the column. Liners for this purpose are sold by the instrument makers.

Relations between Peak Areas. If your chromatograph has a flame ionization detector, compare the detector response for compounds containing oxygen. Using a pipet, measure equal volumes, say 2.0 ml each, of ethyl acetate, *n*-butyl acetate, *n*-butyl alcohol and *n*-amyl alcohol and mix them together. Inject the mixture, taking care that all four peaks are fully resolved and that none of them go off the chart paper. Measure the areas of the peaks by triangulation and compare the ratios of peak areas with the ratios of masses of carbon injected. The latter can be calculated from the densities of the pure liquids (which are, respectively, 0.901, 0.882, 0.810, 0.814 g/ml), their chemical formulas and formulas weights. You will find that the esters, which contain more oxygen than the alcohols, have relatively smaller peak areas.

Results of experiments like this should always be presented in tabular form. A suggested form is the following, which shows data obtained with a DEGS column and flame ionization detector:

Compound	Density, g/ml	Mol. wt.	Mmol per ml.	Atoms C in molec.	Mg-atom C per ml.	Carbon ratio, butanol = 1	Peak area, sq.cm	Area ratio, butanol = 1

C_4H_9OH	0.810	4	10.9	4	43.6	1.000	15.7	1.000
$C_5H_{11}OH$	0.814	5	9.25	5	56.2	1.06	16.5	1.05
$CH_3COOC_2H_5$	0.901	4	10.2	4	40.8	0.94	4.9	0.31

Experiments with a Nonpolar Column. Apiezon L, a hydrocarbon oil, is a good stationary phase for this experiment. Silicone oil DC-200 can also be used, as it is only slightly polar. Substances to be injected may include hexane, cyclohexane and heptane, benzene and toluene. Follow the same plan as with polar compounds, that is, make experiments at three or four temperatures, record retention times, and plot retention times (corrected for void volume) against the number of carbon atoms in a homologous series, like n-pentane - n-hexane - n-heptane. Measure plate numbers, and analyze a mixture provided by your instructor.

An interesting experiment is to compare cyclohexane with an ordinary commercial grade of hexane or low-boiling petroleum ether (which contains pentane and hexane). You will find that cyclohexane gives one single sharp peak, while commercial hexane gives an irregular, broader peak that, if you choose the right conditions, can be resolved into several peaks, each corresponding to one of the five isomers of hexane. Best resolution is obtained at low temperature.

For a real "unknown" examine a sample of cigarette-lighter fluid or fuel sold for camp stoves. Gasoline may be tested, but gasoline contains a great many components, and the higher-boiling ones come out very slowly unless the temperature is raised considerably. Indeed, temperature programming is almost essential for gasolines. An interesting and fairly simple unknown is paint thinner or a paint-removing solvent. These contain alcohols as well as hydrocarbons. Report as fully as you can the composition of your sample, making whatever tests you can devise to identify unknown peaks.

Experiment 12-2

Quantitative Analysis of a Mixture of Xylenes

Materials. Benzene, toluene, commercial xylene (mixture of isomers), two pure isomers of xylene (one of which must be ortho-xylene), and ethyl benzene.

The separation of ortho-xylene from the other two isomers is easy; the separation of the other two isomers is difficult. This fact could be inferred from the boiling points of the three isomers, which are: para-, 138.3°C; meta-, 139.1°C; ortho-, 144.4°C. However, all three can be separated by a good chromatograph equipped with a flame ionization detector. If such an instrument is not available, meta- and para-xylene will be separated together as one peak in the chromatograph.

The stationary phase should be Apiezon L or its equivalent. Silicone oil DC-200 may be used if this is not available. The best column temperature with Apiezon L is about 120°C, but you should make tests to determine the best temperature for your instrument.

First run commercial xylene alone, to find the temperature and flow rate that will give good separation into three peaks. (Be content with two peaks if necessary.) Having established the desired temperature and flow rate, run individual samples of benzene, toluene, ethyl benzene (this may be omitted if it is not available), and ortho-xylene to find their retention times. Now, carefully prepare a mixture of equal volumes of toluene and ortho-xylene, inject this mixture, and measure the ratio of the two peak areas. Repeat two or three times to see if you are getting consistent results, and compare the ratio of areas with (a) the ratio of weights of toluene and xylene, (b) the ratio of the amounts of carbon in each compound. (The densities of toluene and o-xylene are 0.867 and 0.880; their formula weights are 92 and 106.) If you are using a flame ionization detector the ratio of peak areas should equal the ratio of the number of carbon atoms.

Next, identify the three peaks of the commercial xylene. The surest way to do this is to mix with the commercial xylene an equal volume of pure o-xylene, run the chromatograph, and see which of the three peaks is increased in height; then do the same with one of the other two isomers, taking whichever one you have available in pure form. Proceeding in this way you will avoid any doubt you may have about the constancy of retention volumes between runs.

Now that you know which peak corresponds to which isomer, run the commercial xylene by itself, choosing the sample size to give the largest possible peaks without the pen running off the paper and without obscuring the resolution. Measure the individual peak areas as accurately as you can, compare them, and calculate the percentage by weight of each isomer in the sample. You may assume that the peak area is proportional to the mass of material in the same way for all three isomers.

To check your analysis, measure the peak areas for the mixtures of commercial xylene with pure isomer, and see if they are in the expected ratios. Note that commercial xylene may contain ethyl benzene.

Experiment 12-3

Plate Number and Flow Rate: The Van Deemter Equation

Equipment. For this experiment we must compare accurately one flow rate with another. This is hard to do with the flow gauges normally mounted on gas chromatographs. Flow rates can be measured with the simple apparatus shown in Fig. 12-10. This consists of a 10-ml buret with the stopcock removed and replaced by a small rubber bulb containing a soap solution. A side arm is attached to the bottom of the buret, as shown. These devices can be bought commercially, or one can make them oneself. To use the flow meter, the side arm is connected by a narrow-bore rubber tube to the outlet of the chromatograph column. With the carrier gas flowing, a little soap solution is forced up into the gas stream; soap films form and are carried by the gas up the graduated tube. One takes a stop watch and notes the time taken for one of the soap films to traverse the graduations.

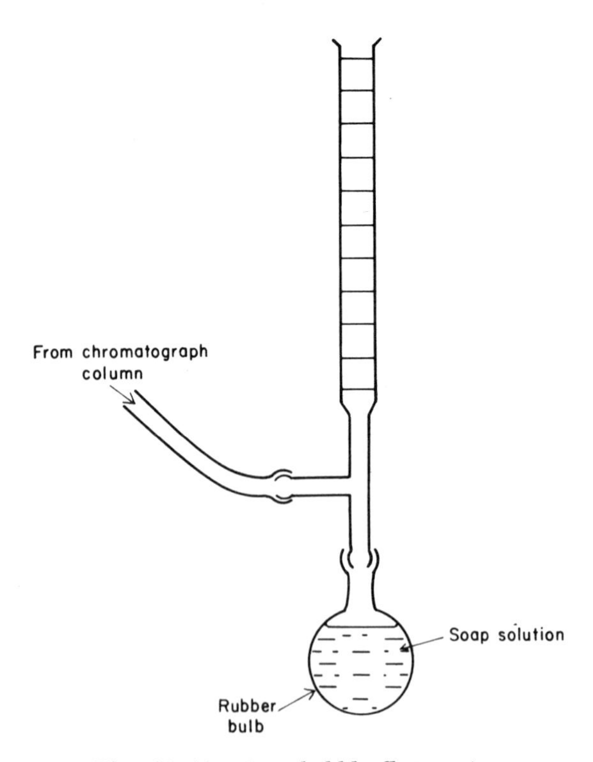

Fig. 12-10. Soap bubble flow meter.

Using a thermal conductivity detector, this measurement can be made with the detector in operation. This cannot be done with a flame ionization detector, as the gas emerging from the tube includes the excess of air that fed the flame. Measure flow rates, therefore, with the detector turned off, and with no air and hydrogen flowing, then note the valve settings and floating-ball meter readings, and calibrate these in terms of the flow rate measured with the soap film.

The next step is to choose any combination of stationary phase and substrate that gives a symmetrical peak with little or no "tailing." The retention volume should be at least three times the void volume. A good choice would be the peak of toluene on a column of Apiezon L, as studied in the previous experiment; you should work at a lower temperature than was used in that experiment, however, to make the elution (retention) volume somewhat longer. A temperature of 100°C would probably be suitable. If the column in your instrument has a polar stationary phase, a good substrate choice might be methyl ethyl ketone or *n*-butanol. Again, choose a temperature such that the retention volume is at least three times the void volume and preferably more.

Now, holding the temperature constant, inject samples of your substance at several different flow rates, from a value as low as possible (remember, if you have a thermal conductivity detector, that you must not overheat the filament; get your instructor's advice) to a value as high as possible. Determine the retention time, or distance on the recorder chart, that corresponds to one void volume at every speed you use. (With a thermal conductivity detector this is easy; inject a little air with your sample and note the "air peak." With a flame ionization detector you may have to rely on your flow-rate calibration and a void volume measured at one particular flow rate; see Section VIII.) Measure the peaks in the manner shown in Fig. 12-7 and calculate the plate number. Then take the reciprocal of the plate number (this is proportional to plate *height*) and plot it against flow rate. The relation should follow the equation of Van Deemter; see Chapter 9, equation (9-7).

Experiment 12-4

Quantitative Analysis with an Internal Standard

Materials Needed. Ethyl alcohol, methyl ethyl ketone, 10-ml volumetric flasks, graduated pipet; a gas chromatograph with flame ionization detector. The unknown may be a distilled alcoholic beverage such as rum or vodka.

The flame ionization detector is desirable because of its insensitivity to water; the unknown is an aqueous solution. The stationary phase should be polar, such as DEGS, diethylene glycol succinate. First, experiment with injections of ethyl alcohol and methyl ethyl ketone to find a suitable temperature. The peaks of these two substances must be well separated and, if possible, symmetrical. Now prepare standards for calibration by carefully measuring known volumes of ethyl alcohol (absolute or 96%, but know which you have!) and methyl ethyl ketone into 10-ml flasks and make up to the mark with water. Suitable volumes are: 1.00 ml methyl ethyl ketone in all cases; 1.00, 2.00, 4.00, and 6.00 alcohol. Inject samples of each of these standards into the chromatograph; the size of the sample need not be accurately known. Measure the areas of the two peaks in every case, and plot a graph of (alcohol peak area/MEK peak area) against (milliliters of alcohol per 10 ml sample). This is your calibration curve. Now take an empty, dry 10-ml volumetric flask and add to it 1.00 ml methyl ethyl ketone. Make up to the mark with the "unknown" alcoholic beverage sample. If there is no volume change on mixing, the flask will contain 9.0 ml sample. Mix, run the chromatogram, measure the peak areas, and take their ratio; from the graph, read off the number of milliliters of ethyl alcohol in your 9-ml sample, and calculate the alcohol content of the unknown in grams of alcohol per 100 ml of beverage. For a check, make another experiment in which you pipet 5.00 ml of sample into a 10-ml flask, add 1.00 ml methyl ethyl ketone, make to the mark with water, run the chromatogram, and compare peak areas as before. To convert the measured volumes of alcohol into weights you will need the following data:

Per cent ethyl alcohol by weight:	90.0	95.0	100.0
Density, grams per milliliter at 20°C:	0.8182	0.8045	0.7895

QUESTIONS

1. Of the three types of detector discussed in this chapter, which would you use for the following conditions:

   (a) When you wanted to collect the substrate after it had passed through the detector; that is, the detector must not destroy the substrate;

   (b) For the greatest possible range of proportionality between the detector response and the amount of the substrate;

(c) For the quantitative analysis of different mixtures of benzene, methanol, and acetone;

(d) For the analysis of aqueous solutions of organic compounds;

(e) For the detection and measurement of pesticides containing chlorine?

2. Suppose that, after you had made several injections, the septum started to leak; how would the recorder tracing show that it was leaking?

3. Suppose you ran a mixture of benzene, cyclohexane, acetone, and methanol, first on a column packed with "Poropak" (Section III), then on one packed with silicone oil, then on one packed with Carbowax. How would you expect the three curves to differ?

4. What is the Clausius-Clapeyron equation, and how can one apply it to gas chromatography?

5. How might one analyze a mixture of sugars by gas chromatography?

APPENDIX

TABLE I

Ionization Constants of Acids and Bases[a]

Acid		pK_a	Acid		pK_a
Acetic		4.76	Glycine	K_1	2.35
Ammonium ion		9.25		K_2	9.87
Anilinium ion		4.60	Hydrocyanic		9.31
Benzoic		4.20	Hydrofluoric		3.45
Boric		9.23	Hydrogen sulfide	K_1	7.05
Carbonic:	K_1	6.37		K_2	14.0
	K_2	10.25	8-Hydroxyquinoline		
Cinnamic,	*cis*-	3.89		K_1	4.91
	trans-	4.44		K_2	9.8
Citric	K_1	3.13	Lactic		3.86
	K_2	4.76	Maleic	K_1	1.92
	K_3	6.40		K_2	6.23
Cystine	K_1	7.85	Malonic	K_1	2.83
	K_2	9.85		K_2	5.69
Ethylene diammonium			Nitrilotriacetic	K_1	1.65
ion	K_1	6.85		K_2	2.95
	K_2	9.95		K_3	10.28
Ethylene diamine			Nitrophenol, *o*-		7.17
tetraacetic acid			*p*-		7.15
	K_1	2.0	Oxalic	K_1	1.23
	K_2	2.6		K_2	4.19
	K_3	6.85	Phenol		9.89
	K_4	9.95	Phosphoric	K_1	2.12
Formic		3.75		K_2	7.21
Fumaric	K_1	3.02		K_3	12.38
	K_2	4.44			

TABLE I (continued)

Acid		pK_a	Acid		pK_a
Phthalic	K_1	2.89	Sulfuric	K_2	1.92
	K_2	5.51	Sulfurous	K_1	1.76
Picric		0.4		K_2	7.19
Pyridinium ion		5.22	Tartaric	K_1	3.04
Selenious	K_1	2.46		K_2	4.35
	K_2	7.30	Vanillin		7.40
Succinic	K_1	4.21			
	K_2	5.63			

[a]The constants are presented as their negative logarithms, or pK_a values. The strengths of bases are shown as the pK values of their conjugate acids.

For example: Acetic acid, $pK_a = 4.76$, $K_a = 1.75 \times 10^{-5}$
Anilinium ion, $C_6H_5NH_3^+$; $pK_a = 4.60$, $K_a = 2.5 \times 10^{-5}$;
therefore, for aniline, $pK_b = 14.00 - 4.60 = 9.40$;
$K_b = 4.0 \times 10^{-10}$.

Values are quoted for 25°C and are thermodynamic constants.

TABLE II

Solubility Products at 25°C[a]

Substance	K_{sp}	Substance	K_{sp}
$AgCl$	1.6×10^{-10}	$Fe(OH)_3$	1×10^{-36}
$AgBr$	7.7×10^{-13}	HgI_2	3×10^{-29}
AgI	1.5×10^{-16}	Hg_2Cl_2	1.3×10^{-18}
$AgCNS$	1.2×10^{-12}	HgS	10^{-53}
Ag_2S	2×10^{-49}	$Mg(OH)_2$	1.8×10^{-11}
$Al(OH)_3$	4×10^{-15}	$MgNH_4PO_4$	3×10^{-13}
$BaCO_3$	8×10^{-9}	$Mn(OH)_2$	1.9×10^{-13}
$BaCrO_4$	2.4×10^{-10}	$Ni(OH)_2$	6×10^{-18}
$BaSO_4$	1.1×10^{-10}	$PbCrO_4$	2×10^{-14}
$CaCO_3$	8.7×10^{-9}	$PbSO_4$	1.6×10^{-8}
$CaSO_4$	1.2×10^{-6}	PbI_2	7×10^{-9}
CaC_2O_4	4×10^{-9}	$SrSO_4$	3.8×10^{-7}
CaF_2	4.0×10^{-11}	$TlCl$	1.7×10^{-4}
$CuCl$	1.0×10^{-12}	$Zn(OH)_2$	1.2×10^{-17}
CuI	5.0×10^{-12}	ZnS	1×10^{-23}
$CuCNS$	5×10^{-15}		

[a]Many of these values are uncertain, especially those for sulfides.

TABLE III

Standard Reduction Potentials

Half-Reaction	Potential, V
$F_2 + 2e = 2F^-$	2.87
$O_3 + 2H^+ + 2e = O_2 + H_2O$	2.07
$MnO_4^- + 8H^+ + 5e = Mn^{2+} + 4H_2O$	1.49
$PbO_2 + 4H^+ + 2e = Pb^{2+} + 2H_2O$	1.46
$Ce^{4+} + e = Ce^{3+}$ (in 1 $\underline{M}$ H_2SO_4)	1.44
$Cl_2 + 2e = 2Cl^-$	1.36
$O_2 + 4H^+ + 4e = 2H_2O$	1.23
$MnO_2 + 4H^+ + 2e = Mn^{2+} + 2H_2O$	1.21
$Br_2 + 2e = 2Br^-$	1.07
$Cr_2O_7^{2-} + 14H^+ + 6e = 2Cr^{3+} + 7H_2O$ (in 1 $\underline{M}$ HCl)	1.00
$VO_2^+ + 2H^+ + e = VO^{2+} + H_2O$	1.00
$AuCl_4^- + 3e = Au + 4Cl^-$	0.99
$Hg^{2+} + 2e = Hg$	0.85
Sb(V) + 2e = Sb(III) (in 6 $\underline{M}$ HCl)	0.82
$Ag^+ + e = Ag$	0.800
$Hg_2^{2+} + 2e = 2Hg$	0.792
$Fe^{3+} + e = Fe^{2+}$	0.771
$Hg_2SO_4 + 2e = 2Hg + SO_4^{2-}$	0.616
As(V) + 2e = As(III) (in 1 $\underline{M}$ HCl)	0.58
$I_2 + 2e = 2I^-$	0.535
$I_3^- + 2e = 3I^-$	0.534

TABLE III (continued)

Half-Reaction	Potential, V
$Fe(CN)_6{}^{3-} + e = Fe(CN)_6{}^{4-}$	0.46
$VO^{2+} + 2H^+ + e = V^{3+} + H_2O$	0.337
$Cu + 2e = Cu$	0.337
$BiO^+ + 2H^+ + 3e = Bi + H_2O$	0.32
$AgCl + e = Ag + Cl^-$	0.222
$BiOCl + 2H^+ + 3e = Bi + H_2O + Cl^-$	0.16
$Cu^{2+} + e = Cu^+$	0.16
$Sn^{4+} + 2e = Sn^{2+}$ (in 1 $\underline{M}$ HCl)	0.14
$S_4O_6{}^{2-} + 2e = 2S_2O_3{}^{2-}$	0.09
$2H^+ + 2e = H_2$	$\pm$0.000
$Pb^{2+} + 2e = Pb$	-0.126
$Sn^{2+} + 2e = Sn$	-0.136
$V^{3+} + e = V^{2+}$	-0.26
$Co^{2+} + 2e = Co$	-0.28
$Tl^+ + e = Tl$	-0.336
$PbSO_4 + 2e = Pb + SO_4{}^{2-}$	-0.356
$Cd^{2+} + 2e = Cd$	-0.403
$Fe^{2+} + 2e = Fe$	-0.41
$Cr^{3+} + e = Cr^{2+}$	-0.41
$S + H_2O + 2e = HS^- + OH^-$	-0.48
$Cr^{2+} + 2e = Cr$	-0.56
$Cr^{3+} + 3e = Cr$	-0.74
$Zn^{2+} + 2e = Zn$	-0.763
$SO_4{}^{2-} + H_2O + 2e = SO_3{}^{2-} + 2OH^-$	-0.92
$Al^{3+} + 3e = Al$	-1.66

TABLE III (continued)

Half-Reaction	Potential, V
$Mg^{2+} + 2e = Mg$	-2.37
$Na^{+} + e = Na$	-2.71

Standard reference electrodes: (25°C)

Saturated calomel	+0.2412
Calomel, 0.10 M KCl	+0.3337
Ag-AgCl, 0.10 M KCl	+0.2881
Quinhydrone, 1.0 M HCl	+0.696

INDEX

D